Coal Combustion and Conversion Technology

Other titles published in this series

Hydrogen and Energy Charles A. McAuliffe
Fuel Cells Angus McDougall

ENERGY ALTERNATIVES SERIES
Series editor: C. A. McAuliffe

Coal Combustion and Conversion Technology

D. Merrick

National Coal Board
Coal Research Establishment
Stoke Orchard
Cheltenham

MACMILLAN

First published 1984 by
Higher and Further Education Division
MACMILLAN PUBLISHERS LTD
London and Basingstoke
Companies and representatives
throughout the world

Printed in Hong Kong

ISBN 0 333 32449 8

CONTENTS

PREFACE

By the middle of the twentieth century, coal, although still the dominant world energy source, had come to be regarded as the product of a traditional industry nearing the end of its useful life-cycle. The prospect of cheap and plentiful energy supplies from oil and nuclear power promised to make the early decline of the coal industry inevitable. During the 1970s, however, attitudes changed dramatically. Oil price increases and doubts about the planned expansion of nuclear power led to a widespread recognition that coal had an important and long-term contribution to make. In parallel, there was a revival of interest in coal technology, particularly in methods of using coal with greater efficiency, environmental acceptability and versatility than previously.

This book is intended as a guide to coal utilisation technology, with the emphasis on new systems under development for the future, especially in the short and medium term. The traditional coal utilisation technologies are also discussed in the context of the future use of coal.

The book has been written for students, engineers, planners and policy makers with an interest in coal, rather than the specialist research worker. Inevitably, some scientific background is assumed, but a knowledge of coal utilisation technology is not required. Both the underlying scientific principles and engineering approaches are covered. However, recent developments in research programmes have been rapid, and no attempt is made to provide a complete catalogue of systems currently under investigation. Instead, engineering approaches are dealt with in terms of their general characteristics and classification so that the reader may be provided with a framework against which to assess specific schemes and future concepts.

The first two chapters discuss coal's position as a world energy source, and the historical background to the present technology and developments. The major technologies themselves, fluidised bed combustion, gasification and liquefaction, are considered in the next three chapters, followed by a further chapter describing other lines of technology under investigation. The last two chapters are devoted to the environmental and economic aspects of coal utilisation, respectively.

In general, SI units have been used throughout the book although, for convenience, pressures are expressed in bar and temperatures in °C.

In writing this book, I am indebted to a large number of colleagues at the Coal Research Establishment and other parts of the National Coal Board, including the Coal Utilisation Research Laboratories, Mining Research and Development Establishment, the Economic Assessment Service of IEA Coal Research, Technical Information Branch and Operational Research Executive. I am extremely grateful for their help in providing information and in commenting on drafts. I would particularly like to thank those most closely involved: Mr J. Highley, Dr J. Holmes, Miss S. Lauder, Dr P. F. M. Paul and Mr. A. A. Randell.

I would also like to thank the National Coal Board for permission to publish this book. Any views implied or expressed are, of course, those of the author and not necessarily those of the National Coal Board.

Stoke Orchard, Cheltenham, 1984 DAVID MERRICK

CHAPTER 1

INTRODUCTION

The exploitation of coal provided the first large-scale, low-cost source of energy and made possible the changes of the industrial revolution. Coal remained the dominant world energy source until the rapid growth of crude oil and natural gas production, particularly during the 1950s and 1960s. However, the large reserves of coal world-wide ensure that, in the forthcoming era of energy shortage, coal will remain a crucially important energy source.

Since much of the economic growth in the industrial world over the last three decades has been based on oil and natural gas, the existing infrastructure and environmental standards have been influenced by the technologies available to these fuels. Substitution by coal therefore requires the introduction of new technologies for coal utilisation that can meet a consumer demand equivalent to that which has been created by the availability of oil and gas. The main new technologies for coal combustion and conversion into gas and liquids are described in the following chapters 2-5, the supporting technologies, environmental effects and economics being considered in chapters 6-8.

1.1 Origin and nature of coal

Formation

Coal was formed by the decomposition of the remains of plants in swamps or river deltas. Two stages of formation existed.

Biochemical

In the waterlogged environment, decomposition of the plant material by bacteria and fungi took place, eventually giving rise to the formation of peat.

Geochemical

Intermittent subsidence of the peat took place, followed by sedimentation and, often, further cycles of plant growth and peat formation. In the buried layers of peat, biological activity ceased and much slower chemical changes (termed 'coalification') took place, transforming the peat into coal.

Since coal-forming plant growth has occurred throughout the last 400 million years, a wide variety of coal types exist corresponding to various stages of coalification.

The main factors affecting the degree of coalification appear to be temperature and, to a lesser extent, pressure and time itself. The cycles of subsidence and sedimentation referred to above generally result in the youngest coals being nearest the surface and the oldest coals being deepest. Since the temperature of rock strata increases with depth, the oldest coals have usually been subjected to the highest temperatures and therefore to the greatest chemical changes.

Not all parts of the original plant material are affected to the same degree by coalification. For example, cellulose decomposes readily, lignin less so and spores, leaf cuticles and resins least. Microscopic examination of coal can identify the remains, after coalification, of the various types of plant material originally present. These remains are referred to as macerals and their study is a method by which coals can be identified and classified.

Composition

As mined, coal contains three components.

Moisture

Some of the moisture is termed free moisture and may be removed by drying in air. There remains, however, inherent moisture; the amount varies with coal type but is generally less than 10 per cent.

Mineral matter

The mineral matter in coal is of two types. Mineral matter present as discrete particles (for example, from adjacent strata or intermediate 'dirt bands') is referred to as 'adventitious' and can be largely removed during coal preparation by washing techniques (see below). However, coal also contains 'inherent' mineral matter that is finely divided (the particles are mainly smaller than 0.1 mm) and distributed throughout the coal substance, making its removal by washing possible only to a limited extent. The type and properties of the mineral matter vary with location and mining methods. Inherent mineral matter contents are generally less than 10 per cent although examples of coals with 30 per cent or more can be found.

Dry, mineral-matter-free coal (dmmf coal)

This is a term used to refer to the pure organic coal substance. The dry, mineral-matter-free coal consists mainly of the elements carbon, hydrogen, oxygen, nitrogen and sulphur.

Coal structure

The molecular structure of coal has been the subject of intensive study using a variety of techniques. The resulting picture shows that coal may be regarded as a complex, cross-linked structure of high molecular weight. Unlike crude oil, coal is not a mixture of well-defined individual molecular types and only statistical models of the 'coal molecule' are, therefore, possible.

Although the evidence is not conclusive, the generally held view is that coal consists of condensed aromatic clusters containing, for bituminous coals, on average about three condensed rings. These nuclei are linked together, mainly by hydroaromatic structures and heteroatomic bridges. Short aliphatic side chains (predominantly methyl) are also attached to the aromatic clusters. In a typical bituminous coal, approximately 75 per cent of the carbon is in aromatic structures, 20 per cent in hydroaromatic structures and 5 per cent in aliphatic structures. About half of the aromatic nuclei exists as single layers with the remainder being stacked as two, or occasionally more, parallel-layer planes.

Although no direct representation of the molecular structure of coal is possible, an attempt to illustrate the main features discussed above is made in figure 1.1. In particular, the figure reflects the wide distribution of molecular structures and the randomness of orientation (which extends into three dimensions). The solid state of coal may be attributed directly to the high molecular weight of the structure.

As noted above, in addition to carbon and hydrogen, coal contains the heteroatoms oxygen, nitrogen and sulphur. The heteroatom content varies greatly from coal to coal although oxygen predominates. A considerable proportion of the oxygen in coal exists as phenolic hydroxyl groups with other structures such as ether linkages also known to exist. The nitrogen and sulphur are generally present mainly in heterocyclic form.

Classification of coal types

Most classification systems for coals are based on the dry, mineral-matter-free coal substance. One of the most elegant systems employs the elemental composition and was introduced by Seyler in 1899. A form of Seyler's system is illustrated in figure 1.2, which also introduces the main categories of coal in common use. Lignites are also known as brown coals; sub-bituminous coals, bituminous coals and anthracites are known collectively as hard coals or black coals. These terms reflect the properties and appearance of the coals.

As shown in figure 1.2, the composition of coals generally falls into a relatively narrow band, implying a close relationship between the carbon, hydrogen and heteroatom contents.

The youngest coals (lignites) have the highest heteroatom content (mainly oxygen) and the lowest carbon content. As coalification proceeds, hydrogen and

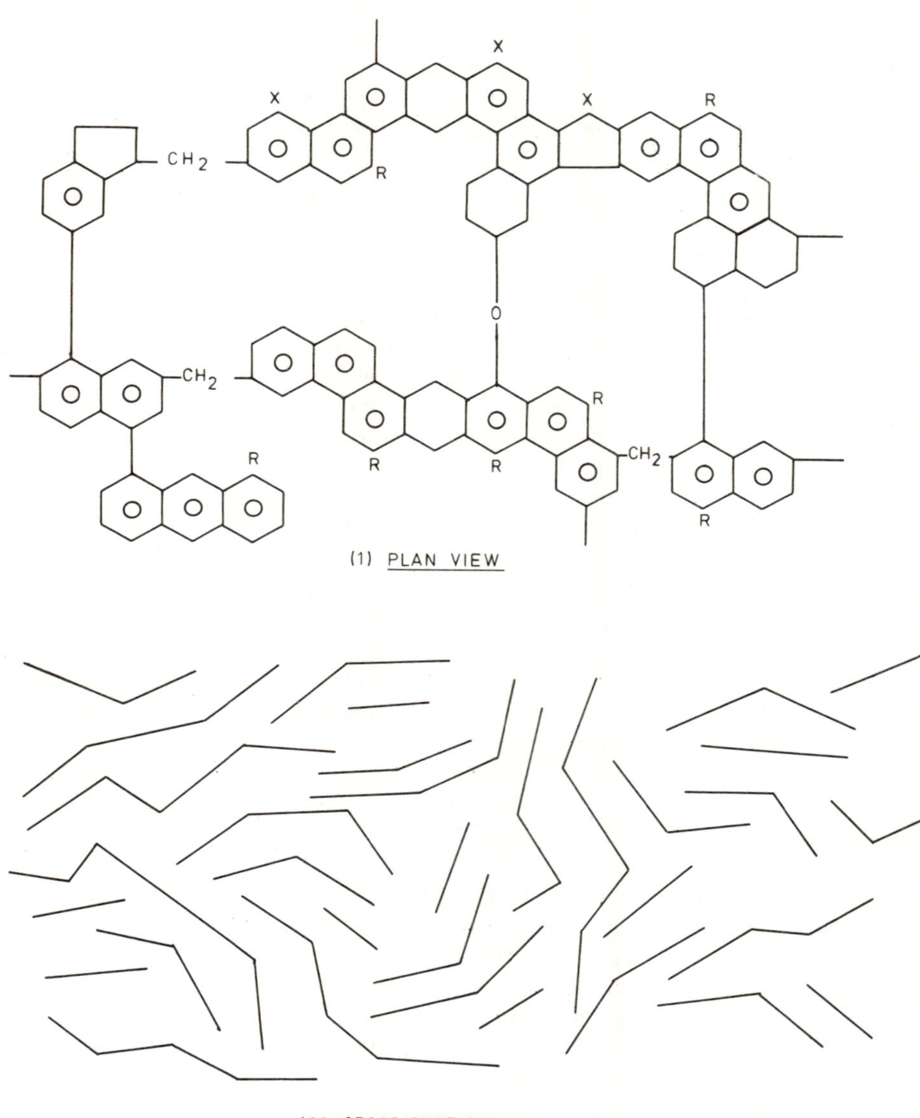

Figure 1.1 Illustration of coal structure. X = Heterocyclic atoms of oxygen, nitrogen and sulphur; R = side groups, mainly methyl and hydroxyl

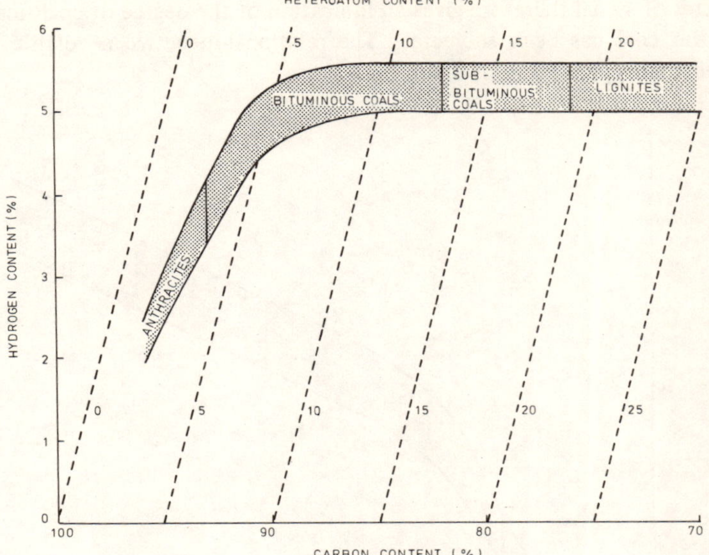

Figure 1.2 Classification of coal types using Seyler's system. The composition is based on the dry, mineral-matter-free coal

oxygen are removed, mainly as methane and water. The methane may become trapped in suitable strata to form a natural gas reservoir. The oldest coals, anthracites, contain least hydrogen and heteroatoms and most carbon. The carbon content of the dry, mineral-matter-free coal may, therefore, be used as a means of classifying coal types and is referred to as the coal 'rank'.

It is interesting to note that, although the loss of hydrogen during coalification results in a reduction in the hydrogen-to-carbon ratio, the hydrogen content remains approximately constant at about 5 per cent throughout the range of lignites, sub-bituminous coals and bituminous coals. The compositions of these coals differ mainly in terms of their heteroatom contents (principally oxygen).

Although the Seyler system provides a comprehensive method of characterising coals, most practical coal classification systems rely on empirical laboratory tests of coal properties.

The simplest such test, and one that is used in all of the current methods of coal classification, relies on measuring the weight loss (referred to as the volatile matter) when the coal is heated. An inert atmosphere is used to avoid combustion and carefully controlled conditions are specified to ensure reproducibility. This test can be thought of as representing, in approximate terms, an acceleration of the coalification process itself. As coal is heated, the hydrogen and oxygen are removed as water, methane and other products (tar and hydrogen for example), ultimately leaving a coke residue comprising mainly carbon. The measured volatile matter of a coal therefore gives an indication of the degree of coalification to which the coal has been subjected. The relationship between volatile matter, rank and coal type is given in figure 1.3.

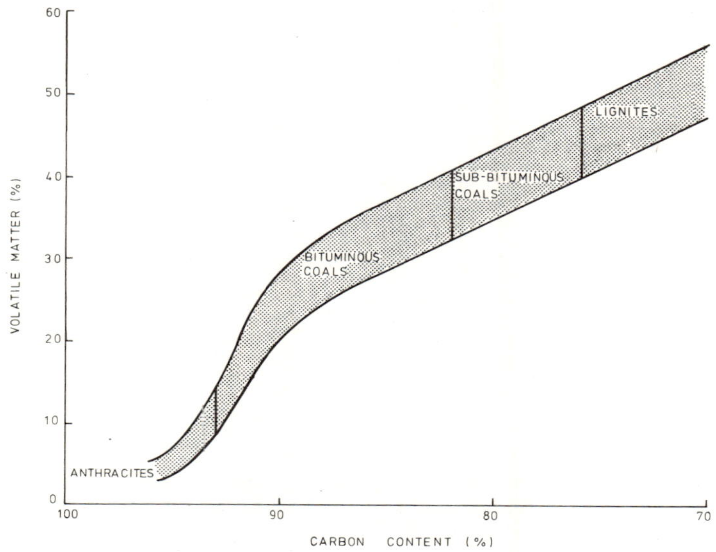

Figure 1.3 Relationship between volatile matter, rank and coal type. The composition is based on the dry, mineral-matter-free coal

It is found, however, that classification by volatile matter alone is insufficient and further tests have to be introduced. These depend on the application and current classification systems are naturally directed towards the present two main markets for coal.

Combustion

For combustion applications the most important parameter is the calorific value of the coal (MJ/kg). This parameter correlates reasonably well with volatile matter and for this reason is excluded from some classification systems.

Coking

When some coals are heated in the absence of air, they pass through a fluid stage during which the particles swell and stick together before forming a solid coke. This behaviour is essential in coals used for manufacturing coke but may cause problems for combustion on a grate. A number of tests based on the caking, swelling and coking properties of coals are used in classification systems.

Although the above characteristics are generally used to describe 'coal types', these may not always provide sufficient information to assess the suitability of a coal for a particular application. Other factors that may be required are as follows.

Minor elements

From an environmental viewpoint, the most important minor element in coal is sulphur because this is released as sulphur dioxide (during combustion) or hydrogen sulphide (during carbonisation). In countries where sulphur-emission controls exist, there may therefore be an economic advantage in using low sulphur content coals in order to avoid the costly processing necessary to reduce sulphur dioxide emissions.

Although organic nitrogen in the coal can contribute to nitrogen oxide formation, so also can atmospheric nitrogen, and probably for this reason the nitrogen content of coals is not as important as the sulphur content.

Small amounts of chlorine are also found in coals and can cause corrosion problems in furnaces if the content exceeds about 0.3 per cent.

Ash characteristics

At high temperatures, coal ash becomes sticky (that is, sinters) and eventually forms a molten slag. This becomes a hard, glassy material on cooling and resolidification. The high temperature behaviour of ash depends on its composition. Ashes with a high silica content and a low content of oxides that act as fluxing agents (iron, calcium and magnesium) require a high temperature to form a mobile slag and vice versa. The ratio of silica to the total content of iron, calcium and magnesium oxides and silica is termed the silica ratio and can be used to predict slag properties. A more commonly quoted parameter, however, is the initial ash fusion temperature as this provides an indication of the onset of sintering. In general, sintering and slag formation are undesirable in combustion processes but are essential in some gasification processes.

Reactivity

Reactivity is less important for current uses of coal but this may change with the introduction of new coal conversion processes such as gasification and liquefaction. The assessment of the suitability of coals for these applications may require that existing characterisation methods are extended to incorporate new tests.

Coal preparation

Coal as mined (that is, 'run of mine' coal) is seldom marketed directly for two reasons.

The ash content is typically up to 40 per cent (estimated United Kingdom value) and such a high ash content is generally uneconomic or unacceptable for combustion markets because the efficiency of the plant or appliance is reduced and the quantity of ash for disposal is increased. For coke manufacture, a high ash content is detrimental to coke quality. A high ash content also increases transport costs because a greater mass of coal has to be moved to supply the same amount of energy. The balance between coal preparation costs to reduce the ash content and transport costs depends on the market and method of transport.

Run of mine coal includes both lump coal and fines but for the domestic and 'premium' industrial markets conventional combustion appliances usually require a sized coal, containing typically lumps ranging in diameter from 13 to 100 mm, to permit the air to flow freely through the firebed. An important advantage is that sized coal is easier to handle than coal containing fines.

The fines can be removed by screening and used in the coal gradings supplied to other industrial consumers and power stations (in the United Kingdom about 70 per cent of run of mine coal is smaller than 25 mm and 20 per cent is smaller than 0.5 mm).

In order to satisfy the economic and other requirements of the markets for coal, the following grades are generally prepared

(1) Part-treated smalls—coal having an ash content of about 17 per cent and a size distribution of 0 to 25 mm and used by power stations and large industrial consumers.

(2) Washed smalls—coal having an ash content of 5 to 10 per cent and a size distribution of 0 to 13 mm and used for coke manufacture and by medium-sized industrial consumers.

(3) Sized coal. The main size gradings are doubles (25 to 50 mm) and singles (13 to 25 mm) with both having an ash content of 5 to 12 per cent.

The size gradings are produced by a complex series of operations in a coal preparation plant. The operations are of four main types: screening to separate out coal lumps of different sizes; crushing to reduce the size of lumps that are too large; washing to separate the coal and mineral matter; and blending to produce saleable grades with the required combination of properties.

Of these operations, probably the most difficult is washing. The principle of washing is to separate coal and mineral matter in an aqueous slurry by relying on the difference in the particle densities; pure coal is typically 1200 to 1400 kg/m^3 and mineral matter is 1800 to 2500 kg/m^3.

In practice, particles exist that are partly coal and partly mineral matter and therefore have an intermediate density. For this reason, in addition to 'clean' coal and reject mineral matter, most washing techniques produce a 'middlings' stream. This may be regarded as coal with a high ash content and is usually incorporated in blends to produce medium ash content coal for power stations and large industrial consumers.

A number of different washing techniques are available, the main types used in the United Kingdom being as follows.

Baum Jig washing

This is the most widely used method of washing. A pulsed flow of water is passed upwards through a bed of raw coal. This causes segregation, the clean coal migrating to the top and the mineral matter migrating to the bottom.

The main advantage of the Baum Jig (and its derivatives) is that it is simple and comparatively cheap. It can handle a full size distribution of feed coal but is sensitive to variations in the properties and size distribution of the feed.

Dense medium washing

Dense medium washing achieves separation of clean coal and mineral matter by allowing the particles to float or sink in a liquid of suitable density, usually about 1500 to 1700 kg/m^3. A slurry of finely ground magnetite in water is most commonly used for this purpose and acts in the same way as a true liquid as far as particle separation is concerned.

Although more expensive to build and operate than Baum Jigs, dense medium wash boxes achieve a more precise separation of coal and mineral matter and have the additional advantage that the separation density can be selected and controlled accurately. Dense medium washing is therefore particularly useful for coals containing a high proportion of middlings particles.

Two designs of dense medium washers exist; dense medium gravity washers are used for large coal (25 mm and above) and dense medium cyclones are used for small coal (0.5 to 25 mm).

Froth flotation

This is the only method used for cleaning coal particles smaller than 0.5 mm.

The technique relies on the surface properties of the clean coal and mineral matter particles. Air is passed through a suspension of the feed coal and water to which has been added a small amount of flotation reagent. The coal particles adhere to the air bubbles and are carried into the froth that forms at the top of the cell. The clean coal is removed with the froth and dewatered by vacuum filtration.

A simplified flow diagram illustrating typical coal preparation processes used to prepare the various grades of coal for the market is given in figure 1.4.

Typically, one or two per cent of the chemical energy in the coal washed is lost as combustible material in the discard. Almost all of these losses are accounted for by low calorific value shale, that is, lumps of material comprising mainly mineral matter but containing a small amount of coal. Losses of high calorific value material (equivalent to the clean coal stream) are generally less than 0.5 per cent of the chemical energy in the coal washed.

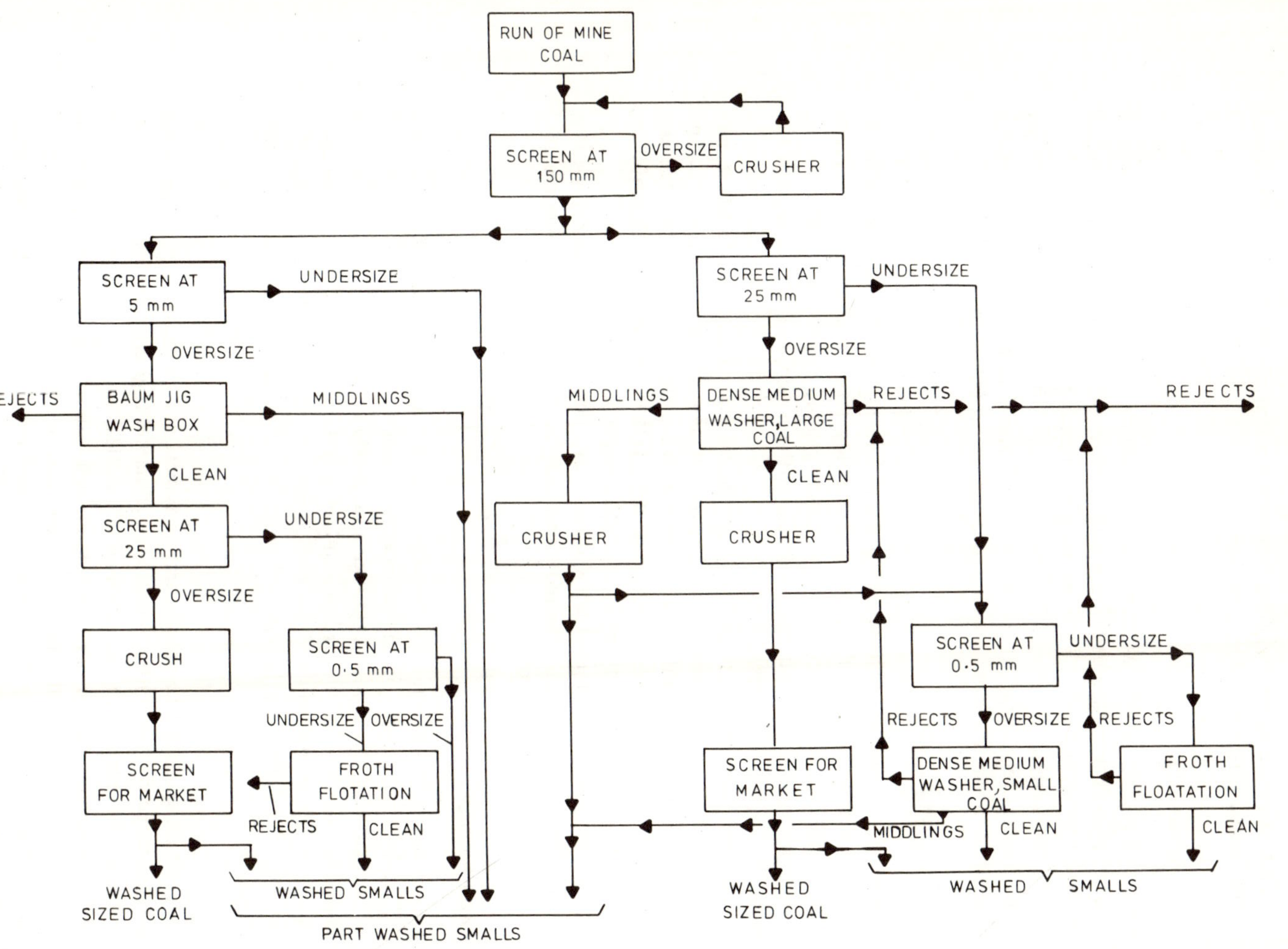

Figure 1.4 Typical coal preparation processes

1.2 The present role of coal

Production

The initial growth in energy demand following industrialisation was met largely by coal and, even as recently as 1950, coal accounted for nearly 60 per cent of world primary energy production.

Traditionally, coal was produced in areas where industry had been introduced at an early stage and where readily accessible reserves of coal existed. In particular, coal production was concentrated at first in North America and Western Europe and in 1950 these two regions accounted for about 65 per cent of total coal production. Since 1950, however, the pattern of coal production world-wide has changed markedly (see table 1.1 and figure 1.5).

In Western Europe, coal production has declined significantly and in North America coal production has remained fairly constant. These two regions now account for only 30 per cent of coal production.

Every other region has achieved a substantial growth in coal production. The increases by centrally planned Europe and centrally planned Asia are especially dramatic and important as these are now the two largest coal-producing regions and together account for nearly 60 per cent of total production.

The combined effect of the above factors has been to produce a modest but steady overall growth in coal production averaging 2 per cent per annum since 1950. However, the growth in total primary energy production has averaged about 5 per cent per annum over the same period with the result that coal's share of primary energy supplies has declined to about 30 per cent.

Consumption

Coal consumption has traditionally been dominated by direct use and coke manufacture and in 1950 these markets accounted for about 60 per cent and 20 per cent respectively of coal consumed.

However, the markets for coal have altered significantly since 1950 (see table 1.2 and figure 1.6) mainly in response to the growth in oil and natural gas availability during the 1960s. Although coke manufacture has grown with coal consumption and has retained its market share, the direct use of coal has remained approximately constant in tonnage quantities and now represents only 35 per cent of total coal consumption. Towns gas production was phased out almost entirely during the 1960s and early 1970s and the manufacture of briquetted fuels has not increased in tonnage quantities.

The main growth area has been electricity generation. In 1950 this accounted for about 15 per cent of coal consumption whereas now 45 per cent of coal produced is used in power stations. There are, however, significant regional differences. In North America and Western Europe more coal is used for elec-

Table 1.1 World coal production[a]

Year	North America	Other America	Western Europe	Centrally planned Europe[b]	Africa	Middle East	Far East	Centrally planned Asia	Oceania	Total production
1950	522	6	486	392	30	3	76	43	21	1579
1951	536	6	511	421	30	3	83	53	22	1665
1952	473	6	523	444	32	3	86	67	25	1659
1953	454	7	519	471	33	4	89	70	23	1670
1954	392	7	526	507	34	4	86	85	25	1666
1955	455	7	531	557	38	4	87	102	25	1806
1956	491	7	538	576	39	4	93	115	25	1888
1957	479	7	541	613	41	5	103	136	25	1950
1958	399	7	531	633	43	5	103	278	26	2025
1959	399	7	511	645	42	5	105	358	26	2098
1960	401	7	499	660	43	4	115	432	29	2190
1961	388	8	494	668	44	4	123	263	31	2023
1962	404	9	498	683	46	5	130	265	31	2071
1963	440	8	496	712	47	5	134	286	32	2160
1964	465	9	500	738	49	6	130	307	35	2239
1965	485	9	480	742	54	6	134	319	39	2268
1966	503	9	452	759	53	6	138	349	41	2310
1967	519	10	429	763	54	6	136	250	42	2209
1968	511	10	414	769	56	6	135	325	47	2273
1969	523	11	396	792	57	6	133	352	52	2322
1970	565	10	381	799	60	6	133	412	55	2421
1971	520	12	379	817	64	7	125	424	53	2401
1972	557	12	340	839	64	7	122	435	64	2440
1973	552	12	340	850	68	7	119	466	65	2479
1974	562	13	316	861	71	8	126	488	69	2514
1975	613	14	331	875	77	8	136	506	74	2634
1976	632	14	308	887	82	8	113	524	80	2648
1977	642	17	306	908	91	8	113	597	82	2764
1978	571	17	304	923	94	8	114	667	85	2783

[a] Sources: *World Energy Supplies* 1950 to 1974 and 1973 to 1978, United Nations. Coal production values are expressed in million tonnes of coal equivalent; one tonne of coal equivalent is defined as 29.3 GJ.
[b] Centrally planned Europe includes the USSR.

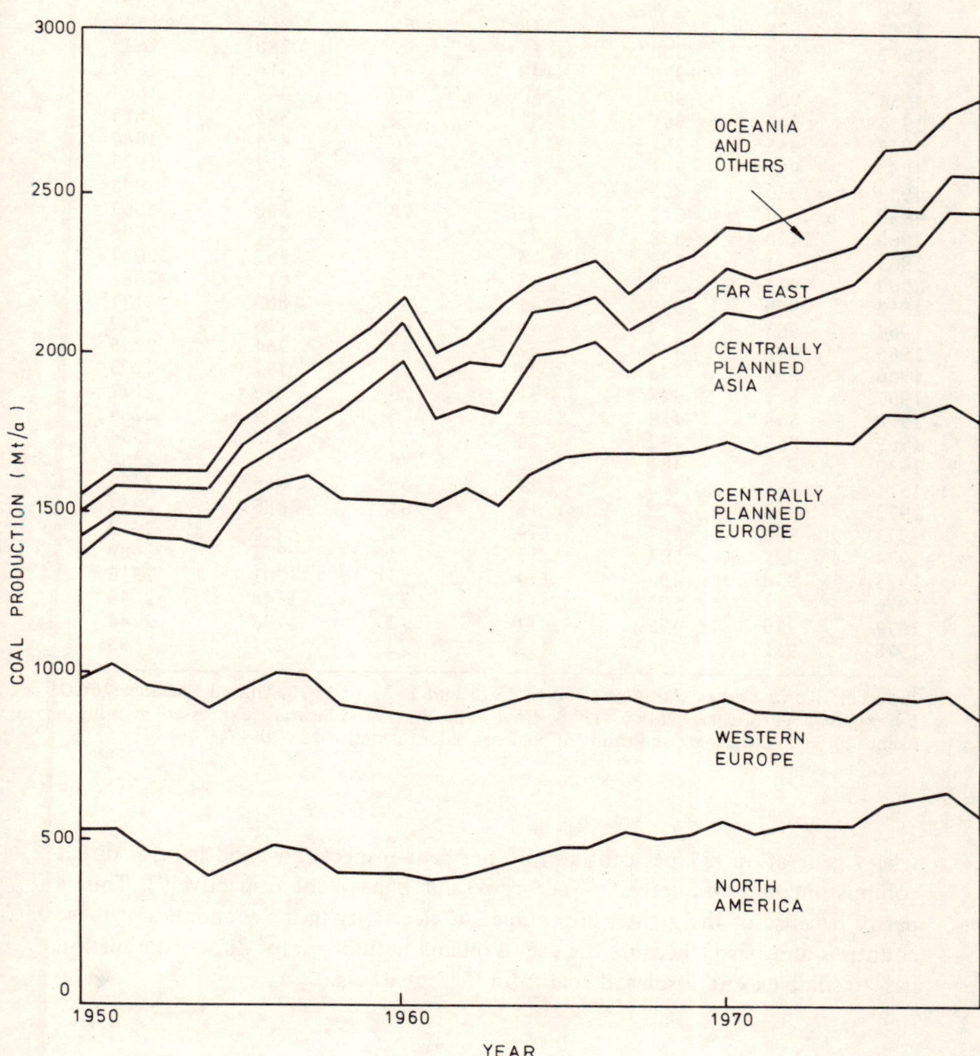

Figure 1.5 Historical trends in world coal production (sources: *World Energy Supplies* 1950 to 1974 and 1973 to 1978, United Nations)

Table 1.2 World coal consumption[a]

Year	Direct use	Coke ovens	Gas works	Briquettes	Electricity	Total consumption
1950	940	264	45	58	227	1534
1951	978	295	48	65	258	1644
1952	939	300	51	67	280	1637
1953	904	325	49	67	314	1659
1954	900	304	50	69	344	1667
1955	949	347	51	72	395	1814
1956	955	367	52	76	434	1884
1957	951	381	51	78	462	1923
1958	1030	365	47	74	479	1995
1959	1040	372	45	70	540	2067
1960	1140	398	45	72	551	2206
1961	971	386	44	74	562	2037
1962	955	390	43	82	611	2081
1963	979	399	44	86	663	2171
1964	961	422	41	82	716	2222
1965	941	436	37	77	764	2255
1966	951	438	34	73	797	2293
1967	811	427	31	70	838	2177
1968	848	438	27	69	904	2286
1969	866	462	24	70	932	2354
1970	932	481	22	70	853	2358
1971	934	470	17	66	865	2352
1972	918	472	15	61	886	2352
1973	899	507	14	74	958	2452
1974	925	504	13	75	972	2489
1975	930	498	14	73	1001	2516
1976	911	503	13	71	1148	2646
1977	970	495	11	72	1196	2744
1978	951	496	12	73	1271	2803

[a]Sources: *World Energy Supplies* 1950 to 1974 and 1973 to 1978, United Nations; *OECD Energy Balance Sheets* 1960 to 1978. Coal consumption values are expressed in million tonnes of coal equivalent; one tonne of coal equivalent is defined as 29.3 GJ.

tricity generation (81 per cent and 66 per cent respectively) and less for direct combustion and briquettes (5 per cent and 8 per cent respectively). This is partly because of the greater importance of electricity in the economies of these countries and partly because the use of oil and natural gas for direct combustion has, to some extent, displaced coal from these markets.

Trade

The 1977 pattern of coal trade is illustrated by the matrix in table 1.3. Each entry in the matrix shows the net amount of coal transferred from one specified region to another. In constructing the matrix, two conventions have been adopted

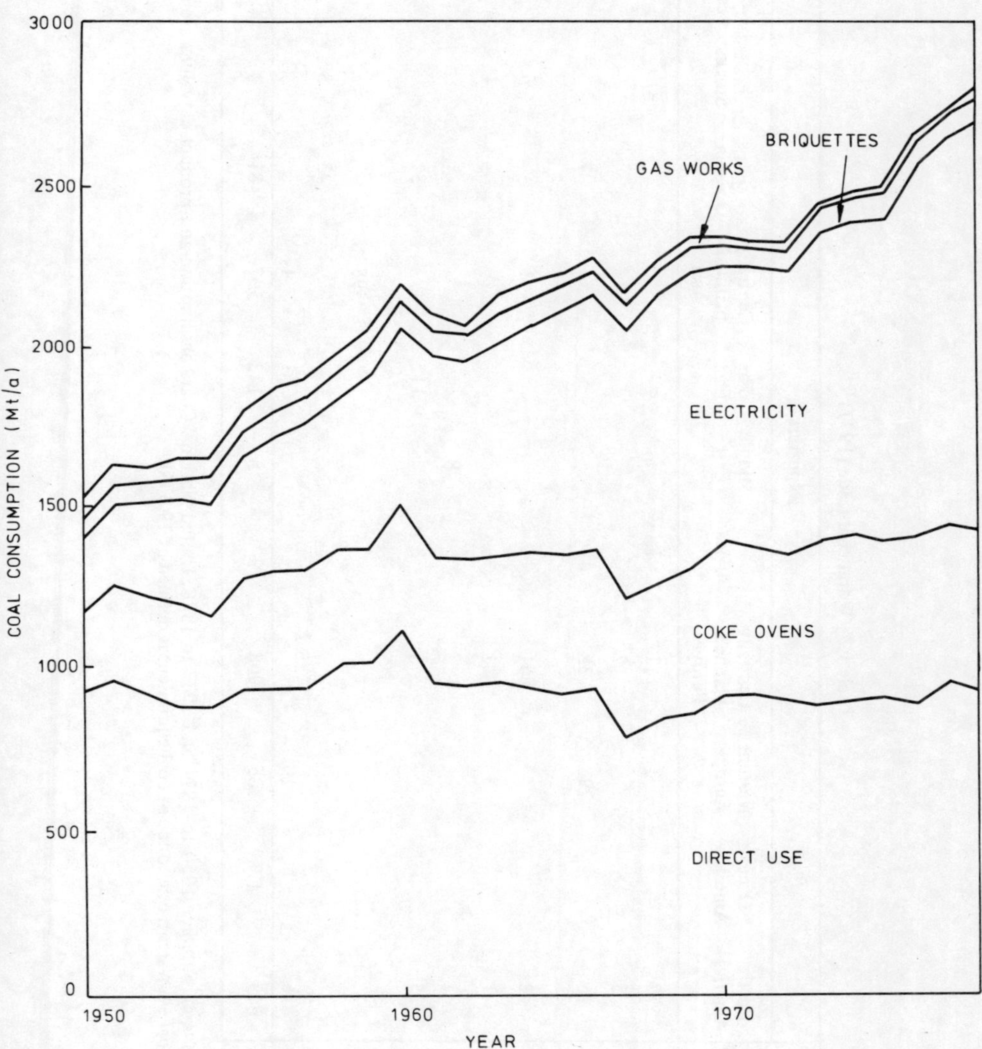

Figure 1.6 Historical trends in world coal consumption (sources: *World Energy Supplies* 1950 to 1974 and 1973 to 1978, United Nations; *OECD Energy Balance Sheets* 1960 to 1978)

Table 1.3 World coal trade (1977)[a]

Importers	Exporters										
	North America	Other America	Western Europe	Centrally planned Europe	Africa	Middle East	Far East	Centrally planned Asia	Oceania	Stocks	Total consumption
North America	577		1	1	1						580
Other America	5	16		2							23
Western Europe	16		295	38	8				7		364
Centrally planned Europe	2			861				1		6	870
Africa	1			1	78					1	81
Middle East	1					8					9
Far East	25			5	3		112		31		176
Centrally planned Asia								595			595
Oceania									43	3	46
Stocks	16	1	10				1	1			
Total Production	643	17	306	908	90	8	113	597	81		2763 / 2744

[a]Sources: *World Energy Supplies* 1950 to 1974 and 1973 to 1978, United Nations. Coal consumption and production values are expressed in million tonnes of coal equivalent; one tonne of coal equivalent is defined as 29.3 GJ.

(1) Diagonal elements represent the coal produced and consumed in the same region.
(2) 'Stocks' can be regarded as an extra region, addition to stocks being similar to 'exports' and retrieval from stocks being similar to 'imports'.

Trade can therefore be regarded as the link between production and consumption, the formal relationship being

production = net trade + consumption + additions to stocks

where

net trade = exports – imports

With the above definitions and conventions, it follows that summing across the rows of the matrix gives the consumption for each region and summing down the columns of the matrix gives the production for each region.

The main trade flows are illustrated in figure 1.7 together with their relationships to production and consumption.

Trade is currently a comparatively minor factor in coal consumption with most coal being used near the point of production. For the breakdown given in table 1.3, coal traded interregionally accounts for only about 5 per cent of total coal consumption. It is, however, generally expected that coal trade will increase significantly in the future, initially to assist substitution by coal for oil in the bulk combustion and steam-raising markets.

The main importing regions at present are Western Europe and the Far East. Imports to Western Europe are mainly from centrally planned Europe and North America whereas the Far East imports mainly from Oceania and North America.

Reserves

The long-term importance of coal as an energy source lies in the extensive nature of the reserves. Recent estimates of world coal reserves are summarised in table 1.4 together with their lifetime at present production rates.

The data in the table refer to reserves that are known to be economically recoverable under present conditions and will almost certainly turn out to be an underestimate for the following reasons

(1) As energy prices increase, known coal reserves that are not economically recoverable under present conditions will become so.
(2) In many parts of the world, known reserves are adequate for the foreseeable future and there is little incentive at present to carry out further exploration to prove new reserves or to increase the size of the resource base. In the future, such exploration may therefore be expected to result in an increase in economically recoverable reserves.

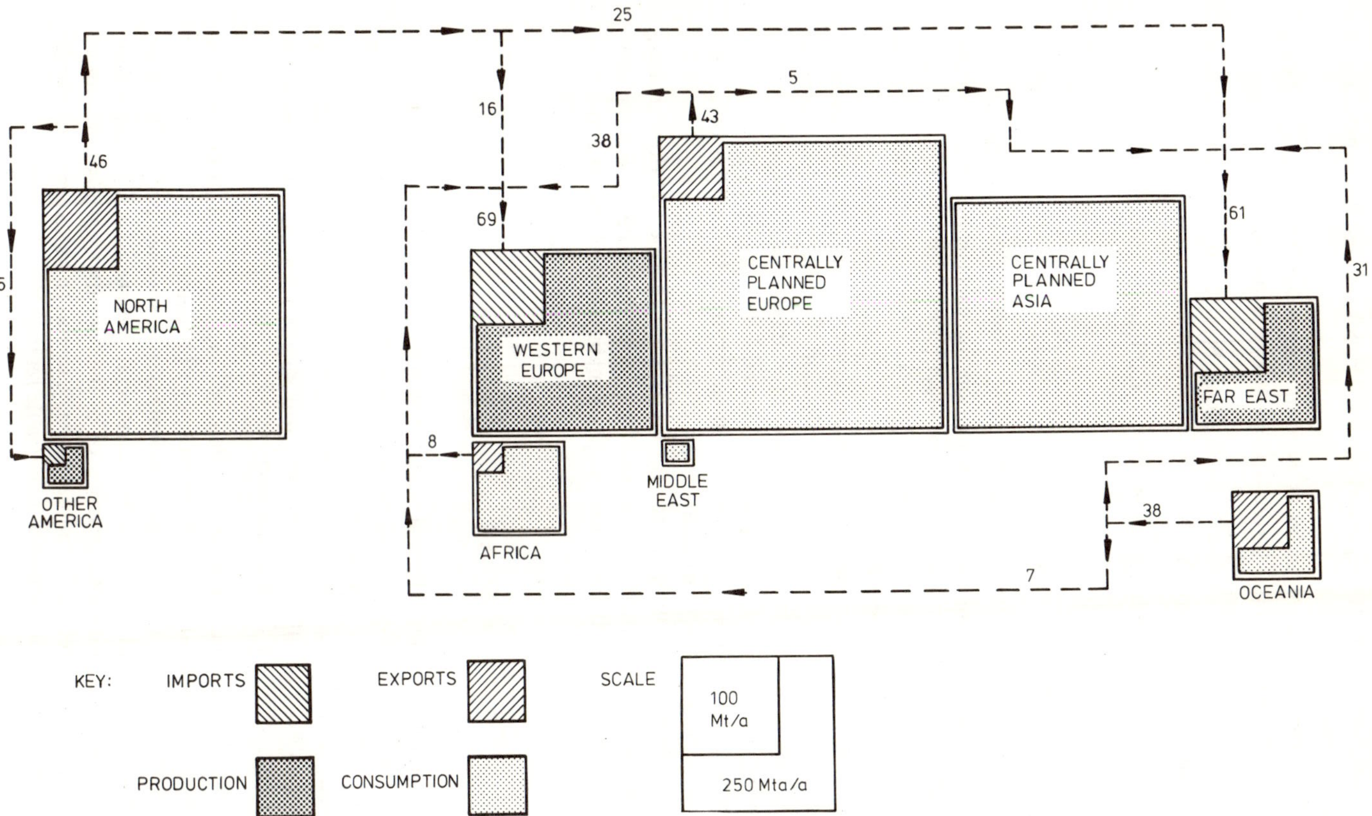

Figure 1.7 World coal consumption, production and main trade flows (in million tonnes of coal equivalent *per annum*, Mt/a) (sources: *World Energy Supplies* 1950 to 1974 and 1973 to 1978, United Nations)

Table 1.4 World coal reserves[a]

Region	Reserves[b] (Gt)	Lifetime[c] (years)
North America	187	327
Other America	6	353
Western Europe	88	289
Centrally planned Europe	175	190
Africa	17	181
Middle East	5	625
Far East	17	149
Centrally planned Asia	112	168
Oceania	28	329
World	635	228

[a] Sources: *World Energy Conference Survey of Energy Resources* 1974, 1976 and 1978.
[b] Reserves are expressed in gigatonnes of coal equivalent; one tonne of coal equivalent is defined as 29.3 GJ. Reserves data refer to coal that is economically recoverable under present conditions.
[c] Lifetimes are calculated from the production rates for 1978 (table 1.1).

As shown in table 1.4, world coal reserves known to be economically recoverable would last over 200 years at present production rates. The global distribution of reserves is roughly paralleled by the current production rates so that lifetimes generally range from 150 to 350 years. In particular, the largest reserves occur in North America and the centrally planned nations of Europe and Asia. The relationships between reserves, lifetimes and production rates are illustrated in figure 1.8 for the main geographical regions.

1.3 Future prospects

A convenient, although somewhat arbitrary, distinction can be drawn between premium and non-premium types of energy demand. Premium energy demands include domestic heating, lighting, etc., road transport and feedstocks for the petrochemicals industry. Non-premium demands are generally for bulk heating applications, for example, in power stations or large industrial boilers. This simple classification clearly has its limitations and some demands, for example, marine transport and the heating of commercial premises, do not fit well into either category. It is, however, useful when considering the future energy market.

Oil and gas can meet both non-premium and premium energy demands and are sometimes referred to as premium fuels. Coal, however, is best suited to bulk

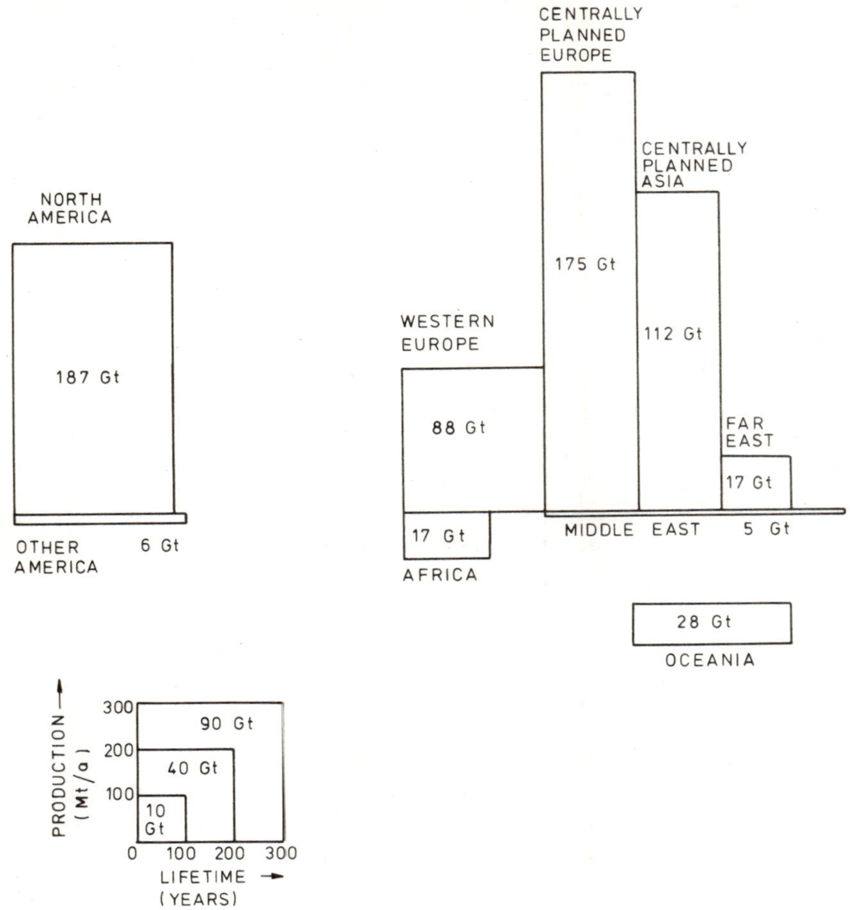

Figure 1.8 World coal reserves and lifetimes at 1978 production rates (sources: *World Energy Conference Survey of Energy Resources* 1974, 1976 and 1978

use in the non-premium sector. In some premium markets, such as road transport, coal does not compete at all because no suitable technology exists whereas in others, such as small-scale heating applications, coal has the disadvantages of being less convenient to transport, store and burn than gas or oil products. These convenience factors can be assessed in economic terms and require coal to be considerably cheaper per unit of energy than the competing premium fuels.

In spite of price increases and relatively low reserves, considerable quantities of oil and natural gas are still used in non-premium markets. For example, about 26 per cent of OECD electricity generation is met by burning oil and natural gas (estimates for 1980).

The first way in which coal can substitute for oil and gas as these fuels become scarcer and more expensive is therefore by direct replacement in non-premium markets. This will enable oil and gas to be diverted from non-premium uses to premium uses so that the same premium demands can be met with a lower total consumption of these fuels. However, the non-premium markets for oil (bulk combustion) generally use heavy fuel oil whereas the premium markets are for 'light' products (for example, gas oil, kerosine, gasoline and LPG). In order for substitution to be effective, therefore, an associated change in oil-refining practice is required.

At present, most of the refineries outside the United States are of the simple 'hydroskimming' type. In these refineries, the complex mixture of compounds present in crude oil is separated into its constituents, giving a high yield of heavy fuel oil, in addition to the light products, as shown in figure 1.9. In the United States, however, the greater demand for transport fuels has led to the use of more complex refineries in which some of the fuel oil is upgraded to light products (also see figure 1.9). Substitution of coal for fuel oil in bulk combustion markets will therefore require the widespread adoption of the advanced refining techniques already being used in the United States, leading to more light products and less fuel oil being produced from the same amount of crude oil. These additional liquid fuels that are made available indirectly by the increased use of coal are sometimes referred to as 'phantom coal liquids'.

The direct replacement of oil and gas by coal can occur in two ways

(1) The installation of new coal-fired combustion equipment instead of oil or gas-fired equipment (either as a replacement for existing equipment or in new facilities). The reintroduction of coal-firing in markets that have enjoyed the low capital costs and high level of convenience associated with liquid and gaseous fuels can be accelerated by improvements in coal combustion technology and this has provided the incentive for an important part of coal utilisation research and development programmes over recent years. An attractive new technology is fluidised bed combustion and is discussed in chapter 3.

(2) Most large, integrated power generation systems have both coal-fired stations and oil/gas-fired stations. Since electricity demand varies diurnally and seasonally, some stations are required to operate most of the time (that is, at a high load factor) while others operate only at times of peak electricity demand (that is, at a low load factor). Clearly, from an economic viewpoint, those chosen to operate most of the time will be those that are cheapest to run. When first built, many oil and gas-fired stations were cheaper to run than coal-fired stations and were therefore used at a high load factor with coal-fired stations being reserved for meeting peak demands. As the relative prices of the fuels have changed, so coal has replaced oil and gas for power generation simply by the duties of the stations being reversed, that is, by coal-fired stations being used at a high load factor and oil and gas-fired stations being reserved for peak demands.

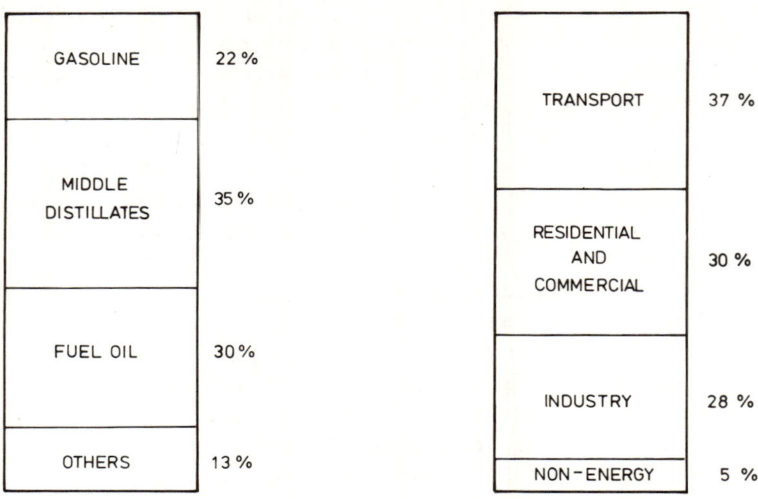

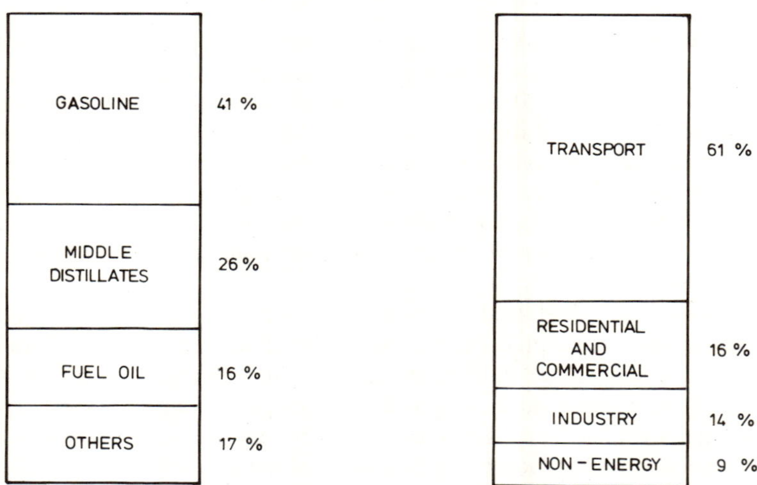

Figure 1.9 Comparison of refinery practice and liquid fuels demand (energy basis) in Western Europe and the United States (sources: *BP Statistical Review of World Energy,* 1981; *Energy Balances of OECD Countries*, 1976 to 1980; *Digest of United Kingdom Energy Statistics*, 1981, H.M.S.O., London)

When direct replacement of oil and natural gas by coal in non-premium markets has achieved saturation, further substitution will be indirect, requiring the manufacture of synthetic liquid and gaseous fuels (referred to as synfuels) from coal. This will enable coal to supplement supplies of oil and gas in the premium markets. While synfuels similar to the natural gas and petroleum-derived liquid fuels will be manufactured, a number of 'unconventional' fuels such as methanol and fuel gas may also find markets. The main options available and the technologies being developed are described in chapter 4 (gasification) and chapter 5 (liquefaction).

As a result of the capital costs of coal conversion plants and their inherent inefficiencies, the cost of liquid or gaseous fuels manufactured from coal is considerably more than the original coal per unit of energy content. At European coal prices, for example, synfuels are expected to cost about three times as much as coal. For this reason, coal can compete with oil and gas more effectively via direct substitution in non-premium markets (particularly in power generation systems where no new equipment is required for the substitution to occur) than through conversion to synfuels for use in premium markets.

It is therefore generally thought that direct substitution for oil and gas by coal will precede indirect substitution. However, the extent of these changes and the time-scale on which they will occur are uncertain.

One of the most important factors is environmental impact and this aspect is considered in chapter 7. A requirement to control pollutant emissions affects the choice between direct substitution by coal for oil or gas (that is, direct combustion of coal at the point of use) and indirect substitution by conversion of coal into clean liquid or gaseous fuels. In general, the larger the plant, the easier and cheaper it is to control pollutant emissions. The world-wide trend over recent years towards increasingly stringent pollution control standards therefore favours the use of clean liquid and gaseous fuels for small plant, with the direct use of coal being reserved for large plant (for example, power stations, large industrial consumers and synfuels manufacture).

A particularly important example is the emission of sulphur dioxide to the atmosphere. In the United Kingdom, a policy of minimising ground-level concentrations by ensuring efficient dispersal of the combustion gases from power stations and industrial combustion equipment has been adopted. Elsewhere, notably in the United States, controls on the rate of sulphur emission from power stations and industrial installations (larger than 25 MW (electrical)) have been introduced.

Environmental considerations in general and the issue of sulphur dioxide emissions in particular have exerted a significant influence on research and development into coal utilisation recently and will be a recurring topic in the following chapters which describe the wide range of technologies that will be required in the future.

The decisions on the direct substitution by coal for oil and natural gas, the introduction of coal conversion processes and the development of new technologies to meet environmental controls will depend on economic factors. Some of the methods of analysis used are described in chapter 8, together with qualitative economic assessments of the ways in which coal can contribute to meeting future energy demands.

CHAPTER 2

HISTORY OF COAL COMBUSTION AND CONVERSION TECHNOLOGY

Coal conversion processes have evolved gradually since the inception of large-scale coal utilisation and even technologies that are now heralded as key elements of the future exploitation of coal have a history that can be traced back over many decades. This history is reviewed briefly in the present chapter for the three mainstream technologies of fluidised bed combustion, gasification and liquefaction. The history of other technologies is considered with the process descriptions in chapter 6.

An overview of the present status of the main coal combustion and conversion technologies is given in appendix 1.

2.1 Fluidised bed combustion

Background

In the traditional methods of coal combustion, a fixed bed of coal is supported on a grate. Air for combustion passes upwards through the coal bed either by natural chimney draught or assisted by a fan.

These early systems gradually evolved into the coal-fired industrial stoker systems used today. In particular, the main designs of stoker can be traced back to the first half of the nineteenth century. The underfeed stoker was introduced in 1816 (John Hawkins and Emerson Davison), the sprinkler stoker in 1822 (John Stanley) and the chain grate stoker in 1841 (John Jukes). The basic designs continued to be improved and refined until the 1960s when oil and natural gas became the preferred industrial fuels.

Pulverised fuel combustion was first used as a means of firing cement kilns; the technique was introduced as early as 1894 by Hurry and Seaman of the Atlas Portland Cement Company. It was not, however, until the 1920s that pulverised fuel combustion was applied successfully to power generation, following the pioneering work of John Anderson on the Oneida Street and Lakeside stations of the Milwaukee Electric Railway and Light Company from 1917 to 1921.

From 1930 onwards, nearly all coal-fired power stations and large industrial boilers have been fired by pulverised fuel rather than by stoker systems because of two principal advantages

(1) Pulverised fuel combustors can accept a wider range of coal types and grades than a stoker.
(2) In practice, stokers are limited to a maximum output of about 30 MW (thermal) whereas pulverised fuel systems can be two orders of magnitude larger than this.

During the 1940s and 1950s, there was considerable interest in the development of cyclone furnaces and a number of units were installed. Like pulverised fuel combustion, the coal is burned in a suspension with air, the main differences being that the furnace geometry is arranged to induce a highly turbulent swirling motion in the combustion air and to facilitate the removal of the coal ash as molten slag. The economics of cyclone furnaces are not generally regarded as favourable and they appear unlikely to have a significant impact on the dominance of stokers and pulverised fuel combustors in the industrial and power generation markets.

Against this background, fluidised bed combustion may be regarded as the first major new coal combustion technology to emerge during the last 50 years.

Early developments

Although the first large-scale application of fluidised bed technology to a chemical process was to catalytic cracking in oil-refineries, fluidisation had been used as long ago as the 1920s in the Winkler coal gasification process. However, it was not until the 1950s that interest in applying the technique to coal combustion began. One of the first such processes was the Ignifluid system in France. This operated at a high temperature so that the coal ash sintered and was removed as a clinker on a travelling grate. Although high combustion rates were attained, few of the other advantages of fluidised bed combustion were exploited.

The development of fluidised bed combustion as known today began in the United Kingdom during the 1960s, the interested parties being the National Coal Board (NCB) and the Central Electricity Generating Board (CEGB). The main market was seen as power generation and the incentive was the development of cheaper power stations to improve the competitive position of coal. By the late 1960s a number of test combustors were operating. Although most of these were atmospheric pressure units, the potential of fluidised bed combustion to drive a gas turbine was recognised and, by 1969, a 3 MW (thermal) pressurised fluidised bed unit was operating at the British Coal Utilisation Research Association (BCURA), Leatherhead, United Kingdom.

The possibility of retaining sulphur in the fluidised bed was identified at this stage and created interest in the United States and other countries where sulphur-emission control standards were beginning to be introduced (for example, West Germany, Canada, Sweden). In the United Kingdom, interest in atmospheric pressure fluidised bed combustion for power generation declined as assessments of market prospects and process economics indicated that, in the absence of sulphur-emission control, there was less incentive than was originally thought. Development of this technology continued, however, in the United States and a 100 MW (thermal) test unit was commissioned at Rivesville, West Virginia in 1977.

Present status

Following the increase in oil prices in 1973, the prospects for coal in the industrial market in the United Kingdom improved dramatically and development effort was concentrated in this area. Until this time, all fluidised bed combustion design concepts had involved crushing the coal to a maximum size of about 3 mm. This did not, however, lead to an entirely satisfactory system for the United Kingdom industrial market and research showed that a fluidised bed could also burn lump coal (for example, singles of 25 to 13 mm or smalls of 13 to 0 mm). Most of the development work in the United Kingdom during the 1970s was directed towards industrial combustors burning lump coal. Applications to water-tube boilers, shell boilers and furnaces are well advanced and, in some cases, equipment is available as commercial units. Work on industrial applications also began in the United States in the mid-1970s and a number of boilers of various designs using both crushed and uncrushed coal are operational.

Pressurised fluidised bed combustion is generally regarded as a more difficult and, therefore, longer-term technology. Development has continued during the 1970s with the construction of test facilities in the United States and the United Kingdom, the largest of which was the International Energy Agency (IEA) collaborative project between the United States, the United Kingdom and West Germany at Grimethorpe, England. The main uncertainty associated with pressurised combustion is whether the combustion gases will be acceptable for use in a gas turbine. Until this has been demonstrated with a commercial-scale turbine, the potential of this technology to provide increased power generation efficiencies cannot be realised.

Recently, interest has grown in a concept called 'fast' fluidised bed combustion. This involves operating in a turbulent fluidising regime at gas velocities of up to an order of magnitude higher than in conventional fluidised bed combustion. The technique is not new, having been employed as long ago as 1942 in the first commercial fluidised bed catalytic cracking plants built in the United

States. Although conventional fluidised beds quickly became established as the standard system for catalytic cracking units, fast beds have been re-introduced in recent developments. Fast fluidised beds have also been used commercially for processes in other fields, notably the Synthol catalytic synthesis gas reactor (first used at the Sasol 1 coal liquefaction plant commissioned in 1955) and the Lurgi aluminium hydroxide calciner (introduced in 1970). However, the application of fast fluidisation to coal combustion is at an early stage of development at present although initial results are encouraging and a number of potential advantages are claimed (see section 3.5).

2.2 Gasification

Early history

The origins of the manufacture of gas from coal date from the end of the 18th century. The incentive for the early work was the production of tar from coal as a substitute for wood tar (see also section 2.3). This process was first practised commercially by Archibald Cochrane, Earl of Dundonald, from 1781 onwards. The possibility of making gas from coal by carbonisation was, however, demonstrated and its advantages for illumination quickly appreciated.

The manufacture of gas from coal was pioneered by William Murdoch who, from 1792, installed gas lighting in a number of factories. The first commercial gas company was founded by Friedrich Albert Winsor in 1812 to supply gas to London.

By the 1820s, gas lighting was available in a number of major cities in Europe and the United States and the industry spread rapidly throughout the remainder of the century. At first, applications were confined to illumination but this market met with increasing competition from electricity after the beginning of the twentieth century. During the first quarter of the twentieth century, gas usage turned from lighting to heating with the result that the gas industry continued to expand.

Early methods of gas production relied on heating coal indirectly (that is, carbonisation), batchwise in cast-iron retorts. The temperature of operation of these retorts was limited to about 600 to 700 °C and their lifetime was often less than a year. In the 1830s, however, horizontal firebrick retorts began to be used and had become widespread by the middle of the century. These had a lifetime measured in years rather than months and permitted operation at temperatures of 900 °C and above thereby increasing the gas yield. In the first decade of the twentieth century, continuous vertical retorts were introduced. Continuous operation resulted in a higher efficiency than the batchwise operation of the horizontal retorts and, for this reason, vertical retorts largely superseded horizontal retorts in the twentieth century. Also at about the turn of the century, the use for domestic gas manufacture of the slot-type ovens developed for metallurgical coke production was adopted in some places and notably in the United States.

Gas manufacture by carbonisation in horizontal or vertical retorts or coke ovens yielded a coke residue that could be used as a solid smokeless fuel and a tar by-product that could be processed further to obtain chemicals or fuels. The yield of gas was low and was typically only 25 per cent of the coal on an energy basis (the yields of coke and tar on an energy basis were about 50 per cent and 5 per cent of the coal, respectively, giving an overall thermal efficiency of about 80 per cent).

The high yield of coke made the economics of gas supply dependent on the value of this by-product. The introduction of cyclic water-gas generators in the 1920s and 1930s enabled the coke by-product to be gasified to supplement the production of gas from carbonisation retorts. Cyclic water-gas generators were essentially batch-type processes in which a fixed bed of hot coke was alternately reacted with air and steam ('blown' and 'steamed'). The 'blowing' part of the cycle heated the coke bed and gave a low calorific value gas called producer gas. This contained nitrogen and was unsuitable for distribution to consumers but could be used for heating the carbonisation retorts. The 'steaming' part of the cycle cooled the bed and produced a nitrogen-free medium calorific value gas called water gas which was mixed with the carbonisation gases for distribution.

In the United Kingdom, where gas coke could be sold as a domestic fuel, comparatively little gas was manufactured using cyclic water-gas generators; their main role was to meet peak gas demands. Elsewhere, notably in the United States, they became the main source of manufactured gas. Integrated complete gasification processes in which coke is transferred hot from a carbonisation retort to a cyclic water gas plant were also developed (for example, Tully, Power Gas, Viag, IBEG) but were not widely adopted.

Continuous gasifiers

The main disadvantages of systems based on retort carbonisation and cyclic water gas generators are low thermal efficiency, low specific throughput and high capital cost. These have led to development efforts in recent decades being concentrated on high efficiency continuous gasifiers. Four types of such gasifier are commercially available and are briefly described below. It is interesting to note that, of these, three (the Winkler, Lurgi and Koppers–Totzek) were developed in Germany, reflecting the strength of that country in this technology.

Fixed-bed gas producers

Atmospheric pressure fixed-bed gasifiers were first developed to manufacture 'producer gas', a low-quality gas used for heating the carbonisation retorts in gas works. From these gasifiers, a number of commercial systems capable of gasifying coal on the scale required for industrial applications have been developed. Examples of such systems include the Wellman (that is, the McDowell-Wellman and Wellman-Galusha), Wilputte, Woodall-Duckham/Gas Integrale, Wellman Incandescent and Foster-Wheeler Stoic.

The Winkler gasifier

This was the first high-throughput gasifier of 'modern' design concept and is based on an atmospheric pressure fluidised bed system. Since its introduction in 1926, seventy Winkler gasifiers have been installed.

The Lurgi gasifier

The Lurgi gasifier, introduced in 1935, is a fixed-bed design and is the only commercially available gasifier operating at elevated pressures (20 to 30 bar). One hundred and forty-two Lurgi gasifiers have been installed or are being installed.

The Koppers-Totzek gasifier

The Koppers-Totzek gasifier, employing an atmospheric pressure entrained-phase reactor, was introduced in 1952. Fifty-four units have been built.

Technical details of the above gasifiers are given in chapter 4.

During the 1950s and early 1960s, considerable research and development efforts were devoted towards new gasifier designs that improved upon the existing systems particularly in the following respects

increased tolerance to coal type and specification
increased throughput
higher overall efficiency (including carbon loss, sensible heat loss and steam utilisation)

With the notable exception of the United States (see below), however, the world's gas industries changed rapidly from manufactured gas to natural gas during the 1960s and early 1970s. As a result of this, interest in coal gasification technology generally declined during that period.

The situation in the United States is somewhat different. The beginnings of natural gas exploitation can be traced from the early part of the nineteenth century and are therefore contemporary with the introduction of manufactured gas. Indeed, as far back as records go, more natural gas than manufactured gas has always been consumed in the United States. However, the two industries grew in parallel throughout the nineteenth century and the early part of the twentieth century and it was only during the late 1940s and early 1950s that manufactured gas was replaced almost completely by natural gas.

Current developments

During the early 1970s, interest in coal gasification revived for the following reasons.

In the United States, the long history of natural gas usage had depleted reserves. This resulted in net annual production exceeding additions to reserves for the first time in 1968 and in net annual production declining from 1973 onwards. Recognition that the reserves to production ratio is only about 10 years, and declining, gave added impetus to plans to develop plants for manufacturing substitute natural gas from coal.

The oil crisis of 1973 brought a general awareness of the limited nature of oil and natural gas reserves but of the much greater reserves of coal. This highlighted the ability of the industrialised nations to reduce their dependence on imported oil by greater use of coal and, in the shorter term, natural gas. Although initially coal would substitute for oil mainly in direct combustion markets, a long-term need to develop coal conversion technology including coal gasification was identified.

Advances in gas turbine technology, coupled with increasing pressure in the United States, West Germany and elsewhere to control pollutant emissions from power stations have led to interest in the development of power generation systems employing gasification as a first stage. This application of coal gasification technology is discussed in more detail in section 4.6.

As a consequence of the above factors, a considerable number of gasification processes are now under development world-wide. The operating principles of the processes and their applications are discussed in chapter 4.

2.3 Liquefaction

Coal liquefaction has a long history that can be traced back to the manufacture of coal tar and the early growth of the chemical industry based on the by-products from gas production. Although these technologies are unlikely to be used in the future, many of the principles on which they are based underlie modern developments and are still relevant to the processes currently under investigation.

Early work on tar production

It has been known since the seventeenth century that liquid tar products can be obtained by heating coal in the absence of air. The Reverend John Clayton of Wigan, for example, reported in a letter to Boyle in about 1664 that tar, liquor and an inflammable gas could be obtained by distilling coal in a retort. In 1681, a British patent for the manufacture of pitch and tar from coal was granted to Becher and Serle. Tar was obtained at that time by condensing the volatile matter evolved when wood was carbonised to produce charcoal. The main use of tar was as a proofing agent for ropes and a protective coating for ships' bottoms. Probably because wood tar was readily available, no immediate application of the early experiments on tar manufacture from coal was attempted.

The first industrial-scale manufacture of tar from coal appears to have been about a hundred years later when the American War of Independence disrupted supplies of wood tar to Great Britain. One of the pioneers was Archibald Cochrane, Earl of Dundonald, who installed a number of coal carbonising plants from 1781 onwards to manufacture tar, ammonia and coke. This and other similar ventures were only of limited financial success and duration.

The effect of gas manufacture

Although this early work was directed towards tar production, it did, however, demonstrate that coal carbonisation could produce an inflammable gas that could be burned for illumination. As discussed in the previous section, the potential of this was recognised by William Murdoch, Friedrich Albert Winsor and others who, from the beginning of the nineteenth century, introduced the manufacture of gas from coal for illumination.

At first, tar was an unwanted by-product from city gasworks and was dumped in rivers or the sea. Soon, however, it was preferred to use it as a fuel (with coke or sawdust) for combustion under the carbonisation retort.

More valuable markets for tar resulted from the introduction of a primitive form of tar distillation in about 1818 following the pioneering work of Accum in London. The crude tar was heated in cast-iron pot stills to obtain a residue of 'refined tar' (or pitch) and a light distillate (naphtha). The 'refined tar' was used instead of wood tar as a proofing agent for ropes and as a protective coating for ships' bottoms. The light distillate was used as a solvent for paints, varnishes and rubber; the latter followed the invention by Charles Mackintosh of the process for waterproofing textiles by impregnating them with a solution of rubber in coal tar naphtha.

Discoveries in the 1830s and 1840s led to the following three new outlets for tar.

The use of coal tar as a binder in road construction was pioneered by J. Loudon McAdam and a number of such roads (known as tar macadam roads) were built in Gloucestershire during the 1830s.

In 1838 John Bethell was granted a patent for a means of preserving wood by impregnating it with creosote, the high boiling part of coal tar distillate. The wooden cross-ties ('sleepers') used in the construction of the rapidly expanding railway network represented an important market for creosote preservatives.

The briquetting of coal fines using the pitch residue of coal tar distillation was first carried out in 1842 by Marsais in France and became the main use of this material.

The birth of the chemical industry

The middle of the nineteenth century marked a change of direction for the coal tar industry. Instead of being principally a source of bulk materials such as creosote, 'refined tar' and pitch, coal tar became the starting point for a wide diversity of high-value chemicals often produced in relatively small quantities.

The foundations were laid by Charles Mansfield who developed a repeated fractional distillation process that was patented in 1847. This made possible the recovery and refining from coal tar distillate of benzole (mainly benzene, toluene and xylene), the material destined to become the basis of the new industry.

The chemical industry can be said to have started in 1856 with the discovery by William Perkin of a purple dye which he called mauvine (mauve). This discovery was quickly followed by others and the manufacture of synthetic dyestuffs grew rapidly to become a major and profitable industry. Other applications including antiseptics, perfumes, pharmaceuticals and explosives, followed throughout the remainder of the nineteenth century.

Although much of the early work was carried out in Britain, Germany took the lead and, by the last quarter of the century, came to dominate the world coal tar chemicals industry. As early as 1880, for example, 80 per cent of all synthetic dyestuffs were manufactured in Germany.

Germany's lead in chemicals production was enhanced by the introduction of a new technology for metallurgical coke manufacture. The traditional method employed beehive ovens. These consisted of a dome-shaped brick kiln in which coal was heated by supplying air to burn the volatiles released. The ovens were not, therefore, suited to the recovery of tar. The problem was overcome by the construction of indirectly heated slot-type ovens for metallurgical coke manufacture. The first such ovens were installed in France by Knab in 1856 but it was not until the ovens were modified by Carves in 1862 that by-product recovery was successful.

The first Knab–Carves by-product recovery ovens were built in Germany in 1881 and, with the incentive of the demand for feedstocks by the expanding chemicals industry, a rapid transformation of coking practice in Germany took place. The introduction of by-product recovery coke ovens elsewhere was much slower. By 1909, for example, 82 per cent of German coke production was in by-product recovery ovens whereas in Britain and the United States the corresponding figures were 18 per cent and 16 per cent respectively.

An alternative to tar distillation as a source of benzole was the gas produced by carbonising coal in retorts at gasworks or in by-product recovery ovens. After removing the tar, more than one per cent (by weight) of benzole remained in the gas as a vapour. A means of extracting this by scrubbing the gas with oil and then distilling the resulting 'wash oil' was patented in 1869 by Caro *et al.* It was found, however, that the removal of benzole from gasworks gas reduced its power as an illuminant and made it unsuitable for distribution. From 1887, however, benzole recovery from coke oven gas (which was used only for heating) was introduced commercially in Germany by Brunck at a plant in Dortmund.

Recent history of tar and benzole manufacture

The onset of hostilities in 1914 underlined the dominant position of the German chemicals industry and emphasised the need for increased benzole production by the Western Allies, particularly for explosives. Benzole recovery from gasworks gas was introduced rapidly as the then widespread use of the incandescent gas mantle (first introduced in 1896) no longer made its presence essential. The replacement of beehive ovens with by-product recovery ovens was also accelerated.

World tar and benzole production continued generally to increase from the 1920s to 1950s reflecting the growth in the coke and gas industries during that period. The traditional markets for tar (for example, road construction, creosote, briquette binders and as a fuel) remained but the main market for benzole from the first World War onwards became the manufacture of motor transport fuels. Since the 1960s, however, world production of tar and crude benzole has remained approximately constant with current levels being about 15.5 Mt/a and 3.5 Mt/a respectively (Mt/a million tonnes *per annum*).

The origins of coal liquefaction

The growth of the petroleum industry during the twentieth century has led to a scale of demand for refined liquid fuels (2270 Mt/a in 1981) that is orders of magnitude greater than that which it is feasible to supply by coal carbonisation. For this reason most of the current interest in coal liquefaction centres on processes that maximise the conversion of the input coal into liquid products. It is, therefore, to the history and status of these processes that most of the remainder of the present section is devoted.

Following the accidental discovery of crude oil by Colonel Edwin Drake in 1859 while drilling for brine in Pennsylvania, production in the United States grew rapidly. The United States became established as the world's major oil producer, a position that it maintained until the expansion of production in the Middle East during the 1950s. In Europe, however, early attempts to find and produce oil were largely unsuccessful and importation was relatively expensive. Under these circumstances, it was natural to consider the plentiful supplies and reserves of coal as a starting point for the manufacture of liquid fuels.

The origins of most coal liquefaction processes at present under development can be traced to work carried out in Germany shortly before the first World War. The incentive to develop coal liquefaction arose mainly from the increasing demand for liquid fuels, particularly for transport applications.

The direct liquefaction route

In 1914, Bergius was granted a patent for a process to convert coal to liquid fuels by hydrogenation in the presence of a solvent oil and hydrogen gas under pressure. The main research work took place during the early 1920s and led to the development of effective catalysts that could be mixed with the coal–oil slurry. The first commercial coal hydrogenation plant was built by I. G. Farbenindustrie A. G. at Leuna and began operating in 1927. The initial capacity was 100 000 t/a of liquid fuels (mainly gasoline) and the feedstocks were brown coal and brown coal tar. The plant was later expanded to a capacity of 650 000 t/a of liquids.

During the period from 1936 to 1943, eleven further hydrogenation plants were commissioned in Germany giving a total synthetic liquids production capacity of more than 4 Mt/a. Various feedstocks were used including brown coal, bituminous coal, tar and pitch. The early plants operated at pressures of up to 300 bar. In order to obtain improved conversions when coal or pitch was the feedstock, the pressure was increased to 700 bar in later plants. The first high pressure plant began operating in 1937 at Welheim.

Although the main developments on coal hydrogenation took place in Germany, experimental work was also carried out in a number of other countries and a commercial-scale plant of 150 000 t/a liquids production capacity was built by ICI at Billingham in Britain. This operated from 1935 to 1939 using coal and creosote as feedstocks. After 1939, the use of coal as a feedstock was discontinued.

The twelve German plants and the ICI plant in Britain were all based on the Bergius technology, involving the direct hydrogenation of a coal–oil slurry in the presence of a catalyst. In 1927 Pott and Broche patented an alternative process that relied on dissolving coal in a solvent and then on filtering the solution to remove the mineral matter and insoluble carbonaceous residue. The mineral-matter-free coal solution could then be hydrogenated for conversion to liquid products using a Bergius-type process or coked to produce special carbons. The only commercial-scale Pott–Broche unit was built at the Welheim plant in Germany (see above). The unit produced 40 000 t/a of coal solution, which was used as a feedstock to the Bergius hydrogenation process.

The indirect liquefaction route

A completely different approach to the production of liquids from coal by hydrogenation (either Bergius or Pott–Broche technology) is the conversion of the coal into a gas followed by the synthesis of the required products using a catalyst.

The development of the gasification/synthesis route started in Germany at about the same time as work on the hydrogenation route; the first patent for synthesis technology was granted to BASF in 1913. The early work on synthesis was carried out at pressures of up to 150 bar in order to obtain a high conversion. However, the resulting product contained mainly oxygenated compounds and was unsuitable as a feedstock for the manufacture of transport fuels. It was found that operation at lower pressures reduced the proportion of oxygenates but also reduced the reaction velocity. More active catalysts were therefore required to make low pressure operation feasible.

During the 1920s, fundamental research on catalysts was led by Fischer and Tropsch who, towards the end of the decade, published a number of classic papers reporting the development of active catalysts. Based on this work, a demonstration plant of 100 t/d (t/d = tonnes/day) capacity was constructed by Ruhrchemie A.G. at Oberhausen-Holten, Germany in 1933.

Successful operation of the demonstration plant led to the construction, between 1933 and 1939, of nine commercial-scale Fischer–Tropsch plants in Germany. These plants had a combined output of 600 000 t/a of liquid products. The gas was manufactured mainly from coke using cyclic water-gas generators (see previous section). The catalyst was cobalt-based and the synthesis reaction was carried out at atmospheric pressure.

During the late 1930s and early 1940s, commercial-scale Fischer–Tropsch plants were also constructed in Japan (four plants producing 260 000 t/a liquids) and France (one plant producing 30 000 t/a liquids). These plants were similar to those built in Germany, using coal or coke as the feedstock for synthesis gas manufacture and a cobalt-based catalyst at atmospheric pressure for the synthesis stage.

Recent developments and current research

Most of the coal liquefaction plants (both hydrogenation and synthesis) built during the 1930s and 1940s were damaged during the second World War. After the war, only the Leuna plant was refurbished. This plant continued to operate on coal until 1959 when it was converted to oil refining.

During the 1950s and 1960s, increasing requirements for liquid fuels were met largely by an expansion of oil production in the Middle East matched by a corresponding growth in the world's tanker fleet and in oil-refining capacity in Western Europe and Japan. The combination of low production costs and economies obtained by the increasing scale of transport and refinery operations led to a steady reduction in crude oil and refined product prices in real terms.

The availability of relatively low-cost oil generally discouraged the construction of coal liquefaction plant and, since the second World War, commercial-scale installations have been built only in South Africa. The original plant (Sasol 1) at Sasolburg was commissioned in 1955 and produces 250 000 t/a of liquid products. A second plant at Secunda (Sasol 2) started operating in 1980 and is producing up to 2 Mt/a of liquid products. A further plant (Sasol 3) of the same design and capacity as Sasol 2 is also now under construction at Secunda. All three plants are based on Fischer–Tropsch technology although they differ in several important respects from those constructed in the 1930s and 1940s. The synthesis gas is manufactured using the standard Lurgi pressurised gasifier and the synthesis reaction takes place at 20 to 30 bar over an iron-based catalyst. The Sasol 1 plant uses two types of synthesis process; a fixed-bed reactor (Arge) and an entrained phase reactor (Synthol). The Sasol 2 and Sasol 3 plants use only the Synthol process.

Although the economic conditions during the 1950s and 1960s did not generally favour coal liquefaction, research interest continued, particularly in the United States, and a number of processes were investigated experimentally. Following the 1973 oil crisis, the research and development programmes gained new impetus and urgency. Many of the less promising processes were shelved so that effort could be directed towards the construction of demonstration-scale plants. Two units of 250 t/d capacity were built in the United States and another unit of similar size in West Germany. Development programmes have also been initiated in a number of other countries including the United Kingdom, Japan, Australia, Canada, Poland and the USSR.

Most of the research and development effort since 1973 has been concentrated on the hydrogenation route. Recently, however, the synthesis route has attracted renewed interest following improvements in catalyst technology such as the Mobil process and greater emphasis is now being given to this option in development programmes. A comparison of the synthesis and hydrogenation processes is included in section 5.11.

CHAPTER 3

FLUIDISED BED COMBUSTION

Fluidised bed combustion (FBC) differs from both conventional stokers and pulverised fuel combustion in being applicable to a wide spectrum of modern coal-fired systems ranging from small industrial boilers and furnaces to large power generation units. The diverse uses of fluidised bed combustion have given rise to a number of different types of combustor design. The main categories are as follows.

Atmospheric pressure FBC boilers

In boiler applications, heat transfer surfaces such as tubes may be placed in the fluidised bed to raise steam or to produce hot water. Further heat may be recovered from the combustion gases using convective heat exchangers.

Atmospheric pressure FBC furnaces

Fluidised bed combustion may be used for furnace applications, for example, to produce a hot gas for drying purposes.

Fast fluidised bed systems

This is a comparatively recent development in which a high combustion intensity is obtained by employing a high air velocity and a method of recirculating the particles entrained in the gas stream. Fast fluidised beds operate in a regime of fluidisation different from the other categories with the bed being highly turbulent but non-bubbling.

Pressurised fluidised bed combustion

A fluidised bed combustor may be operated at elevated pressures, usually of 5 to 15 bar, in order to produce hot pressurised gases that can be used in a gas turbine. Although it has not proved practical to operate a gas turbine with a pulverised fuel combustion system because of erosion, corrosion and fouling of the turbine blades, direct coal firing using a fluidised bed combustor may be possible as a result of the lower combustion temperature.

The fluidisation conditions and chemical reactions for each of the above design concepts are considered together in the first two sections of this chapter.

In subsequent sections, detailed aspects of the combustor design and the prospects for applications of fluidised bed combustion systems are discussed on a case by case basis.

3.1 Fluidisation

A fluidised bed consists of a bed of particles above a grid so that a fluid (air in the case of fluidised bed combustion but, in general, any gas or liquid) can be passed upwards through the bed. The velocity of the fluid is sufficient to separate the particles and to support the bed. If the fluid is a liquid, a uniform separation of the particles occurs whereas, if the fluidising medium is a gas, a bubbling bed is first formed. The remainder of this section refers specifically to gas fluidised bed systems. The main stages of gas fluidisation are illustrated in figure 3.1.

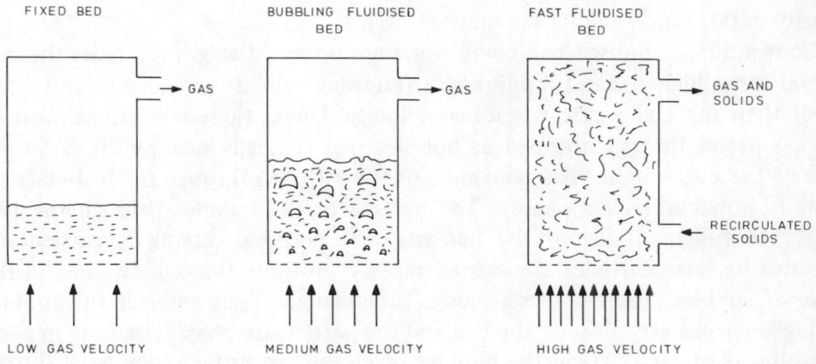

Figure 3.1 Stages of fluidisation

At low gas velocities, the gas flows through the bed without disturbing the particles significantly and the bed remains 'fixed'. As the gas velocity is increased, the force exerted on the particles by the upward passage of the gas increases until a point is reached where the gas stream supports the total weight of the bed. This marks the onset of fluidisation and the gas velocity at this point is referred to as the minimum (or incipient) fluidising velocity.

At gas velocities greater than the minimum fluidising velocity, the excess gas passes through the bed as bubbles and the bed is termed 'fluidised'. The passage of the bubbles through the bed gives the bed the appearance of a boiling liquid. The comparison is more than superficial because the resultant turbulent mixing of the particles provides the bed with some of the properties of a boiling liquid, including good heat and mass transfer and a hydrostatic head. As the gas velocity is increased, the bubbling becomes more violent and considerable splashing above the bed surface occurs.

At still higher gas velocities, the point can be reached where the free-fall velocity of an isolated particle in the gas stream is exceeded. The gas velocity at which this occurs is referred to as the 'particle terminal velocity'. As the gas velocity approaches the terminal velocity, the bed particles become entrained in the gas stream and are carried out of the containing vessel. Steady-state operation of a fluidised bed at velocities approaching or greater than the bed particle terminal velocity is, therefore, possible only if a continuous supply of bed solids is maintained. This is generally achieved by separating the entrained solids from the gas stream using a cyclone and recycling them to the bed. Fluidised bed systems that operate in this mode are known as 'fast' fluidised beds (or, less accurately, as 'circulating' fluidised beds, see section 3.5).

The minimum fluidising velocity and bed particle terminal velocity are dependent on the size of the bed particles, higher velocities being required both to fluidise and to entrain larger particles rather than smaller particles. The relationship between these velocities and the bed particle size is illustrated in figure 3.2 for conditions representative of a fluidised bed combustor (particle density 2500 kg/m^3, air at 1 bar and 850 °C).

Conventional fluidised bed combustors are operated at gas velocities that are several times higher than the minimum fluidising velocity and also several times lower than the bed particle terminal velocity. Under these conditions most of the gas passes through the bed as bubbles that typically occupy 20 to 50 per cent of the bed volume. The remainder of the gas leaks through the bed material (that is, the particulate phase). The passage of the bubbles through the bed causes thorough mixing of the bed material. Vertical mixing is particularly effective because particles are carried rapidly upwards through the bed in the wake of bubbles. There is a continuous interchange of gas between the bubbles during their passage through the bed and the particulate phase, resulting in good gas–solids contacting. When the bubbles reach the bed surface they burst throwing sprays of bed particles into the space above the bed. Some of these particles will be carried out of the system entrained in the gas stream. This process is known as 'elutriation'.

As the gas velocity increases and the operating point moves from minimum fluidisation towards total entrainment, the bubbling becomes violent and the bed expands because more of the bed volume is occupied by bubbles. The preferred operating conditions for a conventional fluidised bed combustor lie in the relatively narrow range denoted as 'good fluidisation' in figure 3.2.

A fast fluidised bed differs from the conventional bubbling bed in a number of respects

(1) The 'bed' fills the containment vessel.

(2) There are no bubbles; instead clusters and strands of particles are turbulently mixed with the gas stream and are continuously breaking up and reforming.

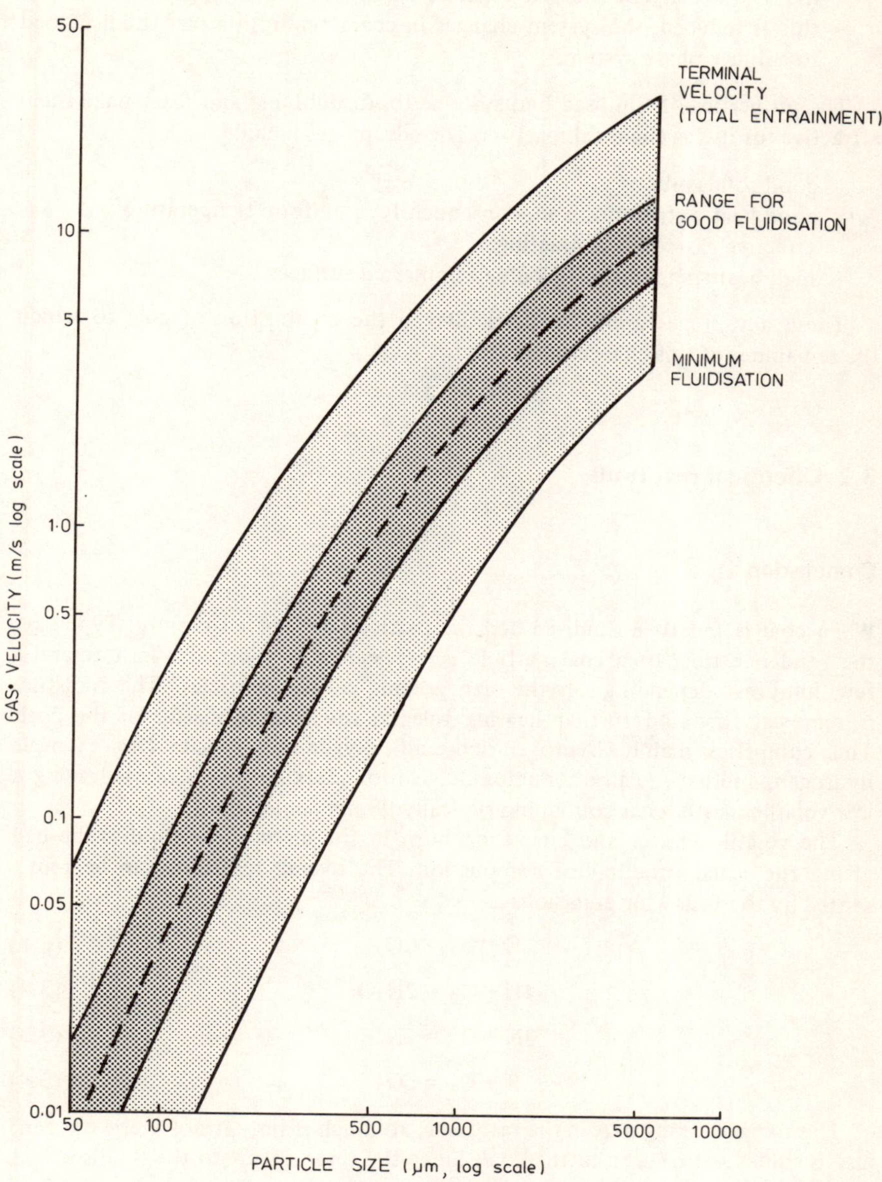

Figure 3.2 Good fluidisation conditions (source: W. V. Battcock and K. K. Pillai, Particle size in pressurised combustors, *Proc. Fifth International Conference on Fluidised Bed Combustion, 1977*, Mitre Corp., Washington D.C.)

(3) The density of the bed depends on the flow rate of recycled solids. If this is reduced, the system changes in character from a fast fluidised bed to a dilute phase system.

The properties of fluidised bed systems (both bubbling and fast) make them attractive for use as chemical reactors. The advantages include

good solids mixing
good heat distribution and, consequently, a uniform temperature
effective gas–solids contacting
high heat transfer coefficients to immersed surfaces

These advantages apply in particular to the combustion of coal to which the remainder of this chapter is devoted.

3.2 Chemical reactions

Combustion

When coal is fed to a fluidised bed the particles are heated rapidly. Typically, the residence time of a coal particle is between a fraction of a minute and a few minutes, depending on the size grading of the particles. The moisture is removed first and further heating releases the volatile matter of the coal. This comprises mainly hydrogen-rich and oxygen-rich species (for example hydrogen, methane, carbon monoxide, carbon dioxide and steam), leaving a low volatile matter char containing typically 98 per cent carbon.

The volatile matter and char then burn in the air used to fluidise the bed giving the usual products of combustion. The overall reactions can be represented by the following equations

$$C + O_2 = CO_2 \tag{3.1}$$

$$4H + O_2 = 2H_2O \tag{3.2}$$

$$2N + O_2 = 2NO \tag{3.3}$$

$$S + O_2 = SO_2 \tag{3.4}$$

The oxygen taking part in the reactions, although mainly atmospheric oxygen, also includes the oxygen in the coal. Since the temperature of the fluidised bed is relatively low (generally 775 to 950 °C, see below), atmospheric nitrogen does not play a significant part in the formation of nitric oxide. In conventional large-scale coal-burning systems (industrial stokers and power station pulverised fuel systems) the peak combustion temperature is considerably higher (usually 1500 to 1800 °C) and under these conditions some atmospheric nitrogen contributes to the formation of nitric oxide. In addition to the products given in the above reactions, traces of other compounds, for example, sulphur trioxide and nitrogen dioxide are also formed.

Under normal operating conditions the volatile matter released as the coal particles heat up will burn completely. Some of the combustion will occur above the bed because part of the volatile matter is inevitably released in a zone where oxygen is not available (near the coal feed point and at the top of the bed).

The reaction in the fluidised bed between oxygen and the char particles remaining after the removal of the volatile matter is constrained by three mechanisms.

(1) The transfer of oxygen from the bubbles to the particulate phase.

(2) Diffusion of oxygen to the particle surface and of carbon dioxide from the particle surface.

(3) Chemical reaction at the particle surface.

For large char particles the rate-controlling step is usually the diffusion of carbon dioxide from the particle surface (mechanism (2)). With fine particles, however, diffusion rates are high and chemical reaction rates (mechanism (3)) become controlling. Typically, the changeover from diffusion to chemical rate control occurs at particle sizes in the range 50 to 100 microns. Since chemical reaction rates increase more with temperature than diffusion rates do, the changeover occurs at a smaller particle size with a higher bed temperature and vice versa.

The bed mainly consists of non-combustible inert material, the carbon concentration in the bed being generally less than 1 per cent if crushed coal is used and less than 5 per cent for larger size gradings. These values provide sufficient surface area for the rate of reaction to be consistent with the rate of oxygen supply. It is the low carbon concentrations in the bed that provide a fluidised bed combustor with its tolerance to the caking properties of the coal. The average temperature at the surface of the burning char particles is up to about 100 °C higher than the bed temperature; this permits the necessary rate of heat transfer from the char to the bed.

As the char particles burn they reduce in size until some become entrained in the gas stream and are removed from the combustor by elutriation. The size of the entrained char particles is mainly between 10 and 100 microns so that most of this material is able to be collected using cyclones. Some loss of char can also occur with the removal of excess bed material to control the bed depth but because of the low char concentration in the bed this loss is small compared with that by elutriation. The total char loss typically represents 2 to 15 per cent of the thermal input of coal to the combustor. However, values above 4 per cent are usually unacceptable commercially and the design must incorporate features to improve the combustion efficiency to an acceptable level (see below).

In general, not all of the char combustion occurs in the fluidised bed. As bubbles burst at the bed surface, sprays of bed particles are thrown into the space above the bed (the freeboard) and combustion of the char continues in this region. A proportion of the heat released by combustion of both char and volatiles in the freeboard returns to the bed as sensible heat in the particles that fall back. The remainder is responsible for heating the combustion gases above bed temperature by as much as 150 °C in some systems. In pressurised combustors, where the gas is cleaned before cooling, combustion may also occur in the cyclones assisted by the good mass transfer properties induced by the high gas velocities.

Sulphur retention

For environmental reasons (see chapter 7) it may be desirable to remove the sulphur dioxide formed from the sulphur in the coal (reaction 3.4) before emitting the combustion gases to the atmosphere.

In conventional combustion systems (stokers or pulverised fuel combustors) sulphur dioxide can be removed only after combustion by scrubbing the stack gases (see section 6.7). The capital costs and efficiency loss associated with such processes can be significant.

Fluidised bed combustion, however, offers the possibility of removing the sulphur dioxide during combustion at comparatively little extra cost by adding limestone or dolomite to the fluidised bed. This retains the sulphur released as calcium sulphate which can be removed with the ash. The course of the reactions involved depends on the operating conditions and in particular on the pressure.

At atmospheric pressure, limestone and dolomite calcine according to the reactions

$$CaCO_3 = CaO + CO_2 \tag{3.5}$$

$$CaCO_3.MgCO_3 = CaO.MgO + 2CO_2 \tag{3.6}$$

The calcium oxide then reacts with sulphur dioxide to form the sulphate, as follows

$$CaO + SO_2 + \tfrac{1}{2}O_2 = CaSO_4 \tag{3.7}$$

$$CaO.MgO + SO_2 + \tfrac{1}{2}O_2 = CaSO_4.MgO \tag{3.8}$$

At high pressures, calcium carbonate is stable at the temperatures used in fluidised bed combustion and does not calcine. As a result, the calcination of dolomite takes the form

$$CaCO_3.MgCO_3 = CaCO_3.MgO + CO_2 \tag{3.9}$$

Limestone and partially calcined dolomite react with sulphur dioxide according to the equations

$$CaCO_3 + SO_2 + \tfrac{1}{2}O_2 = CaSO_4 + CO_2 \tag{3.10}$$

$$CaCO_3.MgO + SO_2 + \tfrac{1}{2}O_2 = CaSO_4.MgO + CO_2 \tag{3.11}$$

As indicated by the above equations, magnesium sulphate does not form when dolomite is used as the sulphur acceptor, either at atmospheric pressure or at high pressures. This is because magnesium sulphate is not stable at fluidised bed combustion temperatures. It is tempting to conclude that limestone would be preferred to dolomite because a smaller mass of material is required to retain a given amount of sulphur and this is indeed true for atmospheric pressure operation.

For operation at high pressures, however, limestone is found to be unreactive and less effective. Since limestone does not calcine at high pressures the porosity of the stone remains low and the gas cannot penetrate easily to the interior of the particles. After the formation of a surface layer of sulphate the reaction effectively ceases because no more active sites are accessible. When dolomite is used the calcination of the magnesium component opens up the pore structure so that the reactivity of the stone is high and sulphation is not limited to the exterior surface of the particles.

In general, therefore, limestone is the preferred sulphur acceptor at atmospheric pressure and dolomite at high pressures. At intermediate pressures of up to about 4 bar, but depending also on the excess air level, some calcination of calcium carbonate will take place and both types of acceptor are equally effective.

Acceptor regeneration

In most fluidised bed combustor designs that incorporate sulphur retention the limestone or dolomite is used on a 'once through' basis so that the calcined and partially sulphated stone is disposed of after removal from the combustion system.

In order to reduce the environmental impact of mining and disposing of acceptor materials and also to save the associated costs, various schemes for regenerating the spent stone to the oxide or carbonate have been proposed, although none have been implemented on a large scale. The main options are as follows.

Thermal decomposition

At temperatures above about 1000 °C calcium sulphate decomposes as follows

$$CaSO_4 = CaO + SO_2 + \tfrac{1}{2}O_2 \tag{3.12}$$

This reaction could, for example, take place in a bed fluidised by air and in which sufficient fuel is burned to maintain the required temperature.

Subsequent processing would then be required to remove the sulphur dioxide from the combustion gases although the gas volume involved is comparatively small.

Reduction

Calcium sulphate can be reconverted to calcium carbonate (which may calcine to calcium oxide) using carbon monoxide. A number of reactions are possible as, for example, indicated below

$$CaSO_4 + CO = CaCO_3 + SO_2 \tag{3.13}$$

$$CaSO_4 + 4CO = CaS + 4CO_2 \tag{3.14}$$

$$CaSO_4 + 4CO + H_2O = CaCO_3 + H_2S + 3CO_2 \tag{3.15}$$

The carbon monoxide required has to be generated externally by a gasification process and the operating conditions have to be chosen carefully to avoid an undesirable mixture of products (for example, the conversion of $CaSO_4$ to $CaCO_3$ and CaO can be hindered by the formation of CaS).

Bed composition

To a first approximation the mineral matter present in the coal can be regarded as an inert in the combustion process. However, some chemical changes do take place and usually result in a weight loss. For example, water of crystallisation is removed and carbonates and sulphides decompose to form oxides. The residue after these changes is referred to as 'ash' to distinguish it from the 'mineral matter' originally present.

The net effect of the combustion and sulphur retention reactions is to provide a bed derived from the following components

(1) coal (present mainly as particles of low volatile char, accounting for less than 1 per cent of the bed material if crushed coal is used and less than 5 per cent for lump coal)
(2) limestone or dolomite (present mainly as calcined and partially sulphated stone)
(3) coal ash (see above)
(4) inert additive (for example, sand)

An inert additive would be used only if insufficient ash, limestone or dolomite were being fed to maintain the design mass of bed material. This occurs, for example, when a washed coal containing only fine inherent ash is burned without limestone or dolomite addition.

3.3 Atmospheric pressure FBC boilers

General description

A schematic diagram of a conventional atmospheric pressure fluidised bed boiler is shown in figure 3.3.

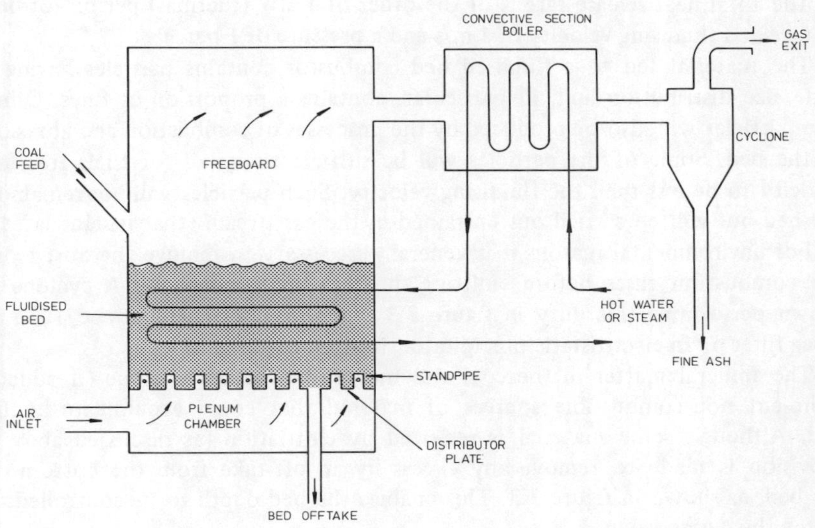

Figure 3.3 Atmospheric pressure fluidised bed boiler

The bed is fluidised by the air required for the combustion of the coal. The air is introduced into the plenum chamber and passes into the bed via the distributor plate. This allows the upward passage of the air but does not allow the bed particles to fall through. Coal is fed to the bed where it burns in the fluidising air. Heat is removed to produce hot water or to raise steam typically by heat transfer tubes immersed in the bed. As a result of the turbulent nature of fluidisation, the heat transfer coefficients between the bed material and the immersed tubes are generally high in fluidised beds as compared with heat transfer coefficients in convective heat exchangers. The bed temperature is maintained in the range 775 to 950 °C and is therefore considerably lower than the operating temperatures of other coal-fired combustion systems.

The combustion gases leave the bed, pass through the freeboard, and exit near the top. As bubbles burst near the bed surface, sprays of bed material are thrown into the freeboard. The purpose of the freeboard is to allow sufficient space above the bed for these particles to disengage from the gas stream and to fall back to the bed.

The gases leave the combustor at approximately the bed temperature; the exact value depends on the degree of combustion and heat transfer in the freeboard. The gases then pass into a convective section where further heat is recovered and the gases are cooled to the required stack exit temperature (usually 150 to 200 °C).

The total heat release rate is of the order of 1 MW (thermal) per m^2 of bed surface at a fluidising velocity of 1 m/s and a pressure of 1 bar.

The material fed to a fluidised bed combustor contains particles having a wide size distribution and, in particular, contains a proportion of fines. Other fine particles will also be produced by the processes of combustion and abrasion in the bed. Some of the particles will be sufficiently small for their free-fall velocity to be less than the fluidising velocity. Such particles will not remain in the bed but will be carried out entrained in the gas stream (that is, elutriated).

For environmental reasons it is generally necessary to remove the dust from the combustion gases before emitting them to the atmosphere. A cyclone is shown performing this duty in figure 3.3 but a more efficient device, such as a bag filter or an electrostatic precipitator, is often required as well.

The mineral matter in the coal and the sulphur acceptor stone (if added) represent non-combustible sources of material that could accumulate in the bed. Although some material is removed by elutriation (as discussed above), provision is made to remove any excess by an off-take from the bottom of the bed, as shown in figure 3.3. This enables the bed depth to be controlled to match the design value.

The main design parameters for a fluidised bed combustor are the bed temperature, excess air level, calcium/sulphur ratio, bed depth, bed particle size and fluidising velocity. These are discussed in subsequent sections in terms of the following parameters that reflect the combustor performance or cost

combustion efficiency
sulphur retention efficiency
in-bed heat transfer surface
bed area
bed pressure drop

The main interactions are summarised in table 3.1.

Table 3.1 Effects of design parameters[a]

Parameter	Combustion efficiency	Sulphur retention	In-bed heat transfer surface	Bed area	Pressure drop
Bed temperature	+	Optimum	–	0	0
Excess air level	Optimum	+	–	+	0
Ca/S ratio	0	+	0	0	0
Bed depth	+	+	0	0	+
Bed particle size[b]	–	–	+	–	0
Fluidising velocity[b]	–	–	+	–	0

[a]Notation: + indicates an improvement or advantage, 0 indicates no change, – indicates a reduction or disadvantage, when a parameter is increased.
[b]It is assumed that the bed particle size and fluidising velocity are changed together (see figure 3.5).

Other aspects of design (coal feeding, air distribution, fines collection, start-up and part-load operation) are discussed in later sections.

Temperature

The operating temperature of a fluidised bed combustor usually lies in the range 775 to 950 °C. The upper limit is determined by the need to avoid ash sintering, that is, the tendency of particles of ash to stick together at high temperatures. The agglomerates thus formed can lead to segregation of the bed and the cessation of fluidisation. Towards the lower limit of the range, the loss of unburned char by elutriation increases as does combustion above the bed. Operation at the highest temperature consistent with avoiding ash sintering is, therefore, desirable in order to maximise the combustion efficiency and the heat release within the bed.

High-temperature operation also has the advantage of increasing the rate of heat transfer to immersed tubes although the amount of heat available in the bed decreases (because more heat is removed by the combustion gases).

The bed temperature also has an important effect on the sulphur retention efficiency. At low bed temperatures, sulphur retention is inhibited by poor calcination of the stone and at high bed temperatures by decomposition of the sulphate. As a result there is an optimum temperature for sulphur capture at about 850 °C. If, therefore, sulphur retention is required and the fuel has a high sulphur content, the retention efficiency is usually the main consideration determining the bed temperature.

Excess air level

The excess air level (usually expressed as a percentage) is defined as the air supplied in addition to the stoichiometric air requirement (that is, the air required just to burn the fuel). Operation at a high excess air level increases the mass flow rate of hot stack gases emitted to the atmosphere and consequently reduces the boiler efficiency (as shown in figure 3.4). The efficiency loss increases linearly with the excess air level. Operation at a low excess air level, however, leads to a high loss of combustible material (principally unburned char entrained in the gas stream although some combustible gases may also be produced as stoichiometric combustion conditions are approached).

The above factors result in the existence of an optimum excess air level for maximum boiler efficiency, as shown in figure 3.4. The position of the optimum varies with the other combustion design parameters (for example, bed temperature, bed depth and fluidising velocity) but usually lies in the range 10 to 40 per cent excess air.

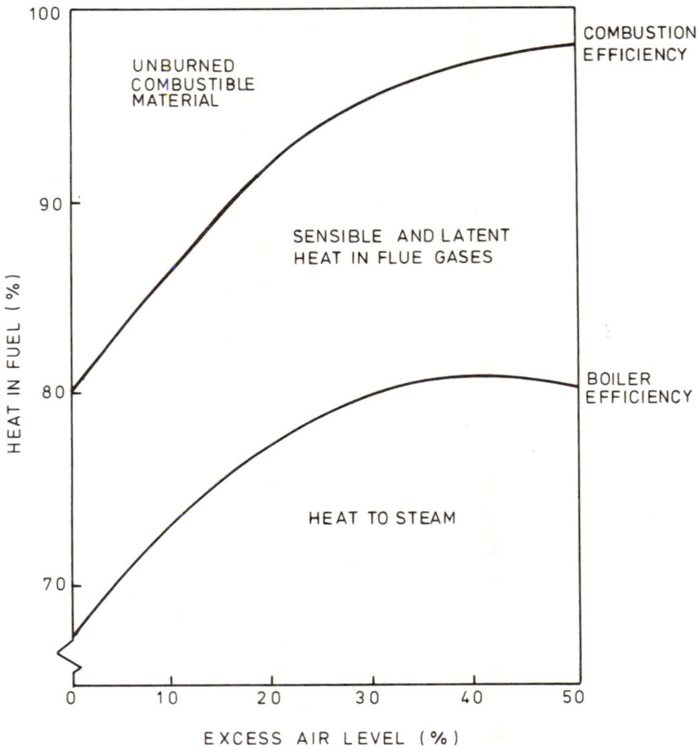

Figure 3.4 Effect of excess air level on combustion and boiler efficiencies

The choice of excess air level affects the requirements for heat transfer surface in the bed. The higher the excess air level, the greater is the amount of heat that is removed from the bed by the combustion gases and the less that is available for transfer to tubes in the bed or to the bed containment. For a system fluidised by air (without preheat) all of the heat would be removed by combustion gases if about 150 per cent excess air were used. At an excess air level of 30 per cent, approximately half of the heat released is therefore removed by the combustion gases and half by heat transfer from the bed.

In industrial boilers, an excess air level of 30 to 50 per cent would normally be used, giving a boiler efficiency of about 80 per cent. For power station applications, however, high efficiency designs are favoured (in spite of their higher capital cost) and an excess air level of 15 per cent would be typical.

Calcium/sulphur ratio

The proportion of the sulphur in the coal that can be retained as calcium sulphate by the addition of an acceptor to the bed depends on the feed rate of the acceptor. The feed rate is usually expressed as the calcium to sulphur mole ratio; a value of 1 corresponds to the stoichiometric calcium feed rate. For limestone, the stoichiometric feed rate is 3.1 per cent of the coal feed rate for every 1 per cent of sulphur in the coal. To retain 85 per cent of the sulphur, for example, would typically require a calcium to sulphur mole ratio of 2 to 4 with the exact value depending on the other design parameters. The reactivity of the limestone is also extremely important and can in itself require a change in the calcium to sulphur mole ratio of a factor of two.

Bed depth

The bed depth for fluidised bed boilers is of the order of 1 m if sulphur retention is required otherwise 0.15 to 0.5 m is usually employed.

The minimum bed depth that may be used in a design is determined by two factors. Firstly, there must be sufficient free space in the bed for lateral mixing to occur in order to avoid temperature gradients and secondly, the bed has to accommodate the tube bundle (if included) without too close a packing being required.

Although a bed depth greater than the minimum may be used, a deep bed adds to the pressure drop and therefore to the fan power requirements. Since fan power is an important operating cost consideration in most boilers, it is usual to design for the minimum bed depth.

As the bed depth is reduced, however, the combustion and sulphur retention efficiencies decrease and the amount of combustion in the freeboard increases. In some cases, it is therefore necessary to design for more than the minimum bed depth specified above in order to achieve the required efficiency targets. In particular, it is not yet known whether effective sulphur retention can be obtained in the shallow beds necessary for use in horizontal shell boilers (although it would inevitably require more limestone than in deeper beds).

Bed particle size

The mean particle diameter of the bed material is usually in the range 0.5 to 1.5 mm, the value depending mainly on the choice of fluidising velocity. A high fluidising velocity requires a coarser bed in order to obtain good fluidising conditions and to avoid excessive carry-over of bed material. The relationship between mean bed particle size and fluidising velocity is given in figure 3.2 and is replotted in figure 3.5.

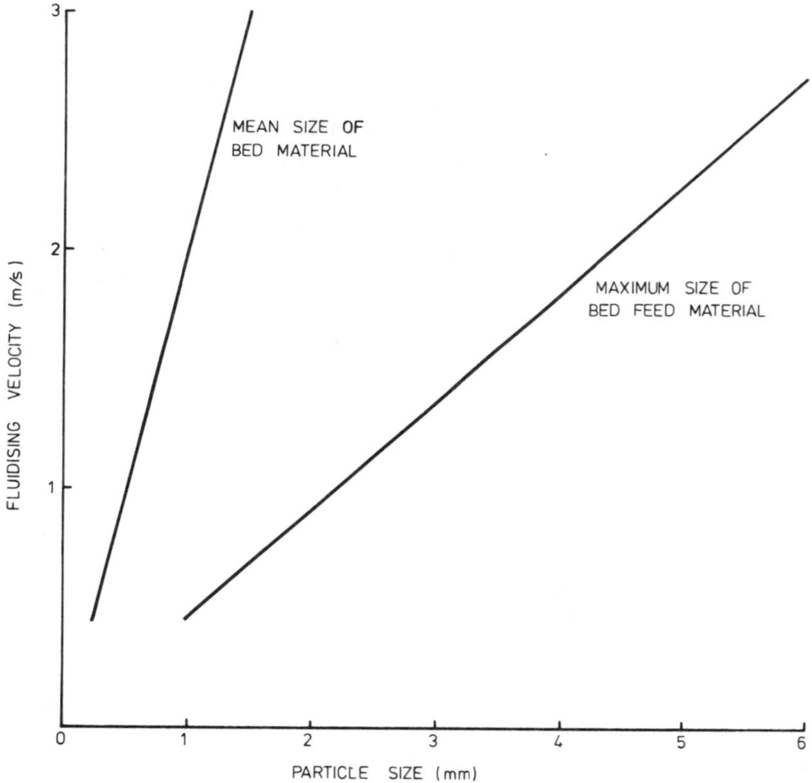

Figure 3.5 Typical fluidised bed combustor conditions

The particle size of the bed material is determined by the particle size of the coal ash, sulphur acceptor or inert additive used to maintain the bed. These feed materials usually have a wide size distribution and, in order to obtain the required mean bed particle size, the maximum size of bed feed material is usually in the range 2 to 6 mm. The relationship between the size of bed feed material and the fluidising velocity is also given in figure 3.5.

The mean bed particle size has a significant effect on the heat transfer coefficient to immersed tubes. After subtracting the radiative heat transfer component (generally small) the remaining convective heat transfer coefficient varies approximately as the inverse square root of the mean bed particle size. Fine beds therefore require less heat transfer surface than coarse beds. Typically, convective heat transfer coefficients of the order of 200 W/m^2 K are obtained with a mean bed particle size of 1 mm.

Fluidising velocity

For most designs of fluidised bed boiler the fluidising velocity lies in the range 1 to 3 m/s.

The choice of fluidising velocity is of fundamental importance as it affects most of the other design parameters. The optimum value may be regarded as a compromise between the capital cost, the bed pressure drop and the efficiencies of combustion and sulphur retention. A high fluidising velocity reduces the bed area required for a given output and thereby decreases the size of the bed containment and distributor plate and the number of coal feed points. Each of these is an important cost factor and capital cost considerations therefore favour a high fluidising velocity. However, a high fluidising velocity results in a high heat release per unit plan area of the bed and necessitates a deeper bed in order to accommodate the additional heat transfer surface. As discussed earlier, this increases the bed pressure drop and therefore increases fan power requirements and operating costs.

A high fluidising velocity also leads to more combustion in the freeboard and increases the carry-over of unburned char and unconverted sulphur acceptor thereby reducing the combustion and sulphur retention efficiencies. If the reduction in performance is unacceptably large, design changes such as refiring the elutriated material may become necessary.

Start-up

Before it is possible to burn coal in a fluidised bed, the bed temperature must be raised to about 600°C by an auxiliary heating system. Two main types of start-up system are used.

Combustion above the bed

In the simplest above-bed combustion system the bed is fluidised with cold air while oil or gas burners in the freeboard direct flames onto the bed surface. Alternatively, gas may be pre-mixed with the fluidising air and lit above the bed. Although such systems have a relatively low capital cost, only a part of the auxiliary fuel is used to heat the bed with the remainder being transferred to in-bed tubes or to the fluidising air. As a result, the rating of the start-up system can approach that of the fluidised bed combustor and auxiliary fuel consumption is high. In this case, the start-up system can, of course, be used as a continuous firing system for the boiler, if required.

Hot gas start-up

Alternatively, initial heating can be provided by passing hot gas through the fluidised bed. This approach is more efficient than above-bed combustion because the gas velocity can be kept sufficiently low to avoid fluidisation until the mean bed temperature has reached 400 °C. Comparatively little heat is absorbed by the in-bed heat transfer surface or is lost with the gases leaving the bed and auxiliary fuel requirements are correspondingly low. However, the design of the plenum chamber and distributor plate has to take into account the high temperatures to which these components are exposed during start-up. For example, the plenum chamber would be refractory lined and the distributor plate either manufactured from heat-resistant steel or water-cooled. These measures can add significantly to the capital cost.

The start-up systems discussed above are needed only when the initial bed temperature is less than 600 °C. It is usually possible to shut down a fluidised bed for several hours before the bed temperature falls from an operating value of, say, 850 °C to 600 °C. Restarting the bed within this period can therefore be achieved without auxiliary fuel and is sometimes referred to as a 'warm start'.

Coal feeding

Depending on its size grading, coal may be fed to a fluidised bed either by a pneumatic transport line to the bottom of the bed (this is the preferred method for feeding crushed coal) or mechanically to the top of the bed (this approach would be used for feeding lump coal, that is, doubles, singles or smalls). The second of these two options is illustrated in figure 3.3.

In the early stages of fluidised bed combustion development, work concentrated on design concepts for firing coal crushed to a maximum particle size of 2 to 6 mm. In such cases the maximum coal particle size corresponds to the desired maximum particle size for the bed material ensuring that both coal particles and ash particles are of a size compatible with the fluidising conditions and that they would neither be removed rapidly by elutriation nor cause segregation and loss of fluidisation. In particular, this system was considered suitable for coals with a high stone content where the crushed stone particles, which are of a similar size to the coal particles, can conveniently form the inert bed material. Crushing, however, inevitably produces many fine coal particles that contribute significantly to the carry-over of unburned char.

In the early 1970s it became recognised that it is also possible to use large coal in vigorously fluidised beds of inert material (for example, sand) of the size required to match the fluidising velocity. Tests demonstrated that the relatively low density of the coal lumps combined with the hydrostatic properties of the fluidised bed and the turbulent conditions created by the bubbles resulted in an even distribution of the coal lumps throughout the bed.

The main advantages of lump coal are as follows.

(1) On-site coal preparation is reduced.

(2) A comparatively small number of low-cost mechanical feeders can be used with lump coal. With crushed coal, however, a large number of coal feed points (typically one per 1 to 3 m^2 of bed area) are required to ensure an even distribution. The practical problems associated with subdividing the coal flow uniformly may be a limitation with this approach.

(3) High combustion efficiencies can be obtained under conditions such as high fluidising velocity or shallow bed that would otherwise lead to a high carry-over of unburned char and would require an efficient fines refiring system.

The greatest improvement in combustion efficiency is obtained with sized coal (doubles of 25 to 50 mm or singles of 13 to 25 mm) rather than smalls because of the absence of fines. However, supplies of sized coal are limited with modern mining methods and smalls must therefore be used for large boilers.

In order to minimise the build-up of large ash particles in the bed washed coal is preferred. Typically, this contains 3 to 10 per cent ash, most of which exists as finely divided 'inherent' mineral matter within the coal lumps, rather than as large stone particles. Even so, a few large stone particles are present and have to be removed from the bed material by purging and sieving otherwise segregation and loss of fluidisation would result.

Although systems employing the mechanical feeding of lump coal to the top of the bed were developed initially for industrial boiler and furnace applications, this system is now also generally favoured for power generation boilers. The main exception to this is for coals (or shales) having a high inherent ash content which masks access to the combustible material; crushing is essential in such cases in order to achieve a high combustion efficiency.

Air distribution

The purpose of an air distributor is to provide a uniform flow of air from the plenum chamber into the bed without allowing bed particles back into the plenum chamber.

A number of designs have been used with one of the most common arrangements employing 'standpipes'. These comprise vertical tubes, sealed at the top and with holes or slots in the side (see figure 3.3), arranged on a uniform pitch of about 100 mm on the base-plate. An advantage of this design is that an unfluidised layer of bed material forms immediately above the base-plate which is thereby insulated from the fluidised bed.

A quite different approach, still in the development phase, is the concept of 'sparge pipes'. These are horizontal pipes placed above the solid base of the combustion chamber and running the width of the bed. The combustion air is introduced via holes or standpipes on the underside of the sparge pipes. An advantage of sparge pipes is that, being fixed only at one end, thermal expansion may be accommodated more easily than with a conventional distributor plate. This is particularly important if the hot gas start-up technique is employed. A further attraction of sparge pipes is that in horizontal shell boiler applications the fire-tube itself can provide the bed containment. In particular, there is the possibility of some direct heat transfer from the bed to the lower part of the fire-tube which would normally be masked by the distributor plate.

Fines collection

As discussed earlier, fine particles present in the feed or produced in the bed become entrained in the gas stream. In order to satisfy environmental requirements, it is generally necessary to install gas-cleaning equipment to remove the particulate material from the gas stream. Cyclones are almost always used as the initial stage of gas cleaning, and may be adequate alone for some small units. For the large boilers and power stations, however, further stages of gas cleaning such as bag filters or electrostatic precipitators are generally necessary in order to meet particulate emission standards.

In simple systems, the fines collected by the cyclone are sent for disposal. However, the fines contain unburned char and unconverted acceptor and may, alternatively, be used as follows.

Recycle

The simplest and cheapest method of utilising the unreacted char and acceptor in the fines is to recycle them to the main combustion bed. In order to avoid refiring the smallest particles, which would have a short residence time, normally only the material collected by the first stage of cyclone gas cleaning would be recycled. In practice, the fines usually have to be cooled before re-injection and a high recycle rate therefore reduces the net heat available within the bed. In some cases, this penalty can be avoided by using 'internal' cyclones (or other dust-collection devices such as grit arrestors or baffles) placed in the freeboard and operating at bed temperature.

Fines burn-up bed

Instead of returning the primary cyclone fines to the main bed they may instead be fed to a separate 'fines burn-up bed'. This bed would usually be designed to operate with a low fluidising velocity and a high excess air level in order to ensure adequate combustion and sulphur retention efficiencies for the fines on a single pass.

The additional capital cost of a fines recycle system or a fines burn-up bed is offset by improvements in the combustion and sulphur retention efficiencies. Other design changes may also be possible, for example, the use of a higher fluidising velocity, lower excess air level and lower calcium/sulphur ratio. The optimisation of a system is, therefore, complex and depends on the application.

Part-load operation

Boilers are typically required to be able to 'turn-down' to about 25 per cent of the maximum continuous rating (MCR) in response to variations in the heat demand. In order to achieve satisfactory part-load operation it is necessary to reduce both the combustion and heat transfer rates.

The combustion rate can be reduced by decreasing the coal and air feed rates. Ideally, these rates would change in proportion to maintain a constant excess air level and therefore a high boiler efficiency. It is not usually possible to reduce the fluidising velocity by more than 60 per cent of the value at maximum continuous rating for steady operation because of the possibility of bed particle segregation and consequent defluidisation.

The main problem in designing a system of part-load operation is reducing the rate of heat transfer to in-bed tube surfaces in proportion to the output so that a nearly constant excess air level can be maintained. One or more of the following techniques may be used to reduce heat transfer as output is lowered.

Bed temperature control

If the bed temperature is reduced during turn-down, the rate of heat transfer to immersed surfaces will be correspondingly lower. However, both combustion and sulphur retention efficiencies are adversely affected. In practice this approach is of limited value with sulphur retention because of the restricted range of operating temperatures and, even without sulphur retention, only about 75 per cent turn-down can be achieved in this way. Dynamic response is also slow because of the thermal inertia of the bed.

Bed depth control

Since the heat transfer coefficient in the bed is considerably higher than in the freeboard, the heat flux can be reduced on turn-down by reducing the bed depth so that some of the tubes are uncovered. This may be achieved either by removing bed material or by utilising the natural variation in bed expansion with changes in fluidising velocity.

Multiple compartments

By sub-dividing the bed into compartments a reduction in the heat demand can be met simply by closing down (slumping) sections of the bed. With sufficient compartments any desired degree of turn-down can be achieved. In practice the compartments in use during a reduced-load period of operation would be varied in order to maintain a 'warm start' capability.

On-off operation

Although full modulating output is necessary for a boiler producing superheated steam, an industrial boiler providing saturated steam can be operated on-off, drawing on the boiler thermal capacity while the bed is slumped. In this way, turn-down to about 20 per cent can be achieved by a single bed.

Effect of coal type

One of the important advantages of fluidised bed combustion is its tolerance to the properties of the feed coal.

Bituminous coals, sub-bituminous coals and lignites can be burned without difficulty. Anthracites can also be burned but their comparatively low reactivity requires special operating conditions to be used (for example, high excess air level, deep bed, high bed temperature) in order to avoid excessive carry-over of unburned carbon. Combustion is not affected by the caking properties of the coal because of the low concentration of combustible material in the bed. The coal may be burned as a graded coal (for example, singles), as smalls or as a crushed coal. However, only in the case of a graded coal can a shallow fluidised bed (less than about 0.3 m deep) be employed. Sulphur emissions arising from the use of a high sulphur coal can be controlled by adding limestone or dolomite to the bed although this requires a combustor design with a bed depth greater than about 1 m.

Coals with high or low ash and moisture contents can also be burned. The combustion of low ash content coals without sulphur retention generally requires an inert additive (such as sand) to be used to provide the bed material. If the inherent ash content is high then the coal would be burned crushed in order to provide good access by the oxygen to the combustible material. The operation of the combustor is not affected by the ash fusion temperature as the temperature of the bed is normally well below that at which ash fusion takes place.

A further feature of fluidised bed combustion is that variability in the characteristics of the coal or ash is acceptable.

3.4 Atmospheric pressure FBC furnaces

The purpose of a furnace is to produce a hot gas, for example, for drying or heating applications. Many of the aspects of combustor design for furnaces are similar to those for boilers and the present section will therefore concentrate only on major points of difference.

Classification of furnace types

If coal is burned at a low excess level under adiabatic conditions the resulting temperature is approximately 1700 °C. In a fluidised bed boiler, heat is removed from the combustion bed to heat water or to raise steam giving a bed temperature of about 850 °C. In a fluidised bed furnace, however, other methods of heat removal are required. Three such systems have been employed and characterise the main types of fluidised bed furnace.

High excess air level operation

The simplest system of heat removal is to operate at a high excess air level (usually about 150 per cent) so that all of the heat generated by combustion is removed from the bed as sensible heat in the combustion gases.

In some applications the presence of coal ash in the furnace gas can be tolerated. This enables a simple furnace design to be used, as illustrated for a drying process in the upper part of figure 3.6. The hot gases from the combustor, mixed with additional air to reduce their temperature if required, pass directly into the dryer. The amount of dilution air is usually kept to the minimum as it reduces the overall efficiency of the unit. The gases at the exit from the dryer are cleaned and separated from the dry product before emission to the atmosphere. Some of the ash entrained in the hot gases is inevitably mixed with the dry product.

For other applications an uncleaned hot gas may not be acceptable but cyclone cleaning of the gas may be sufficient. The system is therefore identical with that shown in the upper part of figure 3.6 except for the introduction of cyclones between the furnace and the dryer/heater. This is referred to here as a 'semi-clean' gas system.

Flue gas recirculation

A disadvantage of high excess air operation is that a reduced overall efficiency is obtained because of high sensible heat losses in the flue gas. One way of avoiding this is to recirculate a side-stream of the flue gases to the combustor for use as part of the fluidising medium. In this mode of operation the recirculating gas stream acts as a heat carrier to remove heat from the combustion bed instead of

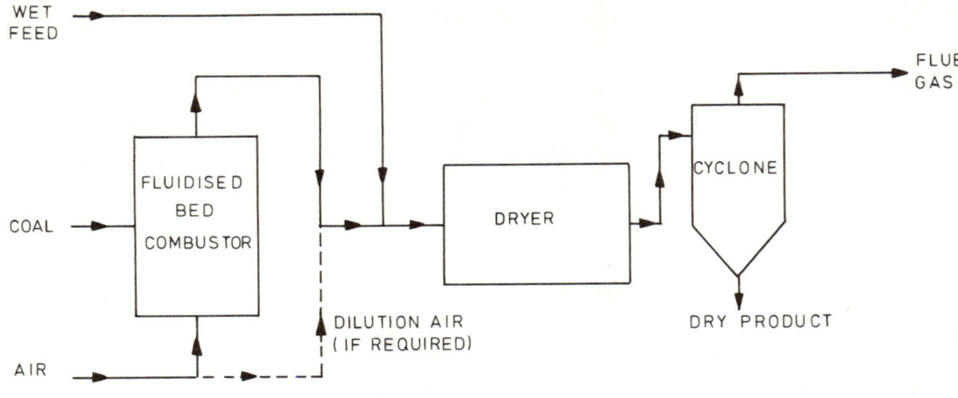

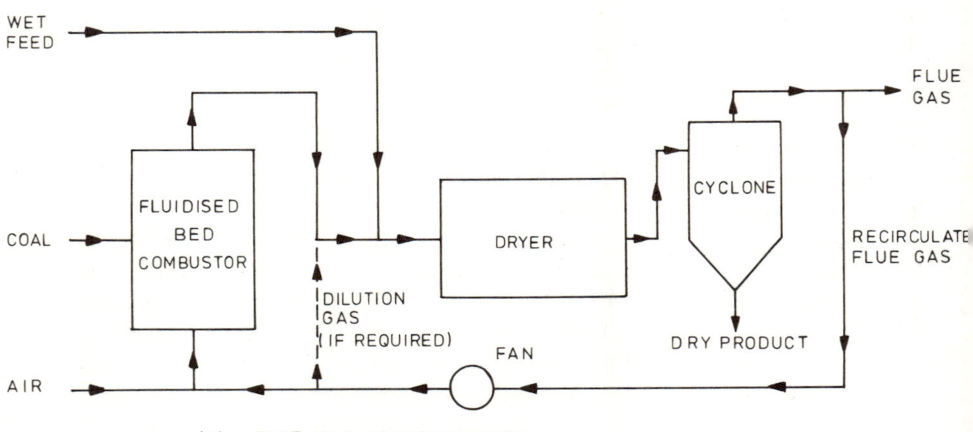

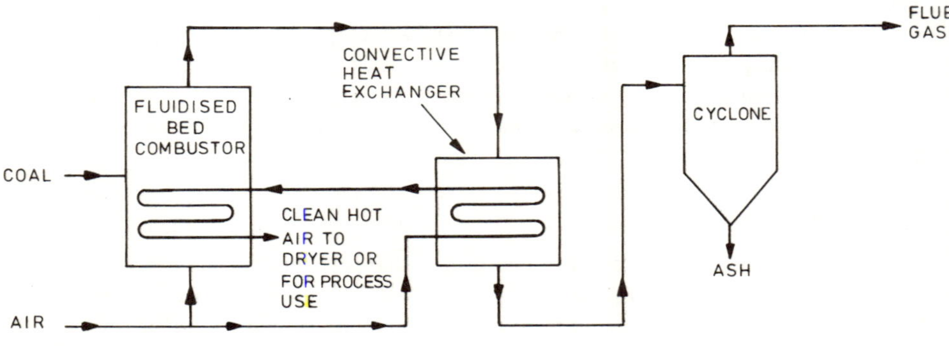

Figure 3.6 Fluidised bed combustion furnaces

the additional air used in the previous case. The system is illustrated in the middle part of figure 3.6 for a simple drying application. The improvement in efficiency obtained by flue gas recirculation is, however, offset by additional capital and operating costs for the fan required to compress the flue gas. This is because compression of the hot flue gas requires considerably more energy than the compression of an equivalent mass of cold air.

An alternative arrangement, intermediate between those shown in the upper and middle parts of figure 3.6 is to recirculate flue gas only for dilution before entry to the dryer and to use air to fluidise the combustor bed.

Indirect heating

In some furnace applications even the small amount of coal ash that passes through cyclones would be unacceptable in the hot gas and indirect heating of clean air has to be employed. The system is illustrated in the lower part of figure 3.6. The combustor is operated at a low excess air level so that clean air can be heated indirectly by passing it through tubes immersed in the bed. Heat exchange to the clean air also occurs in a convective section. Limitations on the working temperature of the heat exchanger materials are likely to restrict the hot gas temperature to 600 °C although this is sufficient for some applications.

Design considerations

The main differences between the design of a fluidised bed furnace (other than the indirectly heated type) and a boiler arise from the absence of tubes in the bed. In particular, part-load operation and the design of the control system are simplified by the absence of heat transfer considerations. Also, because the bed does not have to accommodate a tube bundle, shallower beds can be used subject to satisfactory combustion and sulphur retention efficiencies and solids mixing.

Application to boilers

In some applications, it may be advantageous to employ a furnace of the high excess air level or flue gas recirculation type to produce hot gas for a convective boiler. Although such a boiler would be larger than the convective section of a conventional design (with tubes in the bed), there are two advantages in this approach. Firstly high-quality water treatment is required for conventional boiler designs because of the high heat fluxes encountered by in-bed tubes. This is avoided in the furnace/convective boiler approach. Secondly, as noted above, part-load operation and control system design are simplified in furnace-type combustors that do not contain tubes in the bed.

Alternatively, and for similar reasons, heat transfer from the fluidised bed may be limited in some boiler designs to water-cooled combustion chamber walls. In particular, for a low-grade fuel with a high moisture content, the number of in-bed tubes would be small and little loss in efficiency results from their elimination.

3.5 Fast fluidised bed systems

As discussed in section 3.1, fast fluidised beds operate with relatively high gas velocities and fine particle sizes, and involve a different fluidising regime from that in conventional fluidised beds. Because the maintenance of steady-state conditions in a fast fluidised bed requires the continuous recycle of particles removed by the gas stream, the system is sometimes referred to as a 'circulating' bed. However, the term 'circulating' bed is also used more generally to include fluidised bed systems containing multiple conventional bubbling beds between which bed material is exchanged, and also bubbling beds containing a draught tube. For clarity, therefore, the term 'fast' bed will be used here only for systems that operate in the high gas velocity/fine particle fluidising regime described earlier.

General description

A typical fast fluidised bed combustion boiler is illustrated in figure 3.7. The main combustion and sulphur retention reactions occur in a highly turbulent, non-bubbling fluidised bed usually at least 10 m deep. The fluidisation conditions in this 'fast' fluidised bed reactor are created by a combination of a high fluidising velocity (usually about 10 m/s) and fine bed material (mean particle size about 150 microns). In such a system, the solids rapidly become entrained in the gas flow and are removed from the containment vessel. Recycle of the material is therefore essential to maintain steady-state conditions, the rate of recycle determining the solids concentration in the reactor. The solids-laden gas stream leaving the fast bed passes first to a separation device, usually a cyclone. The solids recovered are then fed to a conventional fluidised bed containing a tube-bundle. This bed extracts heat from the recycling solids to heat water or to raise steam, usually in a series of counter-current stages as shown in the figure. The solids leave the bed cooled, typically to about 100 °C, and are then re-injected into the fast fluidised bed combustor. The recycling solids may therefore be regarded as acting as a heat carrier to remove the heat generated in the combustor, which generally operates at a low excess air level. The cooler is fluidised by air at a low velocity (usually less than 1 m/s), this air then being used

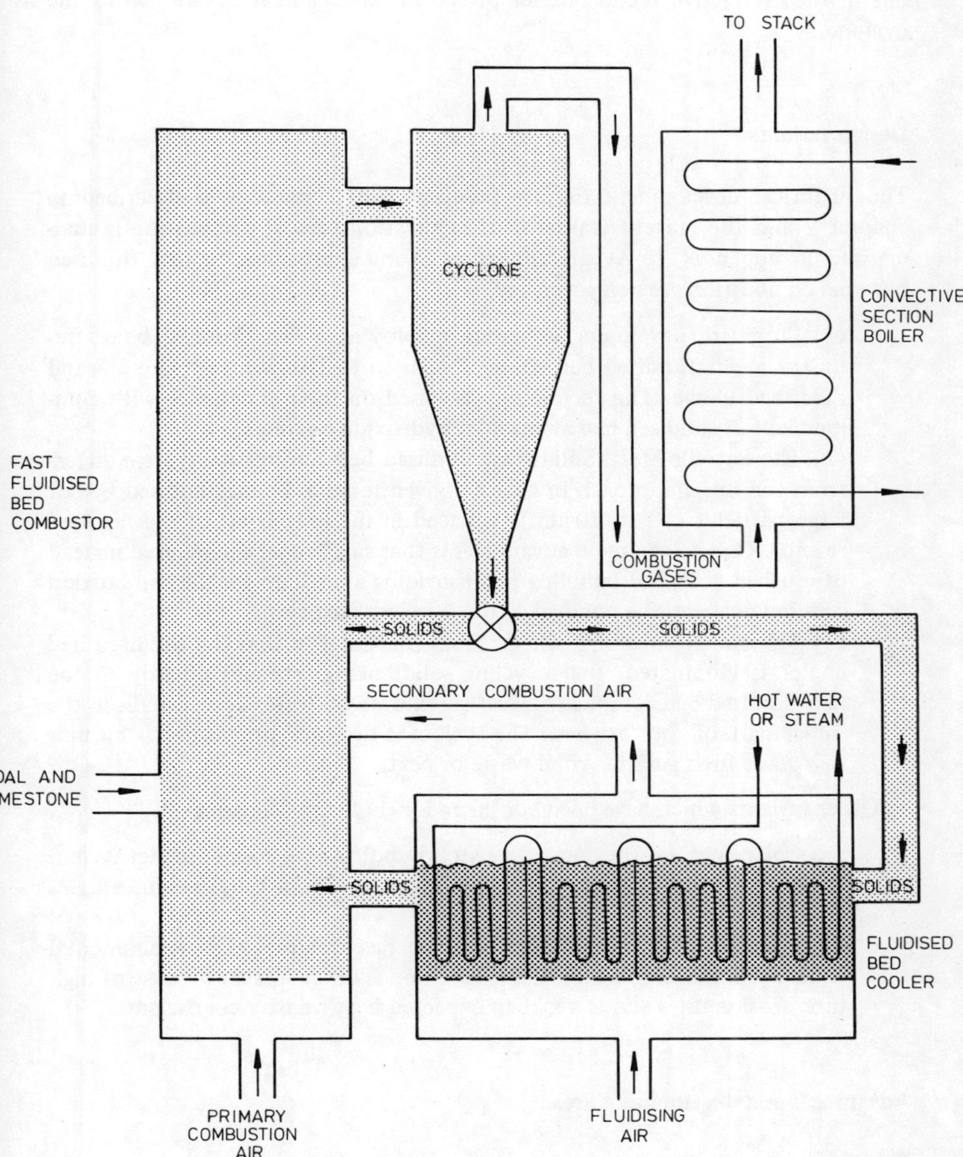

Figure 3.7 Fast fluidised bed boiler

directly as pre-heated combustion air in the fast fluidised bed reactor with no dust removal being necessary. Further water heating or steam raising is carried out in the convective section boiler placed in the clean gas stream leaving the cyclone.

Design variants

The historical development of fast fluidised bed technology is described in chapter 2 and the current status of its application to coal combustion is summarised in appendix 1. At present, three main approaches to fast fluidised bed coal combustion are being pursued.

(1) Lurgi are developing a system employing a fast fluidised bed combustor and a fluidised bed cooler similar to that shown in figure 3.7 and described above. The technology is based on their experience with commercial fast fluidised bed aluminium hydroxide calciners.

(2) The Battelle Multi-Solids fast fluidised bed combustion system differs from the Lurgi approach in that a conventional bubbling fluidised bed of large particles (10 to 20 mm) is placed in the lower part of the fast bed reactor. One of the main advantages is that smalls coal can be used instead of crushed coal, the bubbling bed providing a medium for the combustion of lump coal and the removal of the large ash particles.

(3) The Ahlström design differs from the Lurgi in that the fluidised bed cooler is eliminated, the recycling solids being returned directly to the fast fluidised bed combustor. Heat is removed by water-cooled walls in the upper part of the fast bed. The system is designed primarily for burning low-grade fuels such as wood waste or peat.

Other variants which have been considered include the following:

the replacement of the convective section boiler by a direct contact system using cooled solids from the fluidised bed cooler (a Lurgi patent on this method exists)

the insertion of a tube bundle into the fast fluidised bed—experimental work by Battelle indicates that this option is unfavourable because of high tube erosion rates and lower than expected heat transfer coefficients.

Advantages and development areas

Fast fluidised bed systems may have a number of advantages over conventional fluidised bed systems for coal combustion. These arise from two principal design differences

the use of a high gas velocity/fine particle fluidising regime for the combustor
the use of separate beds for combustion and heat transfer (in the Lurgi and Battelle systems)

The main advantages arising from the use of a fast fluidised bed for the combustor are as follows.

Improved combustion and sulphur retention efficiencies

The use of fine particles, turbulent gas–solids contacting conditions and a high recycle rate makes the fast fluidised bed an extremely effective reactor design with a performance substantially better than that of a conventional fluidised bed combustor with a low recycle rate. For example, a combustion efficiency of more than 99 per cent and a sulphur retention efficiency of more than 85 per cent should be possible with an excess air level of 10 per cent and a calcium/sulphur mole ratio of 1.5.

High throughputs

The use of high fluidising velocities makes the fast fluidised bed much smaller in bed area than a conventional fluidised bed combustor of the same output. This advantage is, however, offset by its additional height and by the size of the fluidised bed cooler which operates at a lower fluidising velocity.

Simple coal feeding system

The small combustor size and turbulent mixing conditions enable the number of coal feed points to be reduced substantially (typically by an order of magnitude) as compared with a conventional fluidised bed combustor, with corresponding cost savings.

Control of nitrogen oxide emissions

Staged introduction of the air into the combustor assists in reducing nitrogen oxide formation (see section 7.3).

The use of separate beds for combustion and heat transfer brings the following additional advantages.

Part-load performance and control

An advantage of the fast fluidised bed system is that turndown may be achieved by decreasing the solids flow rate to the cooler, the solids instead being returned directly to the combustor bed. This reduces the rate at which heat is extracted from the combustor. A simple and effective control system results.

Reduced erosion/corrosion of heat transfer tubes

The tubes immersed in the fluidised bed cooler are subjected to significantly lower gas and particle velocities than are typical in a conventional fluidised bed combustor. Also, oxidising conditions prevail throughout the cooler, whereas reducing conditions exist near the coal feed points in a conventional fluidised bed combustor. These factors combine to reduce substantially the rates of erosion and corrosion of the tubes immersed in the fluidised bed cooler.

High heat transfer coefficients

Because convective heat transfer coefficients in a fluidised bed increase with decreasing particle size, less heat transfer tubing would be required in the fluidised bed cooler than in the bed of an equivalent conventional fluidised bed boiler.

Although, as a concept, fast fluidised bed combustion shows considerable promise, a number of technical questions remain to be answered. Areas of concern include the following:

the pressure drop across a fast fluidised bed is generally greater than for a conventional fluidised bed combustor, resulting in an increase in fan power requirements

because of the large recycle rates, high cyclone collection efficiencies are essential in the recovery of the bed material from the gas stream

the high gas velocities combined with the high particulate loadings could lead to problems of erosion in the combustor, cyclone and associated ducting

Applications

Although considerable experience of fast fluidised bed technology exists in other fields (see section 2.1), the application of fast fluidised beds to coal combustion is at an early stage of commercialisation. In particular, the economics of this approach compared with conventional designs are not yet well established and, as discussed below, it is possible to draw only tentative conclusions regarding the likely future role of the technology.

An important potential market for fast fluidised bed combustors is for coal-fired boilers, the Lurgi and Battelle systems being directed primarily towards this application. Within the boiler market, however, fast fluidised bed systems will probably compete mainly with pulverised fuel combustors (both for industrial and power generation installations) rather than with conventional atmospheric pressure fluidised bed combustors. This is because the additional complexity of a fast bed system is likely to be economic only at the larger boiler sizes where the effects are mitigated by the economies of scale.

A related application which may be attractive is the heating of process fluids, particularly as a special consideration is often the need to limit the heat flux to the fluid in order to avoid thermal decomposition.

Fast fluidised bed combustors can also be used as furnaces or incinerators, either to produce a semi-clean, hot combustion gas (as in the Ahlström design) or possibly to heat clean air indirectly in the fluidised bed cooler. The latter concept could provide the basis for a combined cycle power generation system.

Development work on fast fluidised bed combustors has so far concentrated on atmospheric pressure applications. Although in principle a fast bed system could be used to replace a conventional bubbling bed in a pressurised fluidised bed combustion unit, the combined uncertainties of the technologies makes this a long-term option.

3.6 Pressurised fluidised bed combustion

Direct coal firing of gas turbines

At present, all gas turbine and combined cycle systems are fired by distillate fuel oil or a gas. In spite of considerable research and development in the 1950s and early 1960s, it has not proved possible to operate a gas turbine directly on coal using pulverised fuel combustion. The difficulty lies in obtaining an adequate lifetime for the gas turbine blades when expanding the gas produced by coal firing. Three main problems occur when using a pulverised fuel combustion system for this purpose.

Erosion

As a result of the high temperature in a pulverised fuel combustion flame, the ash particles present in the coal sinter (that is, fuse to form fluid particles) and, on cooling, resolidify to produce extremely hard 'glassy' particles. Even with efficient gas cleaning systems, sufficient particles are ingested by the turbine for blade erosion to be rapid.

Chemical attack

Also as a result of the high combustion temperature, alkali metal salts (for example, sodium chloride) are released as aerosols and can form a liquid film deposit on the blades leading to chemical corrosion.

Deposition

Fine particles of ash assisted by a liquid film of alkali metal salts can stick to the turbine blades. These deposits interfere with the gas flow and may subsequently break off as agglomerates causing catastrophic damage to downstream blading.

These problems result from the high temperature attained in a PF combustion flame. At the temperature in a fluidised bed combustor, however, sintering of the ash does not occur and the release of alkali metal aerosols is limited. For these reasons it is hoped that the combustion gases from a fluidised bed will prove to be acceptable to a gas turbine.

Although the ash entrained in the combustion gas from a fluidised bed combustor is relatively soft, it is still regarded as necessary to remove particles larger than about 10 microns. The main available method of gas cleaning is cyclones and it is generally thought that three stages of cyclones are required. The final stage would be an advanced high-efficiency cyclone. Other methods of gas cleaning (panel bed filters, for example) may be employed if an adequate level of particulate removal cannot be obtained using cyclones.

It has been suggested that the gas turbine inlet temperature may need to be limited to 700 °C or less in order to reduce the concentration of corrosive alkali metal salts in the combustion gases entering the turbine. This does not necessarily imply operating at a low bed temperature, as the gases can be cooled by heat transfer or by dilution before the turbine. Reducing the turbine inlet temperature, however, would result in a lower cycle efficiency than operation at a temperature corresponding to that of the fluidised bed (775 to 950 °C).

Although experimental work on pressurised fluidised bed combustion using static cascades of turbine blades is encouraging, realistic blade lifetimes have yet to be demonstrated using an actual turbine of a size and design representative of commercial power generation practice.

General description

A fluidised bed combustor used to supply hot gas to a gas turbine must operate at the pressure of the turbine, usually in the range 5 to 20 bar. For this reason the bed is much more compact than an atmospheric pressure combustor of equivalent output but has to be enclosed in a pressure vessel. Systems using vertical, horizontal and spherical pressure vessels have been proposed; an example using a horizontal pressure vessel is shown schematically in figure 3.8.

Design concepts for pressurised fluidised bed combustion generally employ crushed coal which is transferred from ambient pressure to the operating pressure by lock hoppers (such as the Petrocarb system). Coal feeding is by pneumatic transport line to the bottom of the bed. Lock hoppers are also used to remove the cyclone fines and excess bed material after cooling.

The combustor bed may be operated at a high excess air level and used simply as an air heater (as in the 'open bed' design shown in figure 3.8) or may contain heat transfer tubes to heat water, to raise steam or to heat clean air. The ways in which a pressurised fluidised bed combustor may be integrated into a power generation cycle are considered in section 3.7.

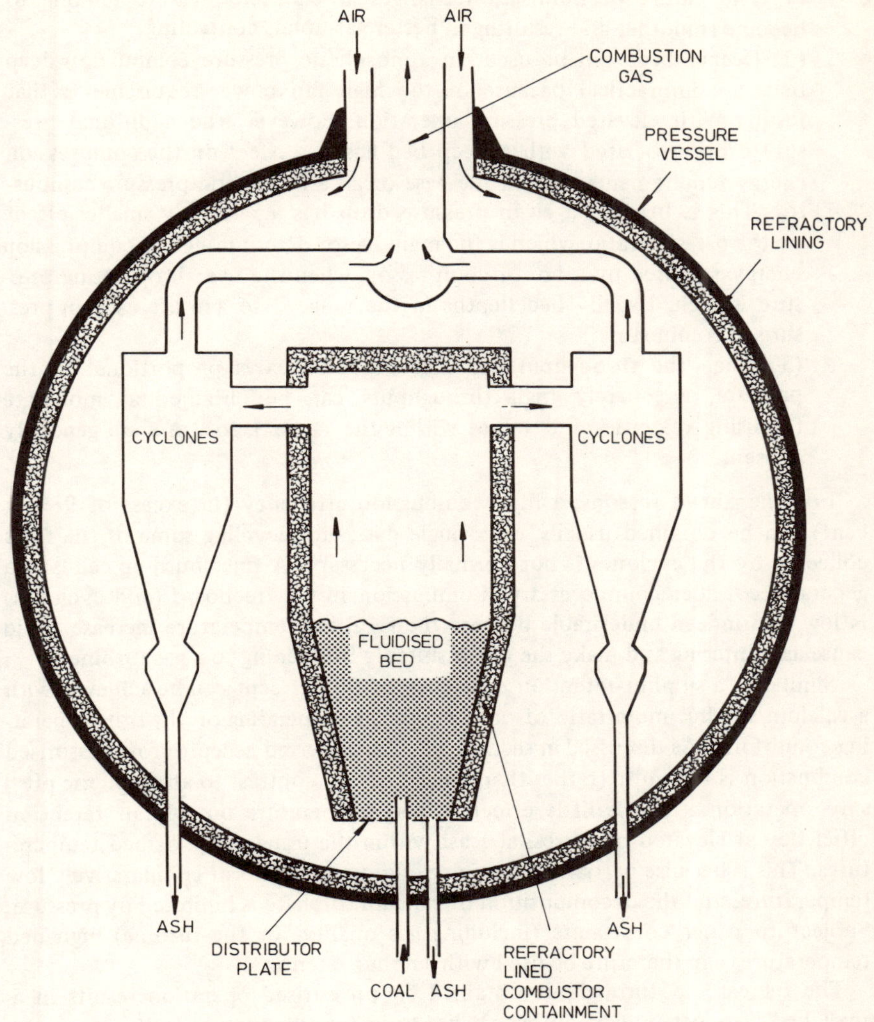

Figure 3.8 Pressurised fluidised bed combustor

Design considerations

There exist a number of important differences between pressurised and atmospheric pressure fluidised bed combustors.

In particular, the efficiencies of combustion and sulphur retention generally achieved in pressurised combustors are considerably better than those for atmospheric pressure combustors. Three factors are relevant:

(1) The nature of fluidisation changes at elevated pressure tending to become smoother and resulting in better gas–solids contacting.
(2) Deeper beds can be used. In atmospheric pressure combustion deep beds are impractical because of the high fan power requirements that result. With elevated pressure operation, however, the additional pressure drop associated with a deep bed has less effect on the compression energy requirements than in the case of an atmospheric pressure combustor. This is because a given pressure drop has a relatively smaller effect on the pressure ratio, which is the main factor determining the compression energy required by a fan or compressor, when the overall operating pressure is high. Usually bed depths in the range 2 to 4 m are used in pressurised combustors.
(3) Since the throughput per unit of bed area is proportional to the pressure, adequately high throughputs can be obtained at moderate fluidising velocities and values within the range 1 to 2 m/s are generally chosen.

For the above reasons, a high combustion efficiency (in excess of 98 per cent) can be obtained usually on a single pass and recycling some of the fines collected by the cyclones is not normally necessary. A fines burn-up cell is also generally considered unnecessary. Combustion in the freeboard (and cyclones) is low and indeed undesirable because the resulting temperature increase could cause ash sintering and make the gas unsuitable for feeding to a gas turbine.

Similarly, a sulphur-retention efficiency of 90 per cent can be achieved with a calcium/sulphur mole ratio of only 1.5 to 2.5 (depending on the other operating conditions). As discussed in section 3.2, the preferred acceptor for pressurised combustion is dolomite (rather than limestone). In contrast to atmospheric pressure operation, there is little effect of bed temperature on sulphur retention efficiency at elevated pressures, at least within the usual range of bed temperatures. This is because partial calcination of dolomite occurs at comparatively low temperatures and the decomposition of calcium sulphate is inhibited by pressure. Subject to other constraints (including gas quality for the turbine) high bed temperatures can therefore be used with sulphur retention.

The increase in throughput obtained by pressurised operation results in a small bed area per unit of output. It has been found experimentally, however, that a coal feed point spacing similar to that used for the pneumatic feeding of crushed coal at atmospheric pressure is satisfactory. The number of feed points required is therefore reduced in proportion to the bed area, and the practical problems of multiple stream splitting to ensure an even distribution of the feed coal are lessened.

With atmospheric pressure operation, the use of lump coal can assist in meeting the conflicting requirements of achieving a high combustion efficiency with a low fluidising velocity, shallow bed and small number of coal feed points. Pressurised operation enables these requirements to be met more easily, so that there is less incentive to use lump coal. For this reason, little work has yet been carried out on the use of lump coal in pressurised combustor designs with present concepts concentrating on the pneumatic feeding of crushed coal.

For beds containing tubes, the high heat release per unit bed area requires a large heat transfer surface (per unit bed area) and therefore a deep bed to accommodate the tube bundle. As discussed earlier, the associated increase in compression energy requirements is relatively small at elevated pressure.

Compression of the air to the operating pressure of the combustor results in the air being pre-heated, typically to 350 °C. At the same excess air level, the proportion of the combustion heat released that is available in the bed is therefore higher than for atmospheric pressure combustion. Conversely, in designs where no heat transfer takes place in the bed (as in figure 3.8, for example), operation at about 300 per cent excess air is necessary for a bed temperature of 850 °C compared with about 150 per cent for a similar atmospheric pressure furnace.

The problems of start-up, part-load operation and load following for a pressurised combustor are generally similar to those at atmospheric pressure with corresponding technical solutions applying. There is, however, one important difference. At elevated pressures, turn-down can be achieved by changing the pressure as well as the fluidising velocity. Indeed, the pressure and final gas temperature are both constrained to follow the operating characteristics of the gas turbine. Varying the pressure should permit a turn-down ratio of more than 5 : 1 for the 'open bed' design although, for beds containing tubes, turn-down will be limited by the extent to which the heat transfer rate can be reduced (as for atmospheric pressure boilers). However, the combination of a gas turbine and a steam turbine makes the plant-control problem more complex.

3.7 Power generation applications

Power generation is the largest market for coal in most industrial countries and it was with this market in mind that fluidised bed combustion was developed first as an alternative to the pulverised fuel system of combustion generally used. The main incentives to develop a new combustion system were to reduce boiler capital cost, to enable coal of low or variable quality to be used and to permit the cheap and effective control of sulphur emissions.

Pulverised fuel power generation

In a pulverised fuel power station the coal is pulverised so that the maximum particle size is about 0.1 mm and is blown into a flame rather like an oil or gas flame. The coal burns at a high temperature (typically 1500 °C) transferring heat by radiation to tubes forming the walls of the combustion chamber. The combustion gases leave the combustion chamber at about 1000 °C. Heat is recovered in convective heat exchangers to raise steam and to pre-heat combustion air before the gases are released to the atmosphere at about 150 °C.

A typical pulverised fuel system is illustrated in simplified form in figure 3.9. The heat released by combustion is used to raise steam for steam turbines. In the cycle shown in the figure the heat available in the combustion chamber is used for secondary superheat and reheat of the steam, whereas the convective section heat exchangers provide primary superheat, evaporation and combustion air pre-heat.

Pulverised fuel combustion is a mature technology capable of operating at a reasonably high efficiency, on a large scale and with a wide range of coal types.

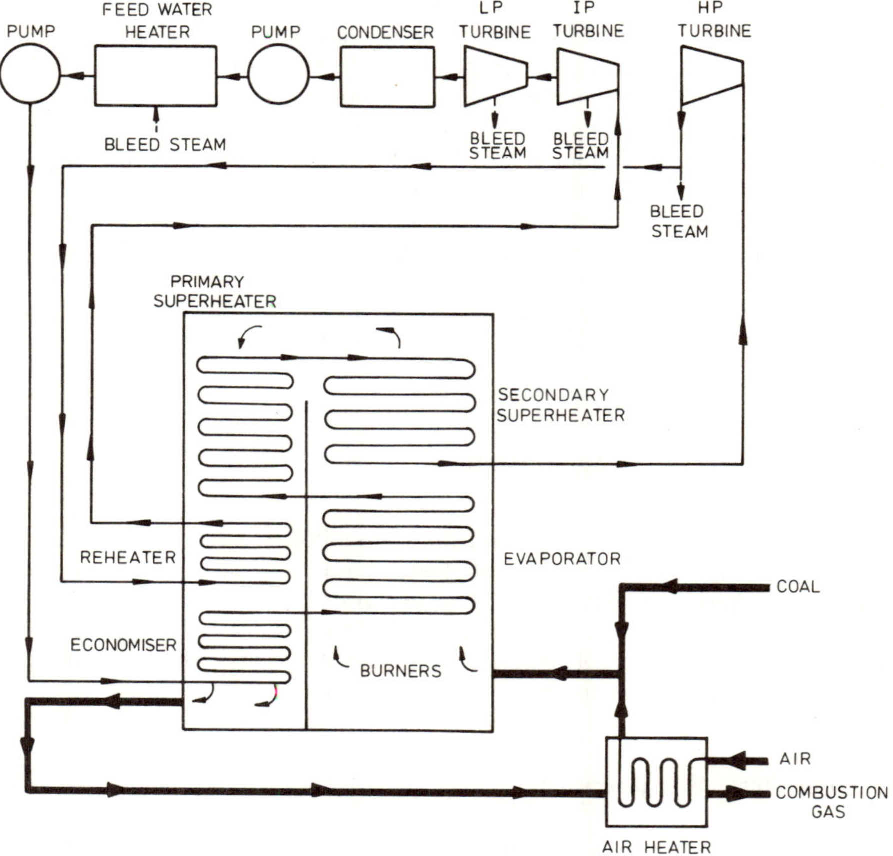

Figure 3.9 Pulverised fuel cycle

If, however, the fuel has a high sulphur content and sulphur emissions have to be limited for environmental reasons, then sulphur dioxide has to be removed from the combustion gases. Although a number of stack gas scrubbing systems are available (see chapter 6), the additional capital and operating costs are significant (see appendix 2).

Atmospheric pressure fluidised bed combustion

A power generation cycle employing atmospheric pressure fluidised bed combustion is shown in figure 3.10.

Separate fluidised beds are generally used for the evaporation, superheat and reheat duties (as shown in the figure) thereby simplifying start-up and load

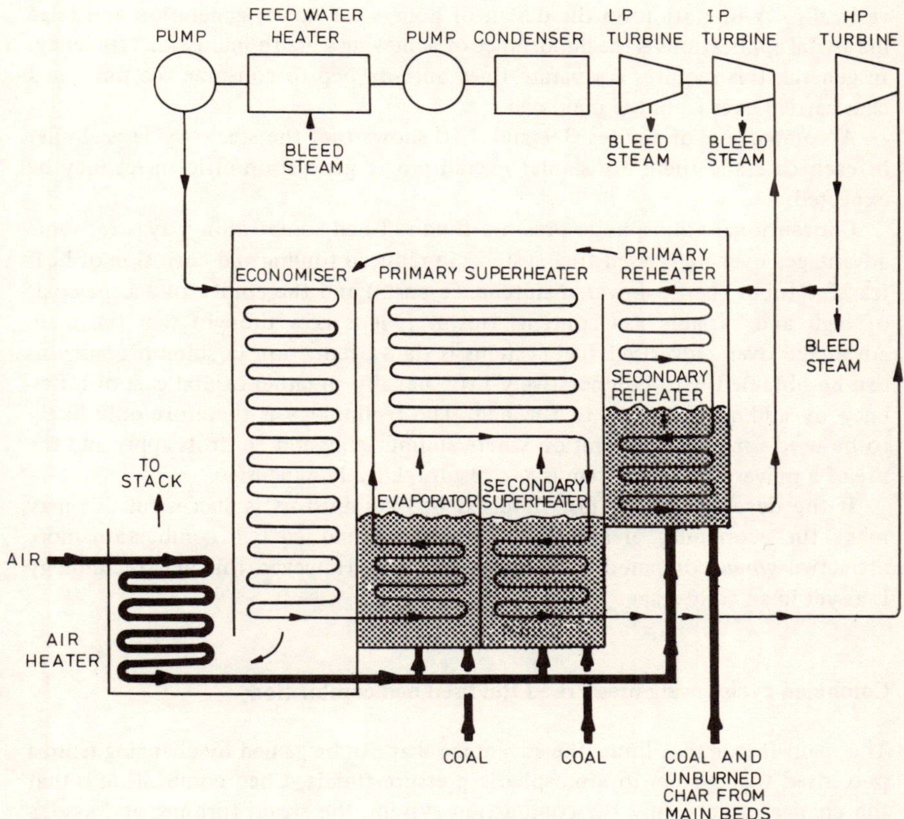

Figure 3.10 Atmospheric pressure fluidised bed combustion cycle

control. Approximately 50 per cent of the heat is transferred to tubes in the bed with the remainder being recovered by heat exchangers in a convective section, usually common to the beds, providing evaporation, superheat and reheat duties. Cost considerations point to using in-bed tubes for the high temperature sections of the steam cycle (superheat and reheat) in order to economise on expensive tube materials, reserving the convective section for low temperature duties (evaporation and air pre-heat). However, other factors such as start-up and load control lead to superheat and reheat being carried out in both bed and convective sections.

Atmospheric pressure fluidised bed boilers are generally fluidised with air velocities of 1 to 3 m/s using a bed of about 1 m deep. For boilers of power station size, a large bed area is required; at 2 m/s, a 2000 MW (electrical) power station would have a bed area of approximately 3500 m^2. Design philosophies are therefore based on multiple beds produced from factory-constructed modules of a transportable size. These can be linked horizontally ('ranch style') or stacked vertically. A limitation on the design of boilers for power generation and large industrial applications is the importance of achieving a high combustion efficiency. In general, this requires a separate fines burn-up bed to consume the unburned char carried over from the main bed.

A comparison of figures 3.9 and 3.10 shows that the steam cycle is similar in each case and therefore similar overall power generation efficiencies may be expected.

Conventional atmospheric pressure fluidised bed combustion may offer some advantages over pulverised fuel systems in reduced fouling and corrosion of heat transfer tubes (hence lower maintenance costs) and the ability to accept coals of high and variable ash content. However, it is now thought that the main advantage over pulverised fuel systems is that a reduction in sulphur emissions can be obtained with comparatively little penalty in either capital cost or efficiency by adding limestone to the bed. The technology is therefore only likely to be used for power generation where sulphur-emission controls apply and the use of a pulverised fuel system is as a result relatively expensive.

If the development of fast fluidised bed combustors is successful this may make the economics of atmospheric pressure fluidised bed combustion more attractive when compared with pulverised fuel. However, this new technology is as yet in an early stage.

Combined cycles using pressurised fluidised bed combustion

The main factor that limits the advantage that can be gained by changing from a pulverised fuel system to atmospheric pressure fluidised bed combustion is that the change affects only the combustion system; the steam turbines and associated equipment remain the same and for this reason the efficiency of power generation is also the same. In order to achieve an improvement in efficiency over the conventional steam cycle, it is necessary to adopt a more advanced power generation cycle. In general, this means using higher cycle temperatures. With a steam cycle, the maximum is less than the tube metal temperature which in turn is limited by materials considerations. Usually, steam cycles operate with a maximum steam temperature of about 565°C and waste heat is rejected from the cycle at the relatively low temperature of about 30 °C. Gas turbines, in contrast, are not so limited, because the blade temperature is normally less than the maximum gas temperature as a result of blade cooling. Typically, gas turbines are designed for inlet temperatures of up to about 1100°C (present technology) and reject heat at approximately 500 °C. By employing both gas and steam turbines in a combined cycle it is possible to use the range of temperatures from 1100 to 30°C, resulting in a high overall efficiency of electricity generation.

Pressurised fluidised bed combustion can be used in a number of ways in advanced power generation systems. These are conveniently classified by the type of power generation cycle employed, the main categories being the simple gas turbine cycle, the waste heat boiler cycle, the exhaust-fired cycle and the supercharged boiler cycle. The main features of these cycles are summarised below.

The simplest application of pressurised fluidised bed combustion is as a direct replacement for the combustion chamber in a distillate or natural gas-fuelled gas turbine. The fluidised bed combustor is used only to heat the combustion air delivered by the compressor to the temperature required for expansion by the turbine and is termed the 'air-heater' design.

An example of an air-heater design operating in a simple gas turbine cycle is shown in figure 3.11. All of the air delivered by the compressor is used to fluidise the bed which therefore operates at a high excess air level (typically 300 per cent) in order to maintain the bed temperature within the design range of

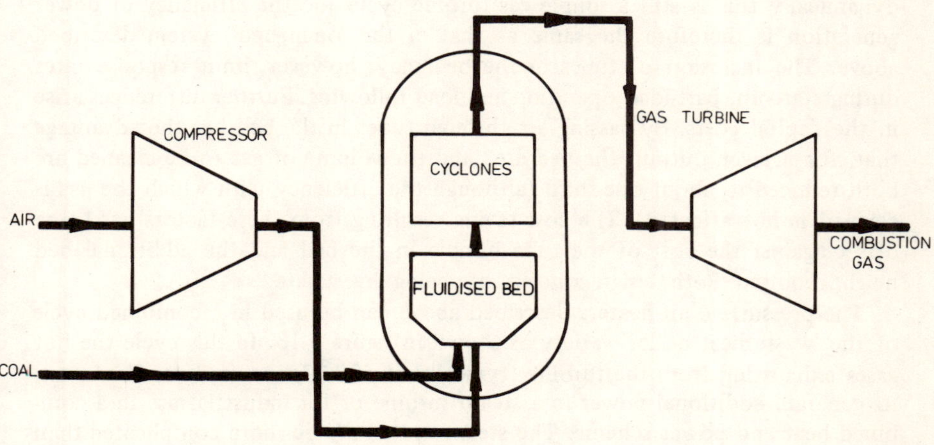

Figure 3.11 Open bed air-heater cycle

775 to 950 °C. Since the fluidised bed does not contain tubes it is referred to as an 'open bed' design. The absence of tubes removes some constraints on operational flexibility and this concept offers the best start-up, part-load and load-following capabilities of the various pressurised combustor options.

A variant of this cycle is shown in figure 3.12 where one-third of the air from the compressor is used for combustion (that is, passed through the bed)

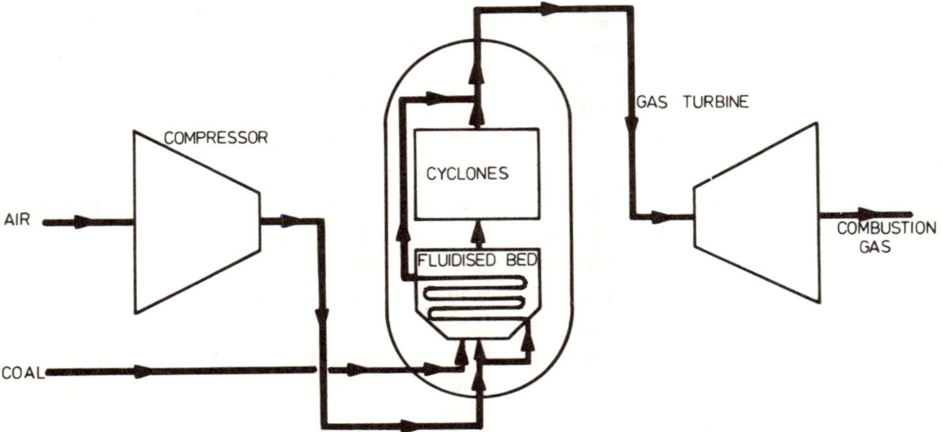

Figure 3.12 Indirect air-heater cycle

with the remaining two-thirds being heated indirectly by tubes immersed in the bed. The hot combustion gas is then cleaned using cyclones and recombined with the indirectly heated air before expansion through the turbine. Thermodynamically this is still a simple gas turbine cycle and the efficiency of power generation is therefore the same as that of the 'open bed' system described above. The inclusion of tubes in the bed may, however, limit response rates during start-up, part-load operation and load following. Further differences arise in the capital costs. By-passing air through tubes in the bed has the advantage that, for a given output, the bed area and the volume of gas to be cleaned are both reduced to about one-third (although the efficiency with which the gas is cleaned is not affected). The cost saving resulting from these factors has to be offset against the cost of the tube bundle in the bed and the additional bed height required. Both design concepts are being investigated.

The pressurised air-heaters described above can be used in a combined cycle of the 'waste heat boiler' variety as shown in figure 3.13. In this cycle the hot gases exhausting from the turbine (typically at 500 °C) are used to raise steam to generate additional power in a steam turbine or for industrial use in a combined heat and power scheme. The steam cycle may be more complicated than shown; twin-pressure or triple-pressure designs have been used in natural gas and distillate-fired systems to improve the utilisation of low-grade heat.

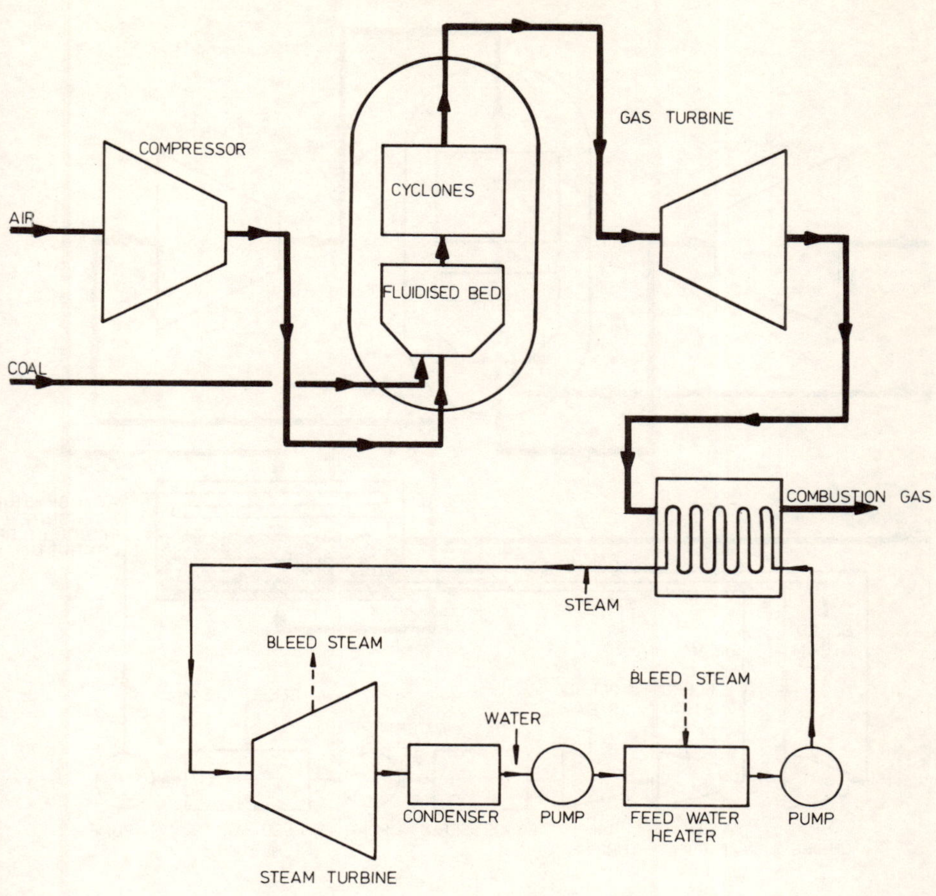

Figure 3.13 Waste heat boiler cycle

The waste heat boiler cycle provides a significant improvement in efficiency over the air-heater cycle but this is limited by the poor steam conditions resulting from the relatively low temperature of the gas turbine exhaust. One option is to increase the temperature of these gases by burning additional coal in them using the excess oxygen present. This is referred to as an 'exhaust-fired' cycle and an example employing both pressurised and atmospheric pressure combustors is shown in figure 3.14. Such cycles appear not to be favoured for pressurised fluidised bed combustion applications and comparatively little work on them has been carried out.

An alternative to the exhaust-fired cycle is to operate the pressurised fluidised bed at a low excess air level and to remove the surplus heat by steam tubes immersed in the bed. This system also makes additional high-temperature heat.

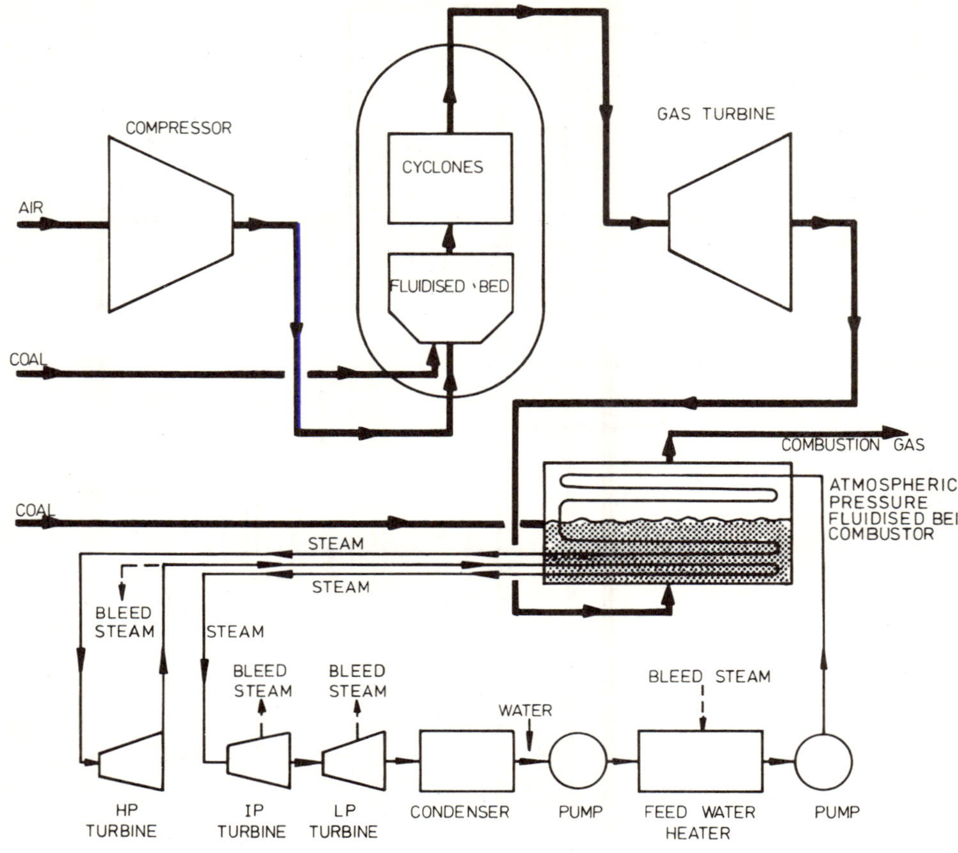

Figure 3.14 Exhaust-fired boiler cycle

available to the steam cycle enabling good steam conditions to be used. The cycle is known as the 'supercharged boiler cycle' and is illustrated in figure 3.15. This is one of the most efficient pressurised fluidised bed combustion cycles although a disadvantage may be poorer part-load and load-following performance than with the other options.

The efficiencies of the main pressurised fluidised bed combustion cycles are summarised in table 3.2 and are compared with those of pulverised fuel combustion and atmospheric pressure fluidised bed combustion. Also shown in the table are the relative contributions of the gas and steam turbines.

Table 3.2 Comparative power generation efficiencies[a]

System	Configuration		Efficiency (%)
	Gas turbine (%)	Steam turbine (%)	
Pulverised fuel			
Without sulphur retention	0	100	38.6
With sulphur retention (regenerable)	0	100	36.7
Atmospheric pressure fluidised bed combustion	0	100	37.9
Pressurised fluidised bed combustion			
Air heater cycle (12 bar, 850 °C)	100	0	22.2
Air heater with waste heat boiler (7 bar, 850 °C)	54	46	35.4
Supercharged boiler cycle (12 bar, 850 °C)	19	81	40.6

[a]Source: process flowsheeting studies using the NCB's ARACHNE system.

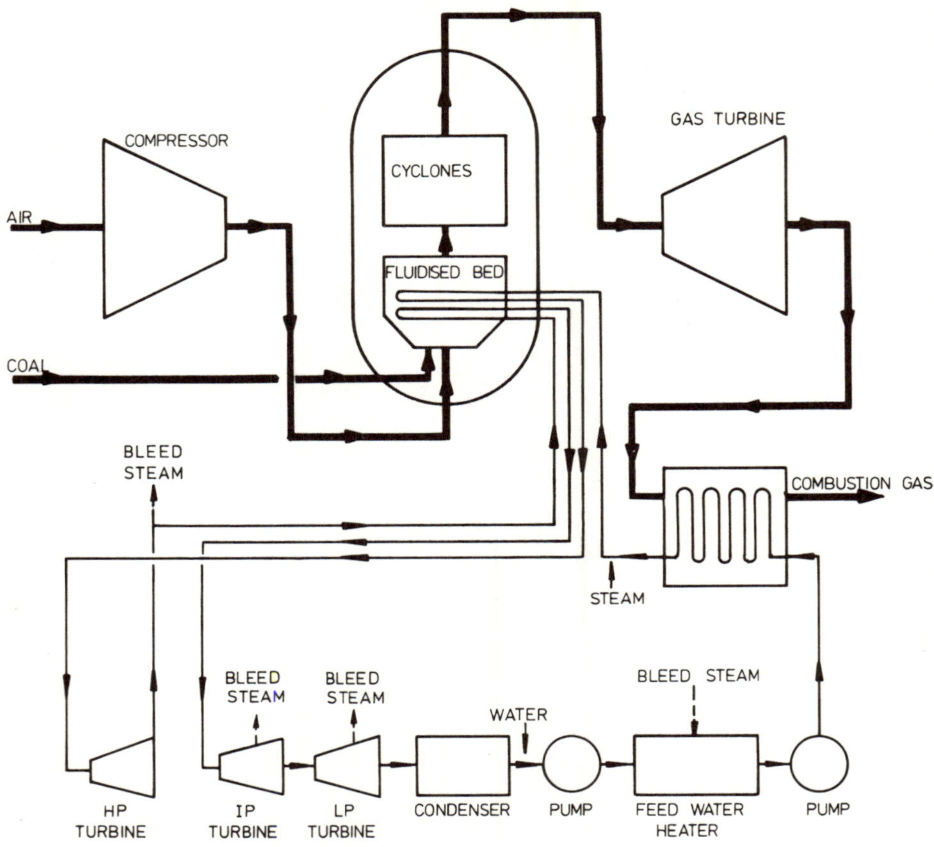

Figure 3.15 **Supercharged boiler cycle**

3.8 Industrial applications

The industrial market covers a wide range of demands both in terms of scale and application. Overall, the market penetration of coal is low, most industrial energy demands being met by oil products and natural gas (see discussion in section 8.5). In many cases, however, coal has a substantial price advantage over the present fuels. For this reason the industrial market is widely regarded as an important future growth area for coal creating an incentive to develop new technology which may make coal more attractive in such applications.

Before assessing the potential of fluidised bed combustion for use in industrial boilers, furnaces and incinerators, it is relevant to consider the characteristics of existing coal-fired combustion systems.

For heat outputs above about 30 MW, pulverised fuel combustors burning part-washed coal (see previous section on power generation) are generally favoured. Below this size, stoker-type systems are used. These burn washed coal (as singles or smalls) in a fixed bed supported on a grate. The main combustion air supply (primary air) is through the grate although secondary air may be required above the bed to burn volatile matter and carbon monoxide.

Types of industrial stoker

Various designs of stoker are available and differ principally in the way in which the coal is supplied to the bed (that is, the method of stoking). Three categories can be identified and are described below.

Overfeed stokers

In overfeed stokers the coal is fed uniformly to the top of the bed and moves vertically downwards, counter-current to the primary combustion air. The main type is the spreader (or sprinkler) stoker, a typical design for which is illustrated in figure 3.16. Since the large coal is thrown to the end of the grate, the moving grate travels towards the sprinkler impeller in order to provide sufficient burn-out time.

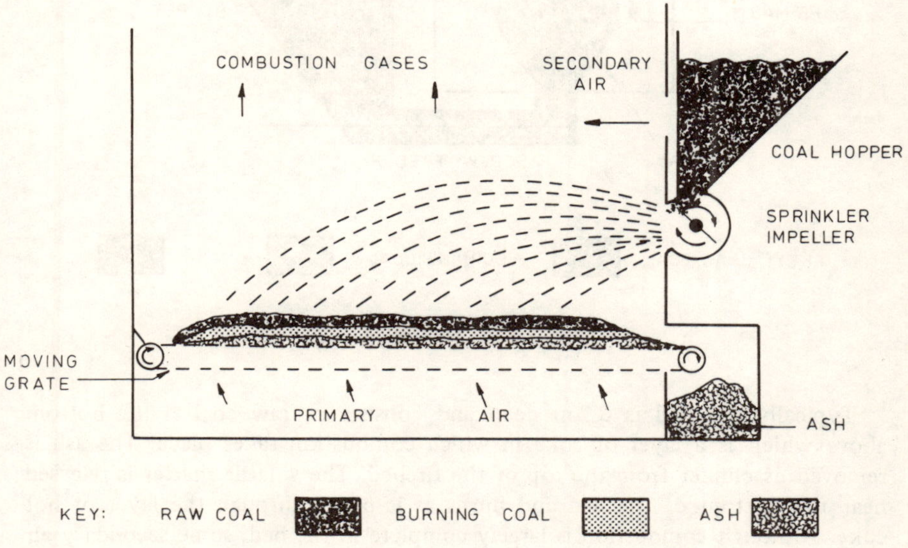

Figure 3.16 Sprinkler/spreader stoker

The bed, which is usually about 0.05 m deep, comprises a layer of coal near the surface, a layer of coke in which combustion takes place and a layer of ash immediately above the grate. The primary air passes upwards through the ash and burning coke to the raw coal on the surface. The volatiles released by the coal are swept out of the bed and burn in the space above the bed where secondary air is injected. Some of the fines in the feed coal also burn above the bed and a proportion may be carried out unburned in the combustion gases. When burning smalls coal it is generally necessary to refire the grits collected from the first stage of particulate clean-up in order to obtain a satisfactory combustion efficiency.

Underfeed stokers

Underfeed stokers use a screw or ram to feed coal from a hopper to the bottom of a retort (see figure 3.17). The bed is forced upwards by raw coal and therefore flows co-currently with the combustion air.

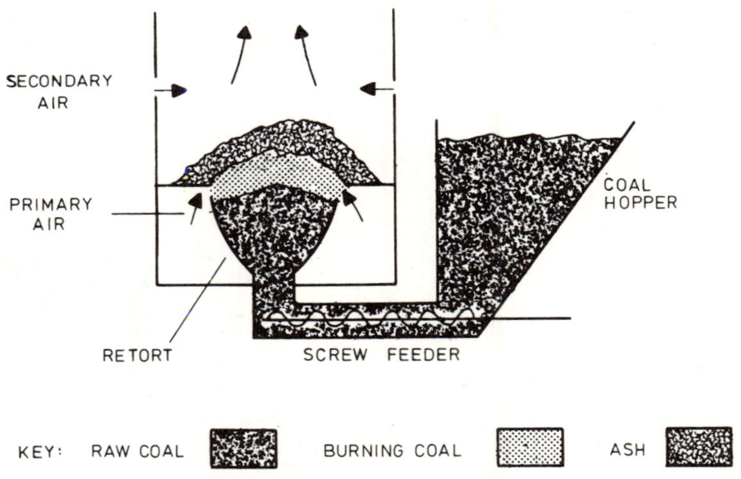

Figure 3.17 Underfeed stoker

Typically the bed is 0.2 m deep and consists of raw coal at the bottom above which is a layer of coke in which combustion takes place. The ash is removed as clinker from the top of the firebed. The volatile matter is released near the bottom of the bed and burns as it passes through the layer of hot coke. Although combustion is largely complete in the bed, some secondary air is injected above the bed to ensure the combustion of any carbon monoxide formed in the bed.

Crossfeed stokers

In crossfeed stokers coal is delivered to one end of the bed which moves continuously across the primary combustion air flow by the action of a grate. One of the main types of crossfeed stoker is the chain grate stoker illustrated in figure 3.18. Coal is fed from a hopper by gravity on to a continuously moving grate that supports the fuel bed moving coal away from the hopper and delivering ash at the far end of the furnace.

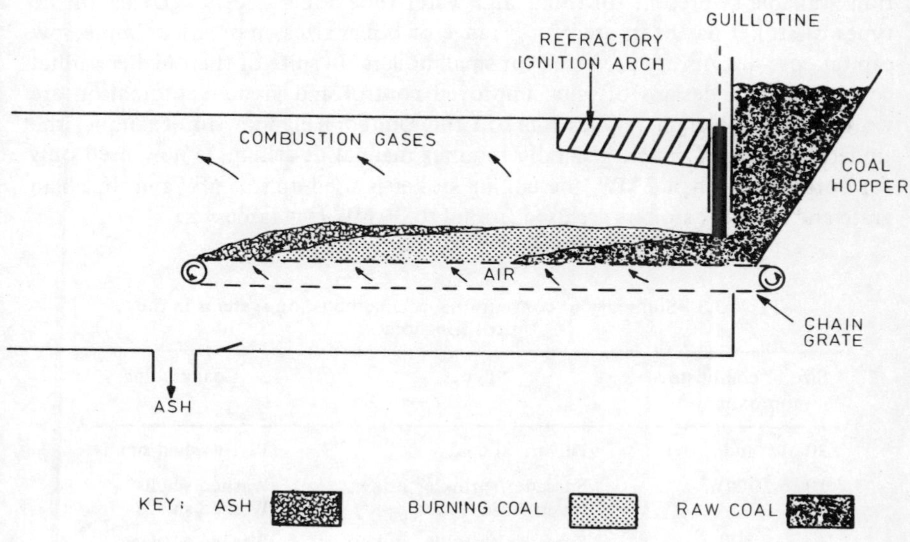

Figure 3.18 Chain grate stoker

The bed is generally about 0.1 m deep; the depth is controlled by the guillotine. Ignition is achieved by radiation from a refractory arch above the entry to the combustion chamber. Combustion proceeds downwards through the bed during the passage of the coal along the furnace (as shown in the figure). The combustion characteristics of crossfeed stokers are therefore more similar to those of underfeed stokers than to those of overfeed stokers. In particular, combustion of the solid fuel occurs principally within the firebed although some of the volatiles burn above the bed (assisted by a supply of secondary air).

Besides the chain grate stoker (and the similar travelling grate stoker), other types of crossfeed stoker include the vibrating grate stoker and the coking stoker. In the vibrating grate stoker coal is fed from the hopper on to an inclined grate which is vibrated to encourage the movement of the coal towards the far end where the ash is removed. In the coking stoker a ram is employed to feed the coal from the hopper. This is necessary because the coal is delivered first onto a static coking plate where the volatiles are released and burn above the bed with the secondary air. The resulting coke passes onto a reciprocating grate arranged to move the bed away from the feed hopper (for example, all of the grate bars move forward together but odd and even bars return separately).

Stokers are the traditional method of coal combustion below about 30 MW and have been used up to about 70 MW. Above about 30 MW, however, pulverised fuel combustion becomes attractive and is the only method of coal combustion available at present for firing large water-tube boilers. Although each of the types of stoker has been used for a range of boiler sizes, in practice simple, low capital cost designs are preferred for small boilers. In spite of their higher capital costs, however, designs offering improved control and greater automation are worthwhile for larger applications. In the United Kingdom, for example, the underfeed stoker, which generally requires manual de-ashing, is now used only for outputs less than 2 MW, the coking stoker is used up to 5 MW and the chain grate and spreader stokers are used from 2 to 30 MW (see table 3.3).

Table 3.3 Summary of conventional coal combustion systems in the United Kingdom

Size of combustor (approximate)	Types	Coal grading
30 MW and above	Pulverised coal	Part-washed smalls
10 to 30 MW	Spreader/sprinkler stoker	Washed smalls
	Chain grate stoker	Washed smalls
5 to 10 MW	Spreader/sprinkler stoker	Washed singles
	Chain grate stoker	Washed smalls
2 to 5 MW	Spreader/sprinkler stoker	Washed singles
	Chain grate stoker	Washed smalls
	Coking stoker	Washed smalls
2 MW and below	Coking stoker	Washed singles
	Underfeed stoker	Washed singles

The grade of coal (singles, smalls, etc.) used at a site is determined by considerations of handling, storage and fuel costs. For ease of handling and storage, singles are preferred for firing small boilers below about 2 MW whereas smalls which are cheaper, are used in larger units. For this reason underfeed stokers in the United Kingdom now generally burn singles although larger units burning smalls have been used in the past. In the case of spreader/sprinkler stokers, adequate height above the grate is required for the combustion of the fine coal entrained in the gas stream. As a result spreader/sprinkler stokers burning smalls are used mainly in water-tube boilers; horizontal shell boilers require singles because of limited headroom. The coal size gradings used in the United Kingdom for the various types of stoker are summarised in table 3.3.

The preferred type of coal for stoker firing is generally a low-rank, weakly caking or non-caking coal (either bituminous, sub-bituminous or lignitic). Anthracites are difficult to burn on stokers because of their comparatively low reactivity; underfeed stokers are the most suitable option for this type of coal. Caking coals, particularly those with a low ash fusion temperature, can cause problems of excessive clinker formation which restricts the primary air flow and inhibits ash removal. Such coals can be burned using crossfeed and overfeed stokers in which sufficient agitation of the bed occurs. In general, the bed of an underfeed stoker is not agitated and caking coals are unsuitable for use on this design.

Since the grate is directly in contact with burning coal in a crossfeed stoker, a minimum ash content of about 5 per cent is required in order to provide an insulating layer and to prevent the grate from overheating. Also, 'conditioning' or 'tempering' of the raw coal (that is, adding water) is common when firing smalls on a crossfeed stoker in order to reduce the loss of fine coal by elutriation or as 'riddlings' through the grate. For an underfeed stoker a low ash coal is preferred to minimise the amount of manual de-ashing. The underfeed stoker is tolerant to a range of moisture contents and the spreader/sprinkler stoker is tolerant to a range of both ash and moisture contents.

The acceptability of coal types for stokers is summarised in table 3.4.

The design of a stoker also determines the response rate, part-load operation and particulate carry-over. The main factor affecting these aspects of performance is the nature of the combustion process, particularly the amount of combustion that occurs above the bed, rather than mechanical features.

In the spreader/sprinkler stoker, volatiles and (when smalls are fired) fines burn above the bed so that the response to load changes is excellent. The more limited above-bed combustion occurring in the underfeed and crossfeed stokers makes the response of these designs slower.

At part-load operation, however, the underfeed stoker has an advantage in that the size of the 'dome' of burning coal reduces permitting the combustion temperature to be preserved while reducing the firebed area. This is not possible for crossfeed and overfeed stokers whose part-load operating characteristics are therefore inferior. The overfeed stoker suffers from the further disadvantage that a large amount of secondary air is required to control the emission of smoke which could result from the incomplete combustion of the volatile matter released into the space above the bed. This can reduce the boiler efficiency markedly at low load operation.

Since some of the fine coal burns in the space above the bed in a spreader/sprinkler stoker, ash is released into the combustion gas stream. Particulate loadings are therefore high and an efficient collection system, such as cyclones, is required to avoid excessive dust emissions. With crossfeed and underfeed stokers, however, combustion of the solid fuel occurs in the firebed and solids loadings in the gas stream are comparatively low. A simple grit arrestor is usually sufficient to control particulate emissions in these cases.

The main operating characteristics of stokers are summarised in table 3.4.

Table 3.4 Characteristics of industrial stokers

Stoker	Preferred coal			Response rate	Part-load operation	Particulate collection
	Type	Ash	Moisture			
Overfeed	Not anthracite	Tolerant	Tolerant	Rapid	Poor	Efficient dust collection required
Underfeed	Non-caking and weakly caking	Not high ash	Tolerant	Slow	Good	Simple grit arrestor only
Crossfeed	Not anthracite	Not low ash	Not low moisture	Moderate	Moderate	Simple grit arrestor only

Types of industrial boiler

The largest sector of the industrial market is boilers producing hot water or steam. The structure of this sector is complex, containing a wide spectrum of boiler sizes and patterns of utilisation. Two main types of boilers are available: water-tube boilers and fire-tube boilers (also referred to as shell boilers).

In water-tube boilers, water or steam is passed through tubes that are heated from the outside by radiation, contact with hot combustion gases or immersion in a fluidised bed. The walls of the combustion chamber are formed by welding together tubes and further heat is extracted from the combustion gases by tube banks which form convective section heat exchangers. This method of construction offers considerable flexibility in design, the size and geometry of the boiler being readily adaptable for different applications and methods of firing.

One of the simplest designs of water-tube boiler, a two-drum 'D-type', is illustrated in figure 3.19. This relies on natural circulation to transfer water from the lower 'mud drum' through the vertical tubes where evaporation takes

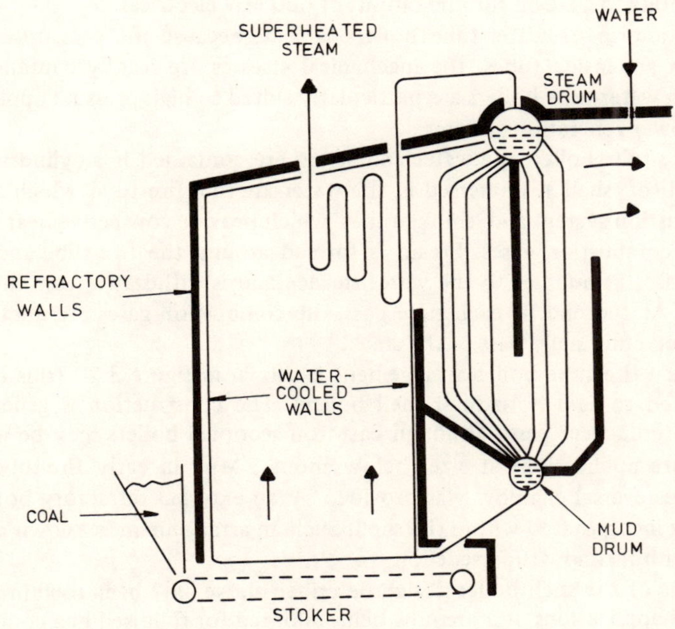

Figure 3.19 Water-tube boiler

place to the upper 'steam drum'. The boiler is designed to deliver wet steam to the steam drum which separates the liquid and vapour phases. Dry saturated steam is drawn off for super-heating and is replaced by a water feed to the steam drum. The feed water, together with that delivered to the steam drum in the wet steam from the evaporator tubes, is transferred to the mud drum by an external 'downcomer' pipe (not shown in the diagram).

Other types of water-tube boiler can involve two mud drums (as in the A-type natural circulation boiler) or pumped circulation (as in large power station boilers).

Water-tube boilers can conveniently be subdivided into two categories: 'packaged' boilers and site-fabricated boilers.

'Packaged' boilers are boilers that are factory fabricated and delivered to site complete with combustion system and ancillaries. The dimensions of packaged boilers are therefore limited by the requirements of road and rail transport. In practice, this means that packaged water-tube boilers are limited to outputs of less than 20 MW for coal and 40 MW for oil or gas (see below). Site fabrication is more expensive than factory fabrication and is therefore used only for boilers that are too large for transport.

No important size limitations have yet been encountered for site-fabricated water-tube boilers; the largest power station boilers are about 2000 MW thermal, corresponding to a steam turbine output of 660 MW electrical.

An advantage of water-tube boilers is that, because the pressurised water and steam are inside tubes, the mechanical stresses are readily contained. For this reason water-tube boilers are particularly suited to high-pressure applications such as power generation systems.

In a fire-tube boiler, the water and steam are contained in a cylindrical pressure vessel (or 'shell'). Immersed in the water are the 'fire-tube' which contains the combustion system and 'smoke tubes' which provide convective heat transfer from the combustion gases. Steam is formed around the fire-tube and smoke tubes. It rises as bubbles to the water surface and is withdrawn from the top of the shell. At the end of each tube pass, the combustion gases are ducted by a 'smoke box' and finally pass to the stack.

A typical fire-tube boiler arrangement is shown in figure 3.20 (this design is also referred to as a horizontal shell boiler). The construction is generally of fabricated mild-steel plate although cast-iron sectional boilers may be used for low-pressure applications at sizes below about 2 MW. In early fire-tube boiler designs the reversal chamber was provided by an external refractory box rather than being incorporated within the shell. Such an arrangement is known as a dry-back boiler but is now superseded.

Variants of the shell boiler design described above have been used in the past for special applications and are now being adopted for fluidised bed combustion (see below).

Most fire-tube boilers are supplied as factory-assembled packages (that is, complete with combustion system and ancillaries). This limits the maximum length of the shell to about 7 m. The maximum diameter of both the shell and the fire-tube is limited by stress considerations. In the case of the shell the maximum diameter is about 4.5 m (depending on the steam pressure) and

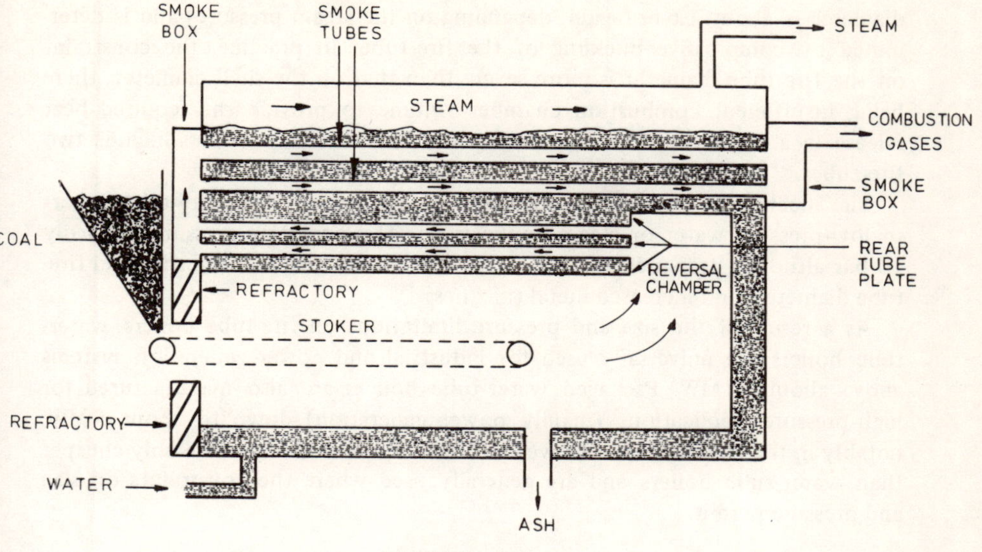

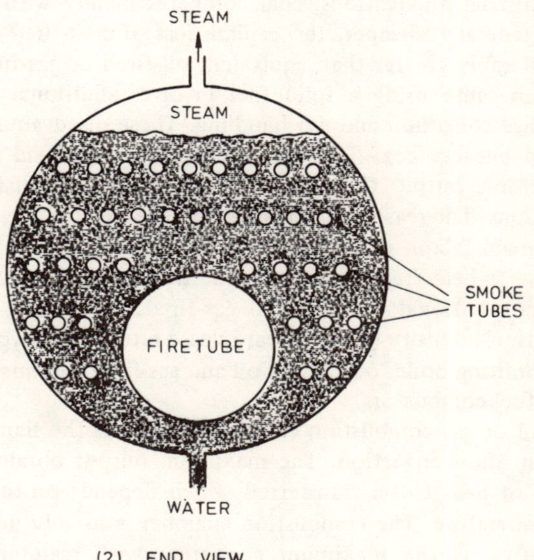

Figure 3.20 Fire-tube boiler

is determined by the stress on the flat end-plates. The maximum fire-tube diameter is about 1.6 m (again, depending on the steam pressure) and is determined by compressive buckling of the fire-tube. In practice, the constraint on the fire-tube diameter is more severe than that on the shell diameter, there being insufficient combustion chamber volume to provide the required heat release in a single fire-tube. For this reason, large fire-tube boilers include two fire-tubes.

The mechanical stresses on the shell and fire-tube restrict fire-tube boilers to low-pressure water or steam applications. The operating pressure is usually 10 bar although pressures up to 20 bar can be used with reduced shell and fire-tube diameters and increased metal thickness.

As a result of the size and pressure limitations on fire-tube boilers, water-tube boilers are universally used for industrial and power generation systems above about 20 MW. Packaged water-tube boilers are also manufactured for high-pressure applications (mainly power generation) down to about 5 MW, notably in the United States. However, fire-tube boilers are considerably cheaper than water-tube boilers and are generally used where the constraints on size and pressure permit.

Oil and gas-fired boilers compared with coal-fired boilers

For industrial applications, coal competes mainly with oil and gas. Although coal is generally cheaper, the capital cost of coal-fired combustion equipment is considerably greater than equivalent oil-fired or gas-fired systems. This arises in part because using a solid fuel involves additional costs in coal handling, particulate collection and ash handling. These disadvantages are, however, compounded because coal-fired systems require larger and more expensive boilers for the same output throughout the range of industrial and power generation applications. The reasons for this are discussed below.

The main factor affecting the output that can be obtained from a boiler of a given size is heat transfer. This is true throughout the range of boiler sizes and for all conventional coal, oil and gas-fired combustion systems. In particular, combustion intensity (that is, heat release rate per unit volume) may be regarded as not limiting boiler output for oil and gas-fired systems, or for stokers and pulverised fuel combustors.

An oil or gas combustion chamber surrounds the flame and cools the gas by radiation and convection. The maximum output obtainable is limited by the amount of heat to be transferred which depends on the combustion chamber exit temperature. The combustion chamber is usually designed so that the exit temperature is the maximum consistent with maintaining the heat transfer surface metal temperatures at the beginning of the convective section within a satisfactory working range. For water-tube boilers the combustion chamber exit temperature is usually about 1200 °C and is limited by considerations of stress and (for oil) corrosion of superheater tubing. For fire-tube boilers there is the additional constraint of the mechanical stress on the 'rear tube plate' (see figure 3.20). This component is subjected to the turbulent flow conditions in the reversal chamber and therefore to high heat transfer coefficients and high metal temperatures. Accordingly, the combustion chamber gas exit temperature is generally limited to about 1100 °C in this case.

The combustion chamber of a coal-fired boiler of any type or method of firing is designed to cool the combustion gases to below the ash resolidification temperature before the combustion chamber exit. This is necessary to avoid the formation of fused deposits on convective section heat transfer surfaces. The amount of cooling necessary depends on the properties of the ash although exit temperatures of about 1000 °C are typical. For pulverised fuel systems the furnace is designed to limit heat transfer coefficients to the combustion chamber walls so that the ash particles cool mainly by radiation. This helps to minimise ash fouling on the walls. A total gas residence time of 2 s is typical compared with a combustion time of about 0.25 s for coal particles. Stokers, whether used in water-tube or fire-tube boilers, suffer from the disadvantage of permitting heat transfer only from above the grate. In a fire-tube boiler, for example, the stoker masks the lower half of the fire-tube for about two-thirds of the length.

Combustion chambers for coal-firing therefore have a lower output per unit size than oil or gas-fired systems for two reasons: more heat has to be transferred to cool the gas to a lower exit temperature and heat transfer rates are lower to avoid ash fouling in pulverised fuel furnaces and because of masking by the grate in stoker systems.

In terms of the overall boiler performance, however, the convective section is at least as important as the combustion chamber. In this respect too, coal is at a disadvantage compared with oil and gas. For both water-tube and fire-tube boilers, the convective section heat exchangers for oil and gas firing are designed for a combustion gas velocity of about 30 m/s (limited by heat fluxes and therefore metal temperatures). For coal firing, the gas velocity has to be reduced to about 15 m/s to avoid excessive erosion of the heat transfer surfaces by the ash particles. The reduction in the velocity decreases the heat transfer coefficient typically to two-thirds that for oil and gas-fired systems. A substantially larger convective section containing more heat transfer surface is therefore required for coal than for oil or gas.

The limitations on boiler output discussed above are summarised in table 3.5. Coal is at a disadvantage in having to adopt more expensive construction techniques (or to supply multiple units) in order to provide the same output as oil-fired or gas-fired systems. This is illustrated in table 3.6 for conditions in the United Kingdom.

For example, at outputs between 20 and 40 MW, coal-fired units are site-fabricated water-tubes whereas packaged water-tubes can be supplied for oil or gas. Similarly, below 10 MW, fire-tube boilers with one combustion chamber

Table 3.5 Limitations on boiler output

Firing system	Combustion chamber	Convective section
Oil/gas	Exit temperature is limited to 1100 °C (fire-tubes) or 1200 °C (water-tubes) to avoid excessive metal temperatures at the convective section entry	Gas velocity is limited to 30 m/s to avoid high heat transfer coefficients (which would give excessive metal temperatures) and excessive pressure drops
Coal (pulverised fuel)	Gas must be cooled to above 1000 °C exit temperature to resolidify the ash mainly by radiation	Gas velocity is limited to 15 m/s to avoid erosion
Coal (stokers)	Exit temperature is limited to about 1000 °C as above; also the grate masks potential heat transfer surface area	Gas velocity is limited to 15 m/s as above

Table 3.6 Preferred United Kingdom boiler designs for conventional combustion systems

Size of boiler	Boiler type	
	Coal	Oil and Gas
40 MW and above	Water-tube boiler (site fabricated)	
20 to 40 MW	Water-tube boiler (site fabricated)	Water-tube boiler (packaged)
10 to 20 MW	Water-tube boiler (packaged)	Fire-tube boiler (two combustion chambers)
5 to 10 MW	Fire-tube boiler (two combustion chambers)	Fire-tube boiler (one combustion chamber)
5 MW and below	Fire-tube boiler (one combustion chamber)	

can be used for oil or gas whereas for coal firing two combustion chambers are required for outputs above 5 MW. These differences contribute to coal-fired boilers having a higher capital cost (by a factor of at least 2 to 3) than oil or gas-fired units of the same output.

Existing coal-fired combustors also compare unfavourably with oil-fired and gas-fired systems in having higher maintenance costs (particularly with stokers that contain hot moving parts), inferior response rates during start-up and load following, greater operator involvement and restrictions on the characteristics and variability of the fuel (for example, caking properties, ash content and size grading).

Fluidised bed combustion offers the possibility of avoiding some of the disadvantages of existing coal combustion systems and minimising others thereby permitting coal to compete more effectively with oil and gas. For this reason, vigorous programmes aimed at developing the technology for industrial applications have been initiated over recent years. These are now reaching the point at which units are beginning to be offered commercially. The opportunities for industrial fluidised bed boilers are discussed in the next section.

Industrial fluidised bed boilers

The principal advantages of using fluidised bed combustion in coal-fired industrial boilers are as follows:

(1) If, for environmental reasons, sulphur emissions to the atmosphere have to be controlled, this can be achieved simply and cheaply by adding a sulphur acceptor (limestone, for example) to the fluidised bed.

(2) A fluidised bed combustor is tolerant to the properties of the feed coal and is particularly advantageous for coals of high, low or variable ash content or of low ash fusion temperature. As discussed earlier, such coals may cause problems with some types of stoker.

(3) There is considerably less fouling of convective section heat exchange surfaces than with pulverised fuel or stoker systems and maintenance-free periods of up to a year may be possible. There are also indications that fluidised bed combustion ash is less erosive than that from conventional coal-fired systems. This may permit higher convective section gas velocities to be used, perhaps approaching those in oil-fired and gas-fired boilers.

(4) At present it appears that, although the boiler is cheaper with fluidised bed combustion than with a stoker, the ancillaries (including the start-up and control system) are more expensive. The result is that there is at present no significant capital cost difference for a boiler of the same output, although further development potential for fluidised bed combustion may remain.

However, the ability to put tubes in the fluidised bed and the improved convective section heat transfer (see (3) above) combine to make a fluidised bed boiler physically smaller than a stoker-fired design. For any given type of boiler (packaged water-tube, fire-tube, etc.) a higher output (up to about 50 per cent) can be provided at the maximum transportable size. This enables coal-fired fluidised bed boilers to compete more effectively with oil-fired and gas-fired systems at the large boiler sizes than stokers, by adopting more favourable construction techniques and by employing a smaller number of units.

The introduction of fluidised bed firing to the industrial boiler market can occur in three main ways

the conversion of existing oil-fired or gas-fired units to coal
new units based on existing boiler designs
novel boiler concepts

Although the lifetime of a boiler is 15 to 50 years, depending on the type (longer lifetimes apply to water-tube boilers), it is necessary to replace the firing system approximately every ten years. There is therefore the opportunity to install new combustion systems in existing boilers before the end of their lifetimes.

If the boiler was originally designed for coal but has subsequently been converted to oil or gas, reconversion to coal may be attractive, particularly if coal and ash handling facilities already exist on-site. However, the number of such boilers is generally small. In any case it is probably preferable to reconvert using the original method of firing (pulverised fuel or stoker) rather than fluidised bed combustion. The main exception to this is for water-tube boilers where sulphur retention is required. Under these circumstances reconversion using fluidised bed combustion appears attractive. At present, however, reconversion of fire-tube boilers to coal may be unattractive if sulphur-emission control is required because of the high acceptor feed rates needed (and, in the absence of sulphur-emission controls, reconversion is more likely to occur using stoker systems because of the extensive changes necessary to adapt an existing facility to fluidised bed firing).

If the boiler was designed for oil or gas, conversion to coal using pulverised fuel, stokers or fluidised bed combustion technology is unlikely to be economic. As discussed earlier, the convective sections of compact oil-fired and gas-fired boilers are designed for higher gas velocities and temperatures than are possible with coal (because of erosion and ash fusion problems). Extensive changes and some derating would therefore be necessary. Furthermore, sites using oil-fired or gas-fired boilers often do not have sufficient space for coal and ash storage and handling facilities to be introduced. The possibilities for conversion using coal-oil mixtures or coal–water mixtures are discussed in section 6.6.

There is clearly an economic advantage in using existing manufacturing facilities by adapting fluidised bed combustion to present boiler designs. The flexibility of the water-tube boiler concept is such that fluidised bed firing can be incorporated into existing designs without departure from accepted principles, except for the provision of tubes immersed in the fluidised bed. In the case of existing fire-tube boiler designs, however, adaptation to fluidised bed firing inevitably involves compromise. The limited headroom available in the cylindrical combustion chamber of a conventional fire-tube boiler involves the use of a shallow fluidised bed which makes it difficult to obtain the most favourable performance characteristics (for example, combustion rate and turn-down capability). In particular, it is not known whether effective sulphur retention can be achieved with shallow fluidised beds.

Although, initially, the application of fluidised bed combustion to industrial boilers was based on the conversion of existing designs, the limitations noted above have led to interest now being directed primarily towards the evolution of novel boiler designs specially for fluidised bed combustion. The main objective of these designs is to provide more bed and freeboard height for the combustor. Several lines of development are being followed.

Vertical shell boilers

Few examples of vertical shell boilers can be found prior to the advent of fluidised bed combustion. A typical fluidised bed design is shown in figure 3.21. In general, the mean velocity of steam leaving the water surface is limited to about 0.05 m/s to minimise splashing and the consequent entrainment of water droplets in the steam. Since there is less water surface area in a vertical shell boiler

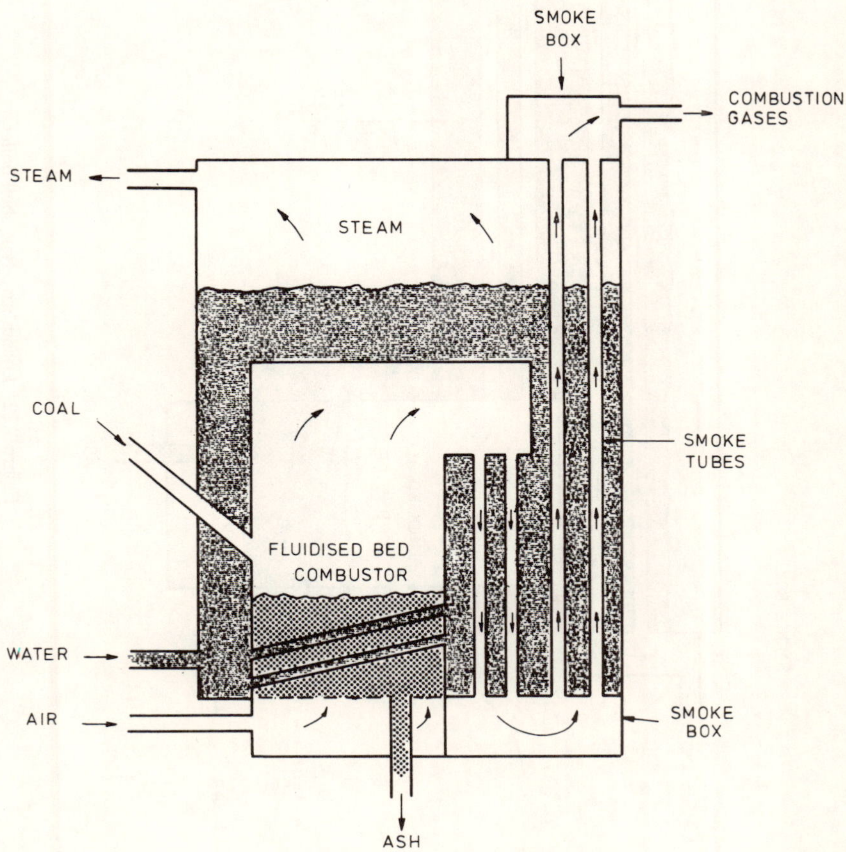

Figure 3.21 Vertical shell boiler

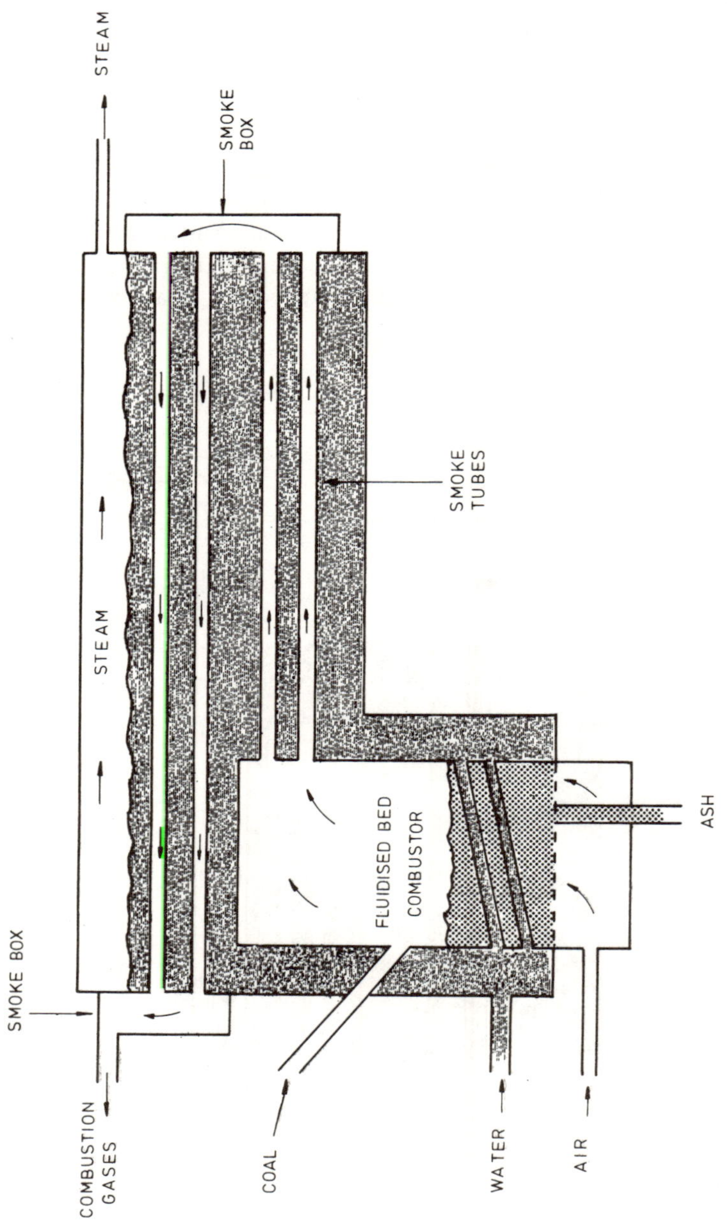

Figure 3.22 Locomotive-type shell boiler

than in the conventional horizontal shell design, the maximum output for steam raising boilers is limited to about 3 MW at the largest shell diameter. For hot water boilers the maximum output is about 5 MW, limited by the fire-tube diameter and therefore the bed area. Several units are in commercial operation in the United Kingdom both for hot water and steam raising.

Locomotive-type shell boilers

This is a traditional boiler design specifically for high-pressure coal-fired applications where the reduced fire-tube diameter does not allow sufficient room for a conventional stoker. As shown in figure 3.22, there is a separate combustion chamber constructed of water-cooled flat plates (with mechanical strengthening) attached to a horizontal shell convective section boiler. For fluidised bed combustion, locomotive-type construction provides the required height of combustion chamber. It is applicable mainly to the boiler sizes intermediate between the vertical shell boilers and packaged water-tube boilers.

Other novel boilers

In addition to the above boilers, there are novel designs for horizontal shells for example, the Wallsend Slipway Engineers' boiler that incorporates an enlarged, non-cylindrical combustion chamber and the Babcock International Ltd boiler that combines a water-tube wall combustion chamber with a horizontal shell boiler convective section. Both of these boilers have been developed in collaboration with the National Coal Board.

For small applications, below about 1 MW, fluidised bed combustion is generally unattractive because of its relatively expensive start-up and control systems. Oil-fired and gas-fired systems are particularly convenient for meeting such demands although fully automated coal-fired systems based on stokers are also available.

The preferred boiler types for coal-fired fluidised bed operation are summarised in table 3.7.

Table 3.7 Preferred boiler designs for fluidised bed firing

Size of boiler	Boiler type
10 MW and above	Water-tube
5 to 10 MW	Novel shell boilers (including locomotive-type)
3 to 5 MW	Novel shell boilers. Horizontal fire-tube boilers
1 to 3 MW	Vertical shell boilers. Horizontal fire-tube boilers
Below 1 MW	Fluidised bed combustion generally uneconomic

The structure of the industrial boiler market is such that the large boiler units operate at the highest load factors (see table 3.8). Large boilers therefore consume proportionately more coal than small boilers and are consequently an attractive target for increased coal sales. This sector also offers the potential for increasing the market penetration of coal because of the concentration of energy consumption in a relatively small number of units (see also table 3.8). There are, as a result, clear incentives to introduce fluidised bed boilers initially in the larger unit sizes and it may be expected that the development of the technology will be directed increasingly towards this objective. However, market penetration is likely to be slow, particularly for large boilers; in the United Kingdom, for example, the average age of water-tube boiler capacity is 20 years whereas, for fire-tube boiler capacity, the average age is 12 years.

Table 3.8 Structure of the United Kingdom industrial boiler market (1980 data)[a]

Size of boiler	Load factor (%)	Energy consumed (%)	Number of boilers (%)
100 MW and above	56	13	Less than 1
50 to 100 MW	42	8	Less than 1
20 to 50 MW	40	13	Less than 1
10 to 20 MW	31	9	Approx. 1
5 to 10 MW	23	16	3
3 to 5 MW	16	23	16
0.5 to 3 MW	12	15	40
0.3 to 0.5 MW	9	3	40
Total	–	100	100

[a] Source: based on the NCB's ORE data bank.

The extent to which coal will penetrate the market for small boilers is uncertain, given also the higher levels of amenity expected in this size range. The introduction of sulphur-emission controls for industrial boilers, while favouring fluidised bed combustion compared with conventional combustion systems, would further reduce coal's competitiveness with low sulphur liquid and gaseous fuels in the small boiler market.

Fluidised bed furnaces

Within the industrial market a significant demand exists for hot gases, for example, for drying and heating processes. Although in the past coal-fired stoker furnaces were widely used for these applications, the convenience and simplicity of oil and gas furnaces has enabled these fuels to capture most of the market. However, considerable work has been carried out to develop fluidised bed furnaces so that coal can take advantage of its competitive price.

For convenience, furnace applications may be subdivided into two categories.

Semi-clean gas

A semi-clean gas can be produced in a simple, low-cost fluidised bed combustor without heat transfer tubes in the bed by using (at most) cyclone cleaning of the combustion gases. Coal generally competes with heavy fuel oil in such applications and indeed is more competitive with heavy fuel oil in this market than for firing boilers. This is because the comparatively low capital cost of simple fluidised bed furnaces enables the price advantage of coal over heavy fuel oil to provide a higher return on capital than for boiler applications.

The use of fluidised bed furnaces for crop and clay drying may now be considered commercial and several other applications are well-advanced.

Clean gas

As described earlier, a clean hot gas can be provided using coal-fired fluidised bed combustion by heating air indirectly, for example, with tubes immersed in the fluidised bed. Although the capital cost of such a system is considerably higher than that of a simple furnace to produce an uncleaned hot gas, the alternatives are to use expensive premium fuels (natural gas or light distillates) or to manufacture an industrial fuel gas from coal (see chapter 4).

Fluidised bed incinerators

Although not directly associated with coal use, the application of fluidised bed combustion technology to waste incineration has developed in parallel with its application to coal combustion.

A wide range of materials can be burned including colliery tailings, sewage sludge and wood waste. The diversity of these materials is testimony to the flexibility of the process. Many of the wastes have too low a content of combustibles to be burned using conventional methods. Some wastes contain more than 80 per cent non-combustibles (usually moisture and mineral matter) and, because they are not capable of self-sustaining combustion, require a supplementary coal feed. The main areas of design development are feeding of the waste materials to the bed and the recovery of heat both of which depend on the properties of the waste and the type of application.

Many industrial sites produce waste materials that are at present discarded but could contribute to their energy needs. The possibility of combining waste utilisation with coal combustion is an advantage of fluidised bed combustion. Similar systems can also be used to incinerate wastes that are potentially hazardous.

Industrial combined heat and power

A possible application of pressurised fluidised bed combustion is in industrial combined heat and power systems. The arrangement that would be used is similar to the waste heat boiler cycle (see figure 3.13) in that the hot gases produced by the combustor are expanded through a gas turbine to generate power and heat is recovered from the turbine exhaust. In combined heat and power applications, however, some or all of the steam raised in the waste heat boiler is used for heating applications.

The potential of combined heat and power systems as a future market for coal is discussed in section 6.7.

CHAPTER 4

GASIFICATION

Gasification is arguably the most versatile of the coal conversion processes having applications in almost every sector of energy demand.

In industrial installations and power generation systems, for example, a low calorific value gas (also called a low BTU gas, producer gas or fuel gas) or a medium calorific value gas (also called a medium BTU gas) may be used. A medium calorific value gas may also be converted into liquid fuels or chemicals and in this case is often referred to as synthesis gas. Finally, a substitute natural gas (commonly abbreviated to SNG but also called pipeline gas, high calorific value gas or high BTU gas) can be manufactured as a direct replacement for natural gas.

In order to meet these diverse needs and to improve on existing systems, numerous new designs of gasifier and gas processing systems have been proposed or are under development. It is, however, likely that only a few of these will be widely used commercially.

4.1 Chemical reactions

Many chemical reactions may occur in a gasifier, the three main types being

pyrolysis reactions
gasification reactions
acceptor reactions

The importance of each type of reaction and the extent of the interactions between them depend on the gasifier design.

Pyrolysis reactions

As coal is heated it decomposes into a char residue consisting mainly of carbon and gases including hydrogen, methane, steam, carbon dioxide, carbon monoxide and tar vapour. This process was the basis for the traditional method of coal gas manufacture in gasworks (see section 2.2) and can also be employed for the manufacture of liquid fuels from coal. The mechanisms occurring are discussed in more detail in section 5.2.

If suitable conditions exist in the gasifier (see below), the gases produced by pyrolysis will form part of the product gas. Alternatively the pyrolysis gases may take part in the gasification or acceptor reactions described below.

Gasification reactions

Combustible gases can be produced by the reaction of the coal, char or volatile matter with oxygen, carbon dioxide, hydrogen or steam. The main reactions are listed below (for simplicity, only reactions with carbon are shown).

Partial-combustion reaction

$$C + \tfrac{1}{2}O_2 = CO \tag{4.1}$$

Boudouard reaction

$$C + CO_2 = 2CO \tag{4.2}$$

Hydrogasification reaction

$$C + 2H_2 = CH_4 \tag{4.3}$$

Water-gas reaction

$$C + H_2O = CO + H_2 \tag{4.4}$$

In regions of the gasifier where oxygen is in excess, combustion may also take place

Combustion reaction

$$C + O_2 = CO_2 \tag{4.5}$$

In gasifiers that are designed to take advantage of the hydrogasification reaction, most of the methane formed appears to result from hydrocracking of the primary tar rather than from hydrogasification of the coal or char. Typically, 80 per cent of the methane is derived from the tars and only 20 per cent directly from the coal or char.

The following gas-phase reactions may also occur.

Shift reaction

$$CO + H_2O = CO_2 + H_2 \tag{4.6}$$

Methanation reaction

$$CO + 3H_2 = CH_4 + H_2O \tag{4.7}$$

The sulphur in the coal is released mainly as hydrogen sulphide and the nitrogen mainly as elemental nitrogen and ammonia. Traces of other compounds including carbonyl sulphide and hydrogen cyanide are also formed.

Acceptor reactions

In some gasifiers, limestone or dolomite may be used to retain the sulphur. If the acceptor is calcined before feeding to the gasifier, carbon dioxide may also be retained. The reactions for calcined limestone are given below.

Sulphur retention

$$CaO + H_2S = CaS + H_2O \tag{4.8}$$

Carbon dioxide acceptor

$$CaO + CO_2 = CaCO_3 \tag{4.9}$$

Retention of sulphur is also possible using dolomite or uncalcined limestone (the analogous reactions for combustion conditions are given in section 3.2).

Heats of reaction

An important factor affecting the choice of reactants and gasifier operating conditions is the heat released or absorbed by the above reactions. This is summarised in table 4.1. As shown in the table, all of the reactions are exothermic (that is, release heat) except the Boudouard and water-gas reactions.

In a gasifier the net heat release has to be just sufficient to bring the reactants to the design operating temperature. Heat therefore has to be supplied to meet the sensible heat requirements and those for the endothermic (that is, heat-absorbing) Boudouard and water-gas reactions. In most designs this is achieved by the combustion and partial-combustion reactions although systems using the methanation reaction, the carbon dioxide acceptor reaction or an external heat source are also under consideration.

The various options for obtaining a heat balance in the gasifier are discussed further in the next section.

Equilibrium considerations

An indication of the effect of the temperature and pressure conditions in a gasifier on the product gas composition may be obtained by considering the theoretical composition if the reactions were allowed sufficient time to reach equilibrium.

In the presence of an excess of carbon, the equilibria for the combustion and partial-combustion reactions correspond to extremely low oxygen concentrations and for practical purposes it may be assumed that the oxygen content of the product gas is zero for most gasifiers (some industrial gas producers may have up to 0.5 per cent oxygen in the product gas).

Table 4.1 Summary of gasification reactions[a]

Reaction	Equation	Equilibrium conditions: Effect of increase in temperature	Equilibrium conditions: Effect of increase in pressure	Kinetics	Heat of reaction
Pyrolysis	Coal = H_2, CH_4, H_2O, CO_2, CO, tar etc.	$\rightarrow$	$\leftarrow$	fast	exothermic (mildly)
Partial combustion	$C + \frac{1}{2}O_2 = CO$		$\leftarrow$	fast	exothermic
Boudouard	$C + CO_2 = 2CO$	$\rightarrow$	$\leftarrow$	rather slow	endothermic
Hydrogasification	$C + 2H_2 = CH_4$	$\leftarrow$	$\rightarrow$	slow	exothermic (mildly)
Water-gas	$C + H_2O = CO + H_2$	$\rightarrow$	$\leftarrow$	moderate	endothermic
Combustion	$C + O_2 = CO_2$			fast	exothermic (strongly)
Shift	$CO + H_2O = CO_2 + H_2$	$\leftarrow$		moderate	exothermic (mildly)
Methanation	$CO + 3H_2 = CH_4 + H_2O$	$\leftarrow$	$\rightarrow$	slow	exothermic (strongly)
Sulphur retention	$CaO + H_2S = CaS + H_2O$	$\leftarrow$		fairly slow	exothermic (slightly)
Carbon dioxide acceptor	$CaO + CO_2 = CaCO_3$	$\leftarrow$	$\rightarrow$	fairly slow	exothermic

[a]Much of the volatile matter is removed rapidly from the coal at temperatures below those at which gasification takes place. For most gasifiers, the oxygen concentration at equilibrium is negligible.

The effect of the hydrogasification, water-gas, and Boudouard reactions on the equilibrium gas composition is illustrated in figure 4.1 for a range of temperatures and pressures. The curves show the conditions under which the conversion of the reactant gas is 90 per cent complete in the presence of excess carbon (starting with hydrogen, steam and carbon dioxide, respectively). In the case of the water-gas reaction, the hydrogen and carbon monoxide products are also allowed to come to equilibrium with the carbon via the other two reactions.

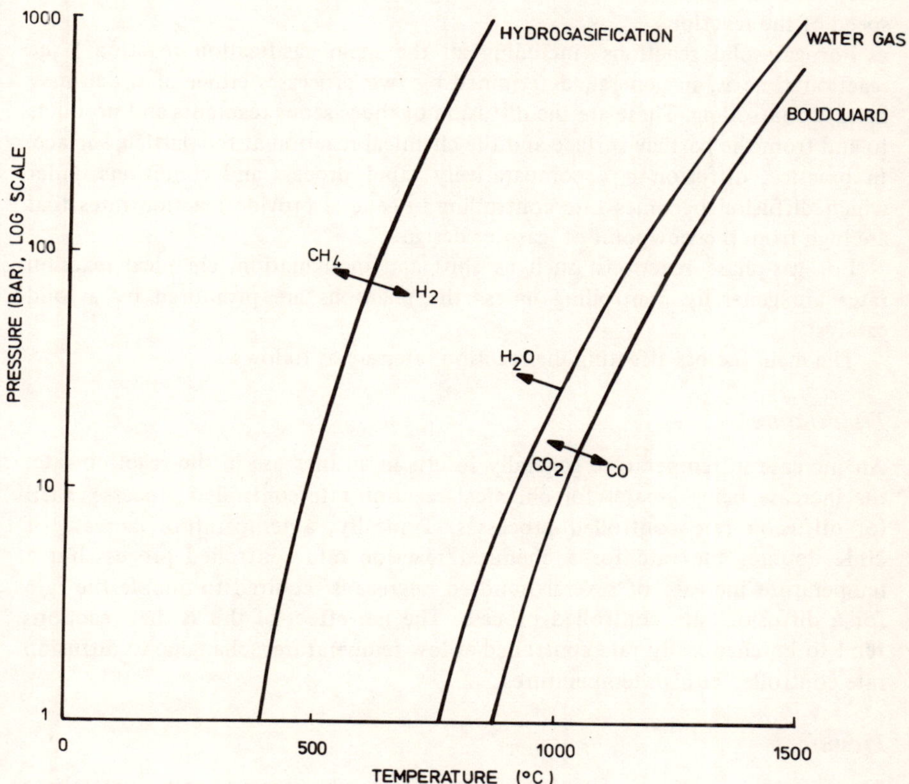

Figure 4.1 Equilibrium conditions for 90 per cent reaction of hydrogen, steam and carbon dioxide with excess carbon. The arrows indicate the direction in which the equilibrium gas composition changes with temperature and pressure

As indicated in the figure, the presence of methane, steam and carbon dioxide in the product gas is favoured by low temperatures and high pressures whereas the presence of hydrogen and carbon monoxide is favoured by high temperatures and low pressures. Equilibrium considerations therefore point to a high temperature, low pressure operating regime for gasifiers except if methane is a desired product.

The effects of temperature and pressure on the equilibria of the main reactions are summarised qualitatively in table 4.1.

Reaction kinetics

The time taken for some of the above reactions to reach equilibrium can, however, be considerable and the design of gasifiers has to take into account the speed of the reactions.

For gas–solid reactions (including all the main gasification reactions) the reaction time is, in general, determined by two processes either of which may be rate controlling. These are the diffusion of the gaseous reactants and products to and from the particle surface and the chemical reaction at the particle surface. In practice, diffusion is a comparatively rapid process and conditions under which diffusion becomes rate controlling in general provide reaction rates that are high from the viewpoint of gasifier design.

For gas-phase reactions, such as shift and methanation, chemical reaction rates are generally controlling unless the reactions are promoted by a solid catalyst.

The main factors affecting the reaction rates are as follows.

Temperature

An increase in temperature generally results in an increase in the reaction rate, the increase being greater for chemical reaction rate controlled processes than for diffusion rate controlled processes. Typically, a temperature increase of 20 K doubles the rate for a chemical reaction rate controlled process but a temperature increase of several hundred degrees is required to double the rate for a diffusion rate controlled process. The net effect of this is that reactions tend to be chemically rate controlled at low temperatures, changing to diffusion rate controlled at high temperatures.

Pressure

The gas–solid reactions occurring in a gasifier can be regarded approximately as first-order chemical reactions so that an increase in the operating pressure results in a proportional increase in the chemical rate constant. The gas-phase reactions are generally second-order (or higher) and the chemical reaction rate therefore increases substantially with pressure in this case.

Diffusion rate controlled processes are little affected by pressure.

Coal/char properties

Both chemical reaction rates and diffusion rates are dependent on the properties of the solid material.

The absolute value of the chemical reaction rate can vary greatly depending on the reactivity of the material. For example, in the case of the water-gas reaction, chars produced under different carbonising conditions can have reaction rates differing by an order of magnitude or more at the same temperature and pressure.

Diffusion rates vary less, being affected mainly by the particle surface area, pore structure and the thickness of the boundary layer across which mass transfer occurs. High diffusion rates are favoured by fine particles and turbulent gas–solids mixing.

Type of reaction

The chemical reaction rates for the combustion and pyrolysis reactions are extremely high, being several orders of magnitude higher than those for the next fastest reactions, the water-gas and shift reactions. The conversion of carbon dioxide to carbon monoxide by the Boudouard reaction is somewhat slower still (typically by half an order of magnitude) with the hydrogasification and methanation reactions being slower than the Boudouard reaction by about a further two orders of magnitude. The relative magnitudes of the chemical reaction rates are summarised qualitatively in table 4.1.

Diffusion rates vary comparatively little with the type of reactants. The main factor is the molecular weight, hydrogen diffusing quicker than the other species.

In practice, for gasification processes using oxygen or air, the oxygen is consumed rapidly at the beginning of the reaction zone and the other reactions occur more slowly as the resulting gases pass through the remainder of the reactor. As a result of the high chemical reaction rates for the combustion reactions, these are generally diffusion rate controlled. Both the combustion and partial combustion reactions can occur at the surface of the coal particles. At non-slagging temperatures the combustion reaction predominates and the production of carbon monoxide depends mainly on the relatively slow Boudouard reaction. The direct production of carbon monoxide by the partial-combustion reaction becomes important at slagging temperatures.

The term 'pyrolysis' covers a variety of reactions that may be either chemical rate controlled, diffusion rate controlled or transitional at non-slagging temperatures. At slagging temperatures, pyrolysis reactions are generally diffusion rate controlled and are comparatively fast.

The other reactions–Boudouard, hydrogasification, water-gas, shift and methanation–are generally chemical rate controlled under non-slagging conditions. At slagging temperatures these reactions may be either chemically rate controlled, diffusion rate controlled or transitional. However, whichever mechanism is rate controlling, the reactions are fast in the context of gasifier design at these temperatures.

The requirements imposed by considerations of equilibrium conditions, the speed of the reactions and practical constraints (for example, solids handling and materials problems) are generally in conflict and a number of compromises exist. The nature of these design options will be discussed in the next section.

4.2 Gasifier designs

A wide variety of gasifier designs has been developed for different applications and types of coal feedstock. In the present review the processes are characterised in terms of the four main design parameters temperature, pressure, reactant gases and method of contacting.

Temperature

Gasifiers can be divided into three categories depending on the physical state of the ash in the gasification reactor.

Dry ash

For most coals, operation at up to approximately 1000 °C enables the ash to be removed 'dry' without sintering or slagging.

Ash-agglomerating

It is also possible to operate at temperatures such that the ash particles become 'sticky', form agglomerates and, with an appropriate reactor design, are removed at a controlled rate to maintain steady-state operating conditions in the gasifier. For most coals, ash-agglomerating conditions occur in the temperature range 1000 to 1200 °C, depending on the composition of the ash.

Slagging

Alternatively, operation above about 1200 °C results in the ash forming a molten slag.

Since dry ash and ash-agglomerating gasifiers are limited by maximum operating temperatures, a number of general features of their performance and characteristics are similar. For simplicity, these two types of gasifier are therefore considered together as 'non-slagging' in most of the following discussion.

The temperature of gasification affects the product gas composition because of its influence on the equilibria and kinetics of the gasification reactions. The raw product gas from gasifiers operating under slagging conditions generally has relatively low concentrations of carbon dioxide and steam and relatively high concentrations of carbon monoxide and hydrogen. The gasification reactions in such gasifiers also produce negligible quantities of methane and tars as these species 'crack' (that is, undergo thermal decomposition) almost completely at slagging temperatures. However, methane and tar may be present in the product gas if the gasifier design is such that devolatilisation occurs separately at a low temperature.

Since methane decomposes at high temperatures all gasifiers based on hydrogasification have to operate under non-slagging conditions in order to obtain an acceptable methane yield.

If steam is used as a gasifying agent under non-slagging conditions it may be necessary to use a considerable excess (in some cases 400 per cent) over that which reacts with the coal. This arises because of the unfavourable kinetics and equilibrium conditions for steam decomposition at low temperatures. The use of excess steam can result in a significant loss in efficiency. However, the efficiency gain arising from the good steam utilisation under slagging conditions may be offset by increased losses from other factors. In particular, high temperatures usually lead to more oxygen being required for gasification and, consequently, a greater energy demand for air separation. Also, depending on the gasifier design, high temperatures can result in more of the input energy of the coal being recovered as steam and less as gas; this may be a disadvantage in some systems.

Gasifier throughputs are generally higher under slagging conditions because the increased reaction rates permit shorter gas and solids residence times and a higher conversion of the reactant gases to product gas. In particular the higher steam conversion obtained under slagging conditions can lead to a significant improvement in throughput.

For operation at slagging temperatures the reaction kinetics are fast and differences in coal reactivity are less important than at non-slagging temperatures. However, ash type and content can be important. The ash type should be such that a sufficiently mobile slag is obtained at the operating temperature. In particular, ashes with a high fusion temperature are generally unfavourable for slagging operation. In some cases, it may be possible to modify the slagging characteristics of the ash by the addition of a fluxing agent such as limestone. It is also important that refractory attack by the slag should not be excessive, as can be the case with a highly mobile slag.

Under non-slagging conditions, the more reactive coals (such as lignites) are generally easier to gasify particularly at high throughputs. For gasifiers employing steam it is usual to operate at as high a temperature as possible (within the limits imposed by the dry ash or ash-agglomerating design) in order to improve the reaction kinetics and equilibrium yields. In this case, coals with a high ash fusion temperature are preferable.

Although gasification at high temperatures has a number of advantages (for example, high rates of reaction and the ability to gasify unreactive coals), the technology is generally regarded as more difficult than gasification at low temperatures. With the exception of the Koppers–Totzek gasifier, which operates at slagging temperatures, all of the designs that are in current commercial use are dry ash gasifiers. The areas in which slagging operation involves special design expertise are the choice of refractory materials to minimise corrosion and erosion by the high temperature slag and the removal of slag from the gasifier (including the control of slag viscosity). Successful operation at ash-agglomerating temperatures depends on control of the rate of agglomeration, the size of the agglomerates and the rate of removal.

Pressure

Gasification processes may be operated either at atmospheric pressure or at elevated pressure.

Equilibrium considerations indicate that operation at elevated pressure tends to discourage the decomposition of carbon dioxide and steam and the formation of carbon monoxide and hydrogen. In practice, however, the effects on product gas composition are small at pressures up to about 30 bar, compared with other factors such as reaction temperature (see figure 4.1).

At higher pressures the formation of methane by the hydrogasification reaction is favoured by equilibrium considerations. Pressures of at least 80 bar are generally regarded as necessary for hydrogasification-based processes.

In most of the large-scale applications envisaged for coal gasification (substitute natural gas, power generation and synthetic liquids), the raw gas from the gasifier passes to downstream units which remove particulates and sulphur and carry out further chemical processing or combustion. Many of these downstream units operate at elevated pressure, usually in the range 10 to 30 bar. In general there are therefore two process-design options; either the reactants are compressed and the gasifier operated at elevated pressure or the gasifier is operated at atmospheric pressure and the product gas compressed for further processing (after cooling and particulate removal). As the product gas volume may be considerably greater than that of the reactant gases (typically 50 per cent greater for air/steam gasifiers and 250 per cent greater for oxygen/steam gasifiers at 0 °C and 1.013 bar), the additional compression energy requirements can represent a significant efficiency penalty for atmospheric pressure gasification particularly with oxygen/steam gasifiers.

Operation at elevated pressure increases the overall reaction rate but the change is generally less than proportional to the pressure because not all of the reactions are chemical rate controlled (for example, the combustion and thermal decomposition reactions are usually diffusion rate controlled—see above). The increase in throughput per unit volume of the gasifier is also less than proportional to the pressure, an empirical square root law having been proposed for a number of gasifiers. In practice the design of pressurised gasifiers often allows for longer gas and solids residence times than for atmospheric pressure gasifiers in order to increase the degree of conversion.

The advantages for the process efficiency and throughput are generally regarded as favouring pressurised operation in large-scale applications. For this reason most of the current development effort on coal gasification is being directed towards elevated pressure systems.

Elevated pressure gasification is a more difficult technology than atmospheric pressure gasification for several reasons. Foremost amongst these is that the coal has to be fed into the gasifier against a pressure gradient. Lock-hopper systems have been widely used for this purpose but other devices, for example coal–water slurry pump systems, are also under development. Similarly, systems for the removal of the ash to ambient pressure are required. The gasifier in an elevated pressure process is contained within a pressure vessel similar to that for a pressurised fluidised bed combustor (see section 3.6). The design of the product gas heat exchangers operating within a high-pressure gas containment system is a further development area. In particular, the construction materials must be selected to withstand the environment of a hot, dust-laden, reducing gas atmosphere.

Reactant gases

The three basic reactants for gasification processes are oxygen, steam and hydrogen. These can be used in a number of ways in practical schemes.

Oxygen/steam

In gasifiers using oxygen and steam the heat absorbed by the endothermic water-gas reaction is provided by the combustion reactions between oxygen and coal giving an overall heat balance within the gasifier.

Air/steam

For applications in which the presence of nitrogen in the product gas is not a disadvantage, air can be used instead of oxygen thereby saving air-separation costs. In this case the steam requirements are lower because more sensible heat is needed to bring the air to the reaction temperature. Indeed, heat-balance considerations indicate that operation with air and steam is usually possible only at non-slagging temperatures (see also below).

Air

At slagging temperatures it is possible to satisfy the heat-balance requirements using air alone as the gasifying agent, the heat released by the combustion reactions being balanced entirely by the sensible heat required to bring the air to reaction temperature. Steam may, however, be required in small quantities for control purposes to maintain a heat balance if the air is preheated or if oxygen-enriched air is used.

Under non-slagging conditions air may be used alone if heat is removed from the process by other than the endothermic steam–carbon reactions (for example, by indirect heat transfer to raise process steam or by rejecting hot char from the gasifier in a partial-gasification process).

Steam

The capital and operating costs of air-separation plant are substantial and this factor has encouraged interest in processes that produce a nitrogen-free gas using steam alone. In this case the heat absorbed by the water-gas reaction has to be provided by a method other than oxidation reactions in the gasifier (see below).

Hydrogen

For gasification processes designed to manufacture substitute natural gas (methane), it is an advantage for the gasifier product gas to contain a high concentration of methane as this reduces the need for further chemical conversion. The production of methane in the gasifier can be promoted by using hydrogen as a reactant, such a system being referred to as a hydrogasifier. The hydrogen is usually obtained from a conventional oxygen/steam gasifier fed with the unconverted char residue from the hydrogasifier.

The choice of reactants is determined primarily by the application and the required properties of the product gas. If low calorific value gas is the final product then air and steam or air alone are used for gasification. To obtain a medium calorific value gas the introduction of nitrogen must be avoided and oxygen and steam, or steam alone, are employed. The absence of nitrogen makes medium calorific value gas suitable for conversion by further chemical processes to liquid fuels and chemicals, hydrogen or substitute natural gas. Alternatively, substitute natural gas can be produced directly with a hydrogasification process by using hydrogen as a reactant.

Processes using steam alone (with an indirect heat supply) are expected to be more efficient than oxygen–steam processes because no energy is required for air separation. For similar reasons, air-blown gasification processes may be expected to be more efficient than oxygen-blown processes. In this case, however, the advantage may be offset by increased losses unless effective use can be made of the greater sensible heat content of the product gas.

For the production of substitute natural gas, direct processes (employing a hydrogasifier) are generally regarded as being potentially more efficient than the manufacture of a synthesis gas followed by conversion to substitute natural gas. This is mainly because the conversion is a strongly exothermic reaction occurring typically at a temperature of 350 °C. Heat can therefore be recovered from this reaction only at a low temperature, introducing an efficiency penalty for the process. Sensible heat and pressure losses encountered in processing the comparatively large gas volumes associated with synthesis gas are also minimised in the hydrogasification route. However, it has yet to be established whether the inherent efficiency advantages of the hydrogasification route can be realised in practical systems.

The main impact of the choice of reactant gases on throughput is found in the comparison between air-blown and oxygen-blown processes. Expressed in terms of coal input or energy output (of gas), the throughput of an air-blown gasifier is reduced approximately in proportion to the amount of diluent nitrogen present. Air-blown gasifiers therefore operate typically at one-third to one-half of the throughput of equivalent oxygen-blown systems. Hydrogasifiers usually have fairly high throughputs, partly because of the high operating pressures (generally 80 to 200 bar).

As discussed earlier, most gasification reactors are designed so that an approximate balance is obtained between the endothermic and exothermic reactions. In particular, with steam–oxygen and steam–air systems, the heat absorbed by the water-gas reaction is provided by combustion reactions. In the case of gasification by steam alone, the heat absorbed by the reaction has to be supplied by another heat source. The three main options are

indirect heat transfer
a parallel exothermic chemical reaction not requiring oxygen
a heat carrier

Examples of all three options exist and are discussed in the next section. As a result of temperature limitations on the use of heat transfer materials and on the conditions necessary for suitable exothermic reactions, only the option of using a heat carrier appears feasible for operation at slagging temperatures.

Similarly, an external heat flow is required in order to obtain a heat balance in gasifiers using air alone at non-slagging temperatures.

Method of contacting

Methods of contacting the solid feed and the gaseous reactants in a gasifier can be considered in four categories.

Fixed bed

Coal is fed to the top of a bed and is heated as it moves downwards by the upward flow of hot gases. The coal passes through a carbonisation zone and then a gasification zone, finally reaching a combustion zone at the bottom of the bed where the reactant gases are injected. The system is illustrated in figure 4.2 and is employed in the commercial Lurgi gasifier and the fixed-bed gas producers.

Fluidised bed

In fluidised bed gasifiers the reactant gases are used to fluidise a bed of particulate material containing the coal. The bed can be regarded as being well mixed (see below). In order to avoid sintering of the ash and the consequent loss of

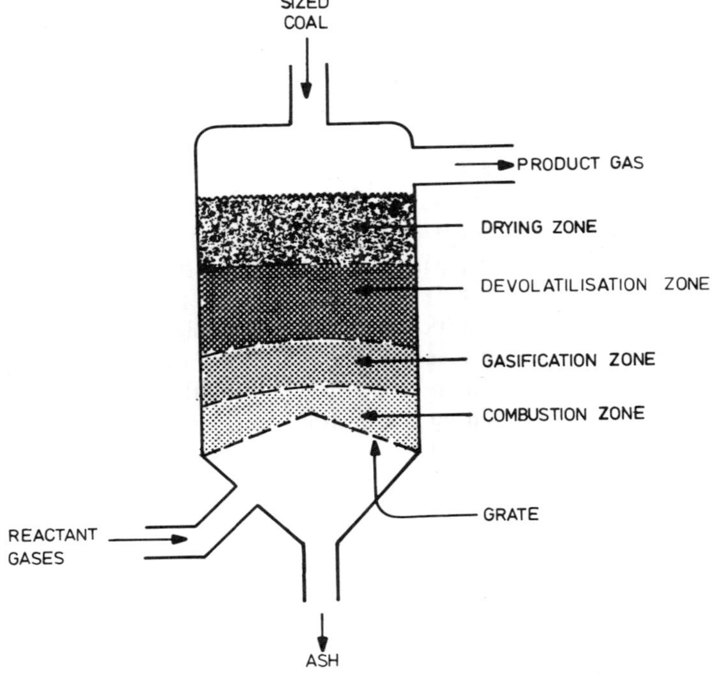

Figure 4.2 Fixed-bed gasifier

fluidisation, fluidised bed gasifiers are restricted to operating at non-slagging temperatures. Fluidised bed gasification is illustrated in figure 4.3 and is used in the commercial Winkler gasifier.

Molten bath

Molten bath gasifiers are similar to fluidised bed systems in that the reactions take place in a well-mixed medium of high thermal inertia. In this case, however, a bath of molten slag, metal or a salt is used. The operating temperature depends on the type of bath; for slag and molten metal baths, a high temperature (1400 to 1700 °C) is necessary but temperatures as low as 1000 °C can be used with molten salts. The reactant gases may be injected from above as jets which penetrate the surface of the bath, as shown in figure 4.4, or may be fed to the bottom of the bath. In either case good gas–solid contacting is obtained.

Entrained phase

In an entrained phase gasifier (see figure 4.5) coal pulverised to less than 0.1 mm is injected with the reactant gases into a chamber where the gasification reactions

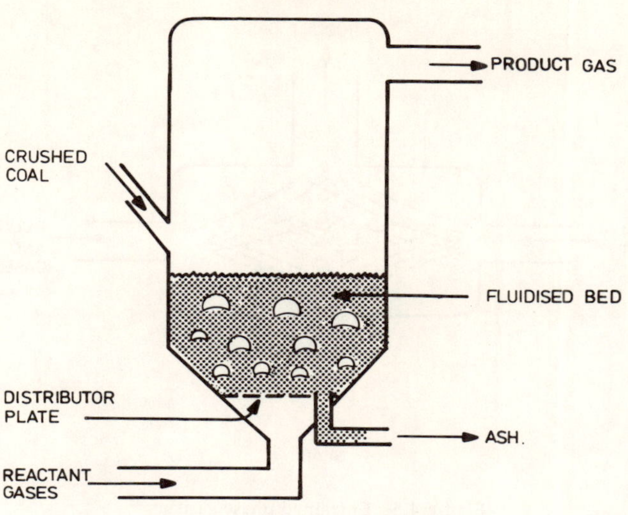

Figure 4.3 Fluidised bed gasifier

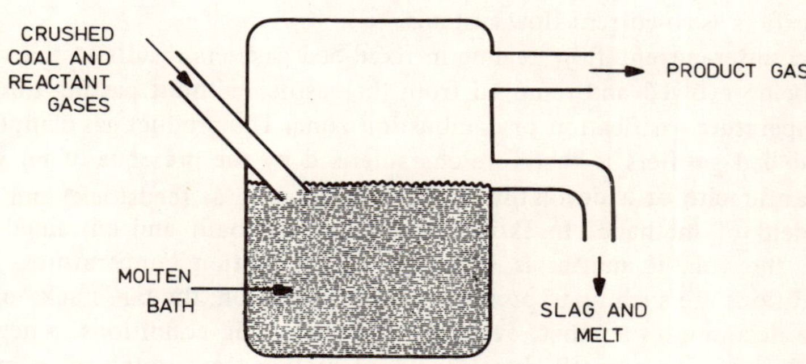

Figure 4.4 Molten bath gasifier

take place in a flame similar to that of a pulverised fuel combustion system. This approach is used in the commercial Koppers–Totzek process. The short residence time of the solid particles in an entrained phase system makes operation under slagging conditions necessary in order to obtain a high reaction rate and hence high carbon conversion. It appears that non-slagging operation in an entrained phase gasifier is attractive only for hydrogasification processes where partial conversion of the coal is acceptable.

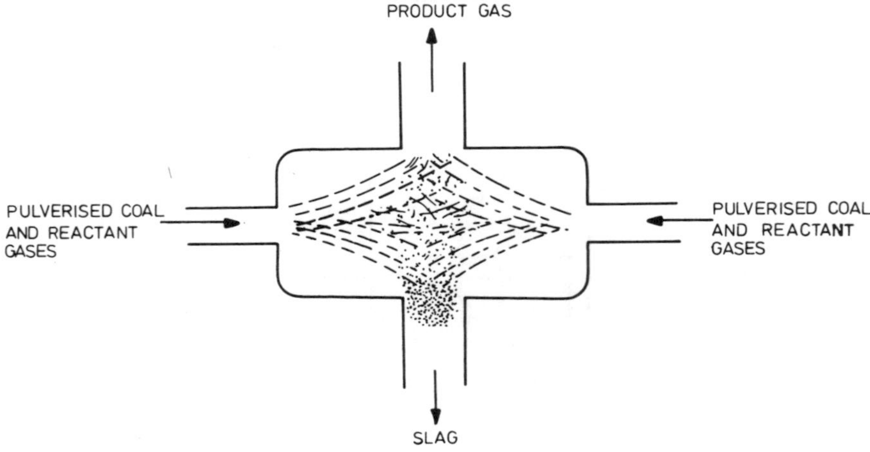

Figure 4.5 Entrained phase gasifier

In terms of the type of gas–solid contacting achieved, fixed bed gasifiers fall into the category of counter-current flow systems, fluidised bed and molten bath gasifiers can be regarded as continuous stirred tank reactors and entrained phase gasifiers as co-current flow systems.

The counter-current flow regime in fixed-bed gasifiers results in the volatile matter being evolved and removed from the gasifier without passing through a high-temperature gasification or combustion zone. The product gas composition for fixed bed gasifiers is therefore characterised by the presence of tar vapour (unless anthracite or a devolatilised char/coke is used as feedstock) and a high direct yield of methane. In fluidised bed, molten bath and entrained phase systems, the volatile matter is subjected to gasification temperatures. Unless the conditions are such as to promote hydrogasification, the tar 'cracks' and the methane decomposes so that, even under non-slagging conditions, a negligible tar yield and only a small direct yield of methane are obtained from these systems. The presence of tar vapour in the product gas is usually regarded as a disadvantage; the presence of methane can be either an advantage or a disadvantage depending on the application (see below).

In all types of gasifier, the gas residence time is of the order of seconds. The longest gas residence times are found in fixed-bed gasifiers where the gas velocity is limited in order to avoid the entrainment of fine coal in the product gas stream. This restricts the throughput of atmospheric pressure, non-slagging fixed bed gasifiers. Operation at elevated pressures or at slagging temperatures (to improve the conversion of the reactant gases) is necessary to make the throughput competitive with that of other systems. The shortest gas residence times occur in entrained phase gasifiers.

The differences in the solids residence time between the various types of gasifier are substantial. In a fixed-bed gasifier the solids residence time is generally several hours and leads to a relatively low carbon loss. The solids residence time in a fluidised bed or molten bath gasifier is generally about one hour. In a fluidised bed gasifier this shorter residence time, together with the restriction to non-slagging temperatures, can lead to a lower carbon conversion than for a fixed bed system. In some fluidised bed systems the unconverted char is collected and fed to a separate gasifier or combustor. With entrained phase gasifiers the gas and solids residence times are similar, each being of the order of seconds. For this reason entrained phase gasifiers, except for hydrogasifiers, generally operate at slagging temperatures in order to obtain a high reaction rate and carbon conversion. The short solids residence time in entrained phase reactors also makes control of the gasification conditions more difficult and consistency of feed coal quality more important.

Summary of classification

The various gasifiers available commercially or under development are classified in table 4.2 using the four main design parameters discussed previously; temperature, pressure, reactant gases and method of contacting. The classification is necessarily approximate but indicates the main characteristics for the examples shown.

It is interesting to note that not all possible combinations of the four variables are represented by existing gasifier designs. In some cases this is because the combination is impractical, for example, hydrogasifiers at atmospheric pressure or slagging temperatures and fluidised bed systems operating at slagging temperatures. Other combinations, however, may be practical but have not been developed, presumably for economic reasons. At atmospheric pressure, for example, there appear to be no non-slagging molten bath gasifiers or fixed-bed gasifiers using air and operating at slagging temperatures.

Existing commercial gasifiers

As noted in chapter 2, commercially available gasifiers are of four types.

Fixed-bed producers

In their simplest form these gasifiers consist of a fixed bed of coke through which air and steam are blown (see figure 4.2). Fresh coke is fed to the top of the bed and ash is removed via a grate at the bottom. The temperature is controlled by the addition of steam to avoid ash slagging.

From the early designs used in gasworks for producer gas manufacture, a number of commercial systems suitable for industrial applications has been

Table 4.2 Classification of processes

Method of contacting	Temperature	Pressure	Reactant gases				
			Oxygen/steam	Air/steam	Air	Steam	Hydrogen
Fixed bed	Non-slagging	Atmospheric	Wellman	Wellman			
			Wilputte	Wilputte			
			Woodall–Duckham/GI	Woodall–Duckham/GI			
				Wellman Incandescent			
				Foster Wheeler Stoic			
				Kilngas			
		Elevated	Lurgi	Gegas			
			Ruhr 100 Lurgi	Steag Lurgi			
	Slagging	Atmospheric	Leuna–Wuerth				
			Thyssen–Galoczy				
		Elevated	BG–Lurgi slagging				
Fluidised bed	Non-slagging	Atmospheric	Winkler	Winkler	Esso CAFB		
				NCB	NCB		
		Elevated	HT Winkler	HT Winkler	NCB	Cogas	Hygas
			U-gas	U-gas		CO_2 acceptor	
			Westinghouse	Westinghouse		Exxon catalytic	
			Synthane	Tri-gas		Agglomerating burner	
				Fast fluidised bed		PNP nuclear heat	
				NCB			
Molten bath	Non-slagging	Elevated	Rockgas		Rockgas		
	Slagging	Atmospheric	Otto–Rummel		Otto–Rummel	Otto–Rummel 2 shaft	
			Humboldt				
			Sumitomo				
		Elevated	Saarberg–Otto		Saarberg–Otto		
			Atgas		Atgas		
			Humboldt				
			Sumitomo				
Entrained phase	Non-slagging	Elevated					CS Rockwell
	Slagging	Atmospheric	Koppers–Totzek		VEW		
					Combustion Eng.		
		Elevated	Texaco		Texaco		
			Shell		Foster–Wheeler		
			Bigas				

developed. All of these processes operate under non-slagging conditions (with gasification temperatures of about 1000 °C) and use air and steam as the gasifying agents although in most cases the processes can be modified to operate using oxygen and steam. The throughputs are comparatively low, being typically only 10 to 20 per cent of those for the Winkler, Lurgi and Koppers–Totzek gasifiers (see below). Two types of system are available—single-stage processes and two-stage processes.

Single-stage processes are so called because they employ a single carbonisation/gasification bed. Steam for gasification is raised by a water-cooled containment section for the gasifier and a mechanical stirrer may be included to reduce the formation of agglomerates when using strongly caking coals. Commercial systems based on this design include the Wellman gasifier (available as the McDowell–Wellman gasifier in the United States and as the Wellman–Galusha gasifier in the United Kingdom) and the Wilputte gasifier.

Two-stage processes differ from single-stage processes in separating the carbonisation and gasification zones. Air and steam are supplied to the bottom of the lower stage in which the gasification reactions take place. At the top of this stage part of the gas is removed to provide an essentially tar-free product gas stream.

The remainder of the gas passes into the upper carbonisation stage to which it supplies sensible heat. A product gas containing tar is obtained from a second gas off-take at the top of this stage. Compared with single-stage processes the two-stage arrangement results in an extended, low-temperature carbonisation zone and has the advantage that tar cracking and the resulting entrainment of soot and pitch in the product gas is minimised.

As a result of the deeper bed, however, none of the present two-stage gasifier designs incorporates a mechanical stirrer and the range of coals that can be used is accordingly limited to non-caking or weakly caking types. Examples of two-stage gas producers are the Woodall–Duckham/Gas Integrale, Wellman Incandescent and Foster–Wheeler Stoic processes.

Fixed-bed atmospheric pressure gasifiers operating under ash-slagging conditions have also been investigated (for example, Leuna–Wuerth and Thyssen–Galoczy) but have not been developed to a commercial stage.

The Winkler gasifier

This is a fluidised bed gasifier (see figure 4.3) that operates at atmospheric pressure. It was originally designed to use lignite but bituminous coal, although less reactive, can also be used. The feed material is crushed to a maximum size of 10 mm and is delivered to near the top of the bed by a screw feeder. The bed is fluidised with steam and oxygen (or air) and ash is removed from the bottom of the bed. The temperature of the bed is maintained at 800 to 900 °C to avoid sintering of the ash. However, at this temperature the gasification reactions proceed slowly and it is necessary to inject additional steam and oxygen (or air) above the bed. The reactions above the bed increase the gas temperature to more than 1000 °C and the ash is therefore cooled to below the resolidification point by radiative heat transfer to a boiler before leaving the reactor. Ash and unconverted carbon are removed from the gas stream by cyclones.

The Lurgi gasifier

The Lurgi is a fixed-bed gasifier (see figure 4.2) and is the only gasifier in commercial use operating at elevated pressure. Coal is fed to the top of the gasifier through lock-hoppers to overcome the pressure differential. The coal moves downwards passing through a carbonisation zone to a gasification and combustion zone where steam and oxygen (or air) are injected. The temperature of the gasification/combustion zone is about 1000 °C. As the hot gases rise through the bed they are cooled by counter-current heat exchange to an exit temperature of about 500 °C. The product gas contains tar vapour produced in the carbonisation zone and is therefore cleaned by an aqueous scrubber. The gasifier can operate only with a sized coal otherwise excessive carry-over of fines and uneven gas distribution through the bed can occur. In order to avoid the formation of agglomerates in the carbonisation zone, the gasifier is restricted to weakly caking or non-caking coals (including lignite).

The Koppers–Totzek gasifier

The Koppers–Totzek gasifier is an entrained phase system (see figure 4.5) operating at atmospheric pressure. Coal, pulverised to a maximum size of 0.1 mm is injected with steam and oxygen into a horizontal, refractory-lined, cylindrical reaction chamber. Usually two burners (one at each end) are used although a four-burner design is now available. The coal is gasified at high temperatures (1500 to 1800°C) in a flame similar to that of a pulverised fuel combustion furnace. The hot gas leaves the reactor via a water-cooled, vertical duct. About half of the ash is entrained in the gas stream as particles of slag that cool and resolidify by radiation to the vessel walls. The remainder of the ash is removed as slag from the bottom of the reactor into a water quench bath. The gasifier can handle most ranks of coal including lignite and strongly caking coals.

Typical performance data for the commercially available gasifiers are summarised in table 4.3.

Gasifiers under development

Gasifiers can be classified according to their state of development; the following nomenclature is widely used.

Table 4.3 Typical performance data[a] for commercial gasifiers[b]

Gasifier	Coal throughput (t/h)	Inputs (kg/kg coal)			Outputs (J/J coal)	
		Air	Oxygen	Steam	Gas	Tar
Winkler	40	–	0.5	0.5	0.7	–
Lurgi	25	–	0.4	1.5	0.9	Up to 0.15
Koppers–Totzek	35	–	0.8	0.2	0.7	–
Wellman	3	3	–	0.6	0.7–0.9	Up to 0.15

[a]The performance data refer to sub-bituminous coals.
[b]Sources: H. F. Hartman *et al.*, *Low BTU Coal Gasification Processes–Selected Process Descriptions*, US DOE/ORNL/ENG/TM-13V2; Martin A. Elliot (ed.), *Chemistry of Coal Utilisation*, Second Supplementary Volume, Wiley, New York, 1981; J. F. Farnsworth *et al.*, Production of Gas from Coal by the Koppers–Totzek Process, *IGT Clean Fuels from Coal Symposium, 1975*; R. A. Ashworth, Coal Gasification Economics–Onsite v. Central Plant and Retrofitting to Low and Intermediate BTU Gas, *Fourth Conference on Coal Gasification, Liquefaction and Conversion to Electricity*, University of Pittsburg, 1977.

First-generation gasifiers

These are the existing commercial systems described above.

Second-generation gasifiers

These are gasifiers that are direct developments of existing commercial processes.

Third-generation gasifiers

These gasifiers are based on new, advanced design concepts. With some exceptions (for example, the Westinghouse, U-gas and Kilngas systems), third-generation gasifiers are generally at an early stage of development.

The present status of the main gasification development projects is summarised in appendix 1. Some examples of the above classification are given in the remainder of this section.

The largest development programmes for second-generation gasifiers are those for Lurgi-based gasifiers and the Texaco gasifier.

A Lurgi gasifier has been adapted to operate on air at the 170 MW (electrical) Steag plant at Lünen, West Germany. A high-pressure version of the Lurgi, the Ruhr 100 gasifier, is under development at Dorsten, West Germany. The object is to operate at pressures of up to 100 bar so that hydrogasification of the tars occurs and increases the direct yield of methane. The Lurgi gasifier has also been modified to operate under slagging conditions by British Gas at their Westfield site. This version, known as the British Gas–Lurgi slagging gasifier, has demonstrated a substantial increase in throughput and improvement in steam utilisation compared with the conventional Lurgi gasifier.

The Texaco entrained phase gasifier qualifies as a second-generation process as it is available commercially as a fuel oil gasification process and is now being developed to operate with coal as the feedstock.

Other developments of existing coal gasification processes centre on pressurised versions of atmospheric pressure systems. These include the High Temperature Winkler gasifier (by Rheinische Braunkohlenwerke) and the Shell gasifier (based on Koppers–Totzek technology).

Third-generation processes are mainly of three types; pressurised fluidised bed gasifiers, hydrogasifiers and slagging gasifiers operating with molten bath or entrained phase systems.

Pressurised fluidised bed processes for producing low calorific value gas include U-gas, Westinghouse, Trigas, fast fluidised bed and the National Coal Board process. These generally use air and steam although the National Coal Board process offers the possibility of using air alone in a partial gasification system. In some cases, for example U-gas and Westinghouse, versions of these gasifiers have been developed to produce medium calorific value gas by gasifying with oxygen and steam. Other developments have been directed towards pressurised fluidised bed gasifiers for producing medium calorific value gas using steam alone. As discussed earlier, provision has to be made in such designs to supply the heat absorbed by the water-gas reaction. One method of doing this is by indirect heat transfer using an external heat source as in the PNP nuclear heat gasification process. Another approach is to use a solid heat carrier that circulates between the gasification reactor where heat is absorbed and a combustor where the heat carrier is reheated. The Cogas and Agglomerating burner processes are based on this principle and use char and ash respectively as the carrier. Finally, an exothermic reaction not requiring oxygen can be employed in the gasifier. For example, in the CO_2 acceptor process, heat is released by the reaction of calcium oxide with carbon dioxide to form calcium carbonate and in the Exxon catalytic process heat is released by promoting the methanation reaction in the gasifier using potassium hydroxide as a catalyst. In the latter case, however, the final product is substitute natural gas.

Hydrogasifiers, for example, Hygas and Cities Service–Rockwell, are also being developed to produce a methane-rich gas directly. In both cases, the conversion is partial and the unconverted char is gasified by conventional means to produce the hydrogen for the hydrogasifier.

A number of gasifiers operating at slagging temperatures have been developed to produce low or medium calorific value gas. The VEW (Vereinigte Elektrizitätswerke Westfalen AG) and CE (Combustion Engineering) gasifiers are entrained phase atmospheric pressure, air-blown systems designed for power generation applications. The Foster–Wheeler gasifier is also an entrained phase, air-blown system but operates at elevated pressure and is a development of the Bigas two-stage (devolatilisation and char gasification) oxygen/steam entrained phase gasifier. Molten bath systems include the Atgas, Humboldt and Sumitomo (molten iron, oxygen/steam), Rockgas (molten carbonate, oxygen/steam or air) and Atomics International (molten carbonate, air).

Effects of coal type

Although specific gasifier designs are usually limited in the range of coals that can be accepted, there is sufficient diversity in the types of gasifier available to allow almost any coal to be used for gasification if the appropriate process is selected.

The main factors affecting the choice of coals for gasification are summarised in table 4.4.

Table 4.4 Choice of coals for gasification

Gasifier type	Coal type			
	Size grading	Caking properties	Reactivity	Ash fusion
Fixed bed	Sized	Non-caking or weakly caking[a]		
Fluidised bed	Crushed	Tolerant[b]		
Molten bath[c]	Crushed	Tolerant		
Entrained phase	Pulverised	Tolerant		
Non-slagging			Reactive coals preferred	High ash fusion temperature[d]
Slagging			Tolerant	Low ash fusion temperature[e]

[a] Fixed-bed gasifiers that incorporate a stirrer (for example, the commercial single-stage industrial gas producers) can accept moderately caking or strongly caking coals.
[b] A fluidised bed gasifier can accept a strongly caking coal only if the appropriate design features (for example, a pre-treatment stage) are included.
[c] The ash properties required for a molten bath gasifier depend on the design.
[d] Although hydrogasification processes operate under non-slagging conditions, a high ash fusion temperature is not an advantage because the operating temperature is limited by equilibrium considerations.
[e] For gasifiers operating under slagging conditions, it is usually also necessary that the slag viscosity remains low over the range of operating temperatures. A fluxing agent, such as limestone, can be added to reduce the slag viscosity, if required.

Commercially available fixed bed gasifiers are restricted to a sized-coal feed although development work on fines injection and briquetting techniques may result in such gasifiers being able to use some or all of the fines in a smalls coal. Fluidised bed, molten bath and entrained phase gasifiers use crushed or pulverised coal and can therefore accept a smalls feedstock.

The high carbon content in fixed bed and fluidised bed gasifiers can result in agglomeration in the bed if a strongly caking coal is used directly.

For single-stage industrial fixed-bed gas producers, designs are available that incorporate in-bed stirrers to reduce the effects of agglomeration and such gasifiers can operate with strongly caking coals. As noted earlier, none of the present two-stage industrial fixed-bed gas producers incorporate a stirrer and suitable coals are accordingly limited to non-caking or weakly caking types. Work on the use of in-bed stirrers with pressurised fixed-bed gasifiers (the Lurgi and derivatives) is being carried out and is expected to lead to these gasifiers being able to accept moderately caking coals.

Most of the fluidised bed gasification systems under development either employ a pre-treatment stage (involving oxidation or pyrolysis) or, equivalently, feed the coal into a dilute-phase zone in the gasifier (for example, a spout or draft tube). In each case, the object is to reduce the caking properties of the feedstock for gasification without forming agglomerates. With such design features, fluidised bed gasifiers can therefore gasify caking coals.

In gasifiers operating under slagging conditions, the reaction rates are sufficiently high for the reactivity of the coal not to be a major consideration. Under non-slagging conditions, however, reactive coals are preferred in order to maximise the conversion and throughput. In particular, reactivity is an important factor for hydrogasifiers, both because the reaction is slow and because equilibrium considerations favour low operating temperatures.

The gasification temperature also determines the required ash properties. For gasifiers designed for non-slagging conditions, a high ash fusion temperature is preferred so that a high operating temperature can be employed in the gasifier to increase the reaction rate while remaining in the non-slagging regime. In the case of gasifiers operating under slagging conditions, however, it is usually important that the molten slag has a sufficiently low viscosity to permit easy removal and a low ash fusion temperature is therefore required. If necessary a fluxing agent such as limestone can be added to reduce the slag viscosity. Excessive fluidity can cause problems of refractory attack, and careful control of the refractory temperature is therefore also necessary.

4.3 Gas processing

The raw gas leaving the gasifier generally contains a number of constituents that have to be removed before the gas is suitable for use. The nature of these constituents depends on the type of gasifier and on the application but may include

particulates
tars and oils
sulphur compounds
carbon dioxide
carbon monoxide

The main gas processing technologies are described below together with representative system configurations and the associated gas specifications.

Cooling

For most applications (with the notable exception of some industrial fuel gas markets) the raw gas can be prepared to the required standard using present technology only if clean-up systems that operate at near-ambient temperatures are employed. For this reason the first stage of gas processing is usually cooling of the raw gas.

Two general approaches to gas cooling exist: direct contact (evaporative cooling with an aqueous spray or quenching with a side-stream of product gas after cooling, cleaning and recompression) and indirect heat exchange (generally to raise steam).

Since effective recovery of heat from the gasifier product gas is an important contributory factor to the overall process efficiency, indirect heat exchange is used as far as possible. The amount of heat that can be recovered by indirect heat exchange depends principally on the raw gas temperature and composition which in turn depends on the type of gasifier. Three main cases arise.

Fixed-bed gasifiers

With fixed-bed gasifiers, the raw product gas leaves the gasifier at about 500 °C and contains tars that can cause fouling of heat exchange surfaces. The initial stage of product gas cooling (where required) is therefore performed by direct contact. The usual method, which also removes the particulates and tars, is evaporative cooling with aqueous liquor in a quench scrubber.

Further cooling (if required) may be obtained by passing the clean gas from the quench scrubber to a waste heat boiler. Special designs of boiler exist for this duty and are able to recover latent heat from the gas by condensing water vapour on the boiler surfaces (the use of evaporative cooling results in a high content of water vapour in the gas). Light oils also condense and the resulting aqueous liquor stream is sent for separation and further processing.

In practice the use of a waste heat boiler after the scrubber is worthwhile only if the gasifier operates at elevated pressure (for example, the Lurgi gasifier and its second-generation derivatives). This is because the temperature at which the latent heat is recovered from the gas increases with the gas pressure and has to be sufficiently high for the boiler to raise steam at a useful pressure.

Entrained phase and molten bath gasifiers

The raw gas produced by entrained phase and molten bath gasifiers is at a relatively high temperature (1400 to 1700 °C) and is essentially tar free. However, the gas generally contains droplets of molten slag (or material from the molten bath) that can cause fouling of heat exchanger surfaces. For this reason, the initial stage of cooling required to reduce the gas temperature so that the molten droplets solidify is accomplished either by a radiant boiler or by direct contact.

In a radiant boiler, the molten droplets are given sufficient time to cool to the solidification temperature before contact with the walls (as in a pulverised fuel furnace). Radiant boilers are therefore large but may be used in both atmospheric pressure gasifiers (for example, Combustion Engineering and VEW) and pressurised gasifiers (for example, the Ruhrkohle Texaco plant).

Alternatively, the gas may be cooled initially by direct contact. Either evaporative cooling or recycled cold gas may be used. The former option is adopted in the existing Koppers-Totzek gasification system although the latter may have an efficiency advantage (the latent heat component of evaporative cooling can be recovered only partially).

After the gas has been cooled to a temperature at which the molten droplets have solidified, further cooling and heat recovery can be achieved by using a conventional boiler. Since the gas temperature is still comparatively high, high-pressure superheated steam is usually produced.

Fluidised bed gasifiers

The raw gas from a fluidised bed gasifier (such as the Winkler or its derivatives) is produced at a temperature of about 1000 °C and contains neither condensible tars nor molten droplets of slag or other materials. The gas can therefore be passed directly to a conventional boiler for cooling, the heat being recovered to raise high-pressure superheated steam.

Cleaning

The main commercially available gas clean-up systems suitable for removing dust or tar fog from a coal-derived gas are as follows.

Cyclones

In cyclones, separation is achieved by imparting a spinning motion to the gas so that centrifugal forces act on the particles. The particles are carried towards the wall where they disengage from the gas stream. The centrifugal forces acting on a particle increase with the mass of the particle and the largest particles are therefore collected more efficiently. The efficiency can be improved, at the expense of pressure drop, by increasing the gas velocity or by using several stages of cyclones in series.

In practice, however, the collection efficiency of cyclones is moderate (typically 80 per cent) but operation at temperatures of 900 °C (or higher) is possible, limited only by materials of construction.

Aqueous scrubbers

Aqueous scrubbers operate by wetting the particles with water or an aqueous effluent. The wetted particles are larger and heavier and can therefore be removed more easily from the gas stream by gravitational or centrifugal forces. In order to wet fine particles effectively, a high relative velocity between the water droplets and the particles is required.

The most efficient designs (high-velocity venturi scrubbers) can remove up to 98 per cent of the particulate matter from a gas stream. However, the operating temperature is limited by the boiling point of water. For example, atmospheric-pressure systems typically operate at 40 °C whereas at 30 bar the operating temperature can be increased to about 160 °C.

Electrostatic precipitators

A precipitator relies on electrostatic forces to separate the particles from the gas stream. The gas is passed between electrodes to which a high voltage is applied and the particles become electrically charged. The charged particles migrate towards and collect on the electrodes from which they are removed mechanically at suitable intervals.

Electrostatic precipitators can collect up to 99.5 per cent of the particulates in a gas stream but present designs can operate only up to a maximum temperature of 375 °C.

Bag filters

If no tar fog is present (for example, where a fluidised bed or entrained phase gasifier is used), it may be possible to use a bag filter for dust removal. These operate by trapping the particles in fabric bag filters. The particles are removed periodically by reversing the air flow and mechanically disturbing the bag. The removal efficiency is similar to that of electrostatic precipitators (about 99.5 per cent) but the operating temperature is usually about 150 °C.

For industrial-scale coal gasification plant (based on fixed-bed gas producers), the gas is generally consumed close to the point of production. There is therefore an incentive to burn the gas hot in order to avoid the costs and losses inherent in a heat recovery system. Where this is possible, either cyclones or electrostatic precipitators are used for cleaning the raw gas depending on the efficiency of particulate removal required. Each of these devices can collect dust particles and tar fog.

For other applications, however, aqueous scrubbers are almost always specified both for existing commercial gasification systems and in specifications for second and third-generation gasification plants. An important advantage is their high particulate removal efficiency and, although the gas has to be cooled for cleaning, in most cases cooling is also necessary for subsequent gas processing stages. In particular, present commercial sulphur removal processes (see below) all operate at a relatively low temperature. Furthermore, in synthesis processes, it is possible to take advantage of the fact that the gas leaves the scrubber at saturation by passing the gas directly to the shift reactor (see below) where the additional steam can be utilised.

In gasification combined cycle power generation systems, as with industrial-scale plant, the gas would usually be consumed on-site and there is therefore a similar incentive to burn the gas hot. For this reason there has been interest in the development of 'hot gas cleaning' systems capable of high particulate collection efficiencies at high temperatures. Cyclones, although expected to be satisfactory for pressurised fluidised bed combustion cycles (see chapter 3), are not thought to provide adequate particulate removal for the advanced, high-temperature gas turbines likely to be used in gasification cycles. Several new gas-cleaning concepts are therefore under investigation, the most promising being granular bed filters and ceramic fibre filters. In both cases, the particulates are trapped in the filter medium which is cleaned and regenerated periodically.

Although hot gas cleaning could improve the power generation efficiency of a gasification cycle by several percentage points, the technology is, as yet, in an early stage of development. A number of problems remain to be solved before a commercial process can emerge. These include

sulphur removal at high temperatures (see below)
control of alkali concentrations
reliable and efficient regeneration of the filters
the control and feeding of a hot gas to a gas turbine combustor

Even if satisfactory technical solutions to the above problems can be found, it is by no means certain at present that an economically competitive system will result.

Waste water treatment

The aqueous effluent from scrubbers contains the particulates and (where present) tars removed from the gas stream. These constituents can be separated by simple gravity methods leaving an aqueous waste stream that may require further treatment before discharge to the environment.

In the case of fixed-bed gasification processes, the main pollutants remaining in the aqueous waste stream are phenols and ammonia. The recovery and removal of these components is described in section 7.6 but, for completeness, the main processes available are summarised below in the order in which they would be used

phenol extraction, by scrubbing with an organic solvent (the Phenosolvan process is often specified for this duty)

ammonia stripping, by distillation with alkali in an ammonia still
destructive biological oxidation of remaining phenols using the activated sludge process

For fluidised bed, molten bath and entrained phase gasifiers, the gasification temperature is high so that negligible amounts of tars and phenols and little ammonia are present. This greatly simplifies the treatment of the aqueous waste which can often be disposed of after settling in a clarifier or lagoon.

For gasifiers that operate at slagging temperatures, the waste water treatment requirements can be reduced substantially or virtually eliminated by recycling the aqueous effluent to the gasifier. Most of the pollutants in the aqueous stream are destroyed by oxidation or thermal decomposition so that the recycle loop can be self-purging. In particular, ammonia decomposes into nitrogen and hydrogen, and phenols are oxidised. Furthermore, the aqueous stream can provide a convenient transport medium for injecting coal fines into the gasifier as a coal–water mixture. Aqueous stream recycle is a standard design feature of the Texaco entrained phase gasifier and is under development for the British Gas–Lurgi slagging gasifier. In the latter case it could simplify considerably the comparatively complex waste water treatment system normally associated with fixed bed gasifiers.

Acid gas removal

Acid gas removal refers to the removal of hydrogen sulphide and carbon dioxide and is carried out downstream of the scrubber. For most coal/gasifier combinations the gas at this stage has a hydrogen sulphide content of 0.1 to 1 per cent (volume basis) and a carbon dioxide content of 5 to 25 per cent (volume basis).

If the gas is to be used in a synthesis process to produce substitute natural gas, liquid fuels or chemicals, removal of the hydrogen sulphide is generally necessary to prevent poisoning of catalysts in downstream conversion units. In such cases, the efficiency of hydrogen sulphide removal has to be extremely high, 99.99 per cent being achievable (giving a hydrogen sulphide content in the sweet product gas of less than 1 ppm, volume basis). Carbon dioxide is also usually removed in synthesis processes as an unwanted diluent.

For combined cycle power generation applications, hydrogen sulphide may be removed in order to comply with sulphur-emission control regulations. In this case, however, a lower sulphur removal efficiency (of the order of 90 per cent) may be acceptable. Carbon dioxide would not be removed from the gas as its expansion through the turbine generates power and improves the overall efficiency of electricity production.

Acid gas removal processes operate by chemical or physical absorption of the hydrogen sulphide or carbon dioxide in a solvent. A simplified flow diagram is shown in figure 4.6. The acid gases are removed by counter-current contact with

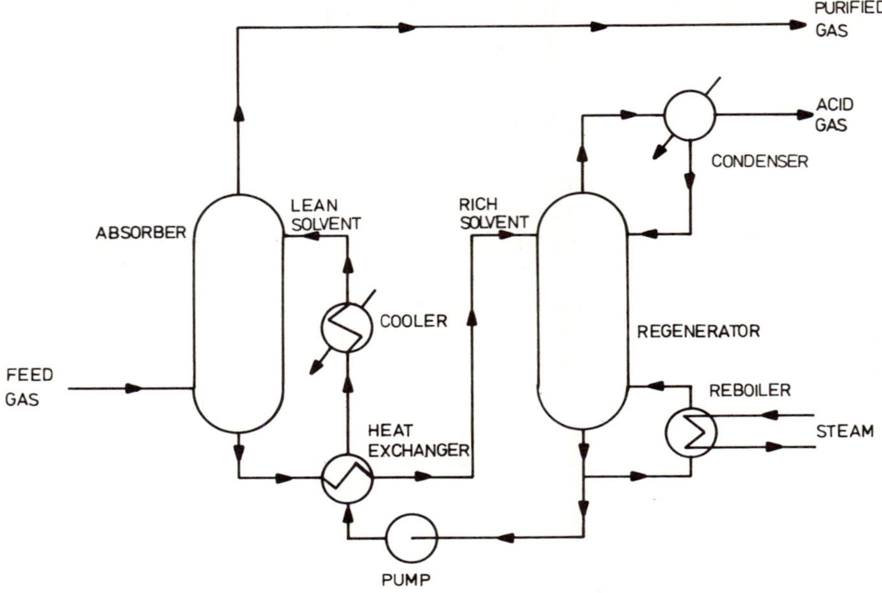

Figure 4.6 Acid gas removal (hydrogen sulphide or carbon dioxide)

the solvent in the absorber tower. The 'rich' solvent then passes to a regenerator where the acid gases are recovered by raising the temperature (or, alternatively, by lowering the pressure). The regenerated 'lean' solvent is cooled (or recompressed) and returned to the absorber tower.

The available acid gas removal processes can be classified according to the type of solvent used.

Amines

Alkanolamines have been used since the 1940s for acid gas removal. The processes are often known by the acronym representing the solvent and include

MEA (monoethanolamine)
DEA (diethanolamine)
TEA (triethanolamine)
MDEA (methyldiethanolamine)
Adip (DIPA, diisopropanolamine)
Econamine (DGA, diglycolamine)

In general, amines offer poor selectivity in the removal of hydrogen sulphide and carbon dioxide.

Compared with the alternative acid gas removal processes, the economics of amine-based processes do not appear attractive for elevated-pressure applications.

A further consideration is that irreversible chemical reactions can occur between the solvent and impurities in the gas (such as carbonyl sulphide, carbon disulphide and hydrogen cyanide) leading to loss of solvent. For these reasons amines are not favoured for acid gas removal in coal gasification systems.

Hot carbonate

The hot carbonate approach to acid gas removal was developed during the 1950s and was commercialised first as the Benfield process; more recent variants are the Benfield HiPure and Catacarb processes.

The solvent in hot carbonate processes is an aqueous solution of potassium carbonate to which catalysts are added. Absorption takes place at about 120 °C.

Selectivity in the removal of hydrogen sulphide and carbon dioxide is poor.

Physical absorption

A number of commercial processes based on the reversible physical absorption of acid gases in a solvent have been developed including

Rectisol (methanol)
Selexol (DMEPEG, dimethyl ether of polyethylene glycol)
Fluor (propylene carbonate)
Purisol (NMP, N-methyl-2-pyrrolidone)
Estasolvan (tri-n-butyl phosphate)
Union oil (methyl cyanoacetate)

Physical absorption processes usually have good selectivity; by the appropriate choice of temperature and pressure, hydrogen sulphide and carbon dioxide may be absorbed and recovered in separate stages (alternatively they may be absorbed together but recovered separately). This results in two off-gas streams; a hydrogen sulphide-rich stream and a carbon dioxide-rich stream.

In addition to absorbing hydrogen sulphide and carbon dioxide, the physical absorption processes also remove many minor impurities from the gas stream, for example, carbonyl sulphide, carbon disulphide and mercaptans. A disadvantage is that hydrocarbons of C_2 and above are also soluble. The processes generally operate at near-ambient temperatures although Rectisol operates at −20 to −60 °C.

An important consideration affecting the suitability of the above acid gas removal processes for a particular application is the choice of method for the disposal of the hydrogen sulphide. The processes available for this purpose are discussed in the following section together with possible system configurations.

In parallel with the development of hot gas cleaning systems for particulate removal (see above), effort has been directed towards the development of high-temperature sulphur removal systems. The main incentive is the same; to permit a coal-derived gas to be used to fire a gas turbine in an advanced power generation cycle without the losses inherent in cooling the gas. A number of systems are being investigated using a range of sorbents including

a fixed bed of iron oxide and fly ash (Morgantown Energy Research Center)
a fluidised bed of dolomite (Conoco)
a molten bath of calcium carbonate (Battelle Pacific Northwest Laboratories)

As with hot gas particulate removal, however, the above high temperature sulphur removal technologies are at an early stage of development and at present the economics are uncertain.

Alternatively, in some fluidised bed gasifiers (Westinghouse, for example) and molten bath gasifiers (Humboldt, for example), there is the possibility of retaining sulphur during gasification by adding an acceptor (limestone or dolomite) to the reactor thereby avoiding the need for a further sulphur removal system.

Sulphur recovery

For both economic and environmental reasons the preferred method of treating the hydrogen sulphide-rich stream is conversion into elemental sulphur. This can then either be sold as a by-product or be disposed of acceptably.

There are two main approaches to sulphur recovery.

The Claus process

This is a two-stage process. In the first stage hydrogen sulphide is burned in a furnace with a limited air supply to give elemental sulphur and sulphur dioxide by the following reactions

$$2H_2S + O_2 = 2H_2O + 2S \qquad (4.10)$$

$$2H_2S + 3O_2 = 2H_2O + 2SO_2 \qquad (4.11)$$

The sulphur dioxide and the remaining hydrogen sulphide are then passed to the second stage where they react at about 200 °C over a bauxite or alumina catalyst to complete the conversion to elemental sulphur

$$2H_2S + SO_2 = 2H_2O + 3S \qquad (4.12)$$

The Claus process can convert up to about 96 per cent of the hydrogen sulphide into elemental sulphur. However, economic operation is possible only if the feed gas contains at least 15 per cent of hydrogen sulphide.

Liquid-phase oxidation

An alternative approach to sulphur recovery is liquid-phase oxidation in which the gas stream is washed with an aqueous solution containing a proprietary mixture of inorganic salts. The hydrogen sulphide is absorbed by the solution and oxidised by the salts to elemental sulphur but not to the oxides. The salts are then regenerated by oxidation with air and the elemental sulphur recovered. The overall reaction may be represented by

$$2H_2S + O_2 = 2H_2O + 2S \tag{4.13}$$

The commercial processes include Stretford (using a mixture of sodium carbonate and sodium vanadate, with catalysts) and Giammarco-Vetrocoke (using an alkaline solution of potassium arsenite and arsenate, with catalysts).

The liquid-phase oxidation processes offer a high efficiency of sulphur recovery (up to 99.99 per cent, corresponding to an effluent gas hydrogen sulphide concentration of 1 ppm), and can operate effectively with a gas stream having a low hydrogen sulphide concentration. Unlike the Claus process, however, the recovered sulphur is impure and is relatively unattractive as a saleable by-product.

The simplest sulphur removal and recovery system comprises a non-selective, hot carbonate process (for example, Benfield) followed by liquid-phase oxidation (for example, Stretford), as illustrated in the upper part of figure 4.7. The gaseous effluent stream may be discharged directly to the atmosphere.

However, it is generally considered that the Claus process is the most economic method of sulphur recovery. For many coal/gasifier combinations, non-selective removal of both hydrogen sulphide and carbon dioxide from the gas stream produces an effluent gas with a hydrogen sulphide concentration less than the 15 per cent required by the Claus process. For this reason, a physical absorption system, selective for hydrogen sulphide, is usually specified for use with the Claus process as shown in the lower part of figure 4.7. A number of arrangements for carbon dioxide removal (if required) are possible with this system including

selective removal/recovery of carbon dioxide in the physical absorption process

application of the selective physical absorption/Claus system shown in the figure to the acid gas stream from a non-selective hot carbonate removal system

downstream removal of the carbon dioxide from the purified gas stream (after further processing) using either a hot carbonate or physical absorption method

As a result of its versatility, the physical absorption/Claus configuration shown in the lower part of figure 4.7 is widely used in design studies for commercial-scale coal gasification plants.

Since the Claus unit recovers only about 96 per cent of the sulphur, the tail gases contain some unconverted hydrogen sulphide and sulphur dioxide. In the past, it has been the practice to incinerate these gases and to vent the products to the atmosphere. Now, however, this may be unacceptable on environmental

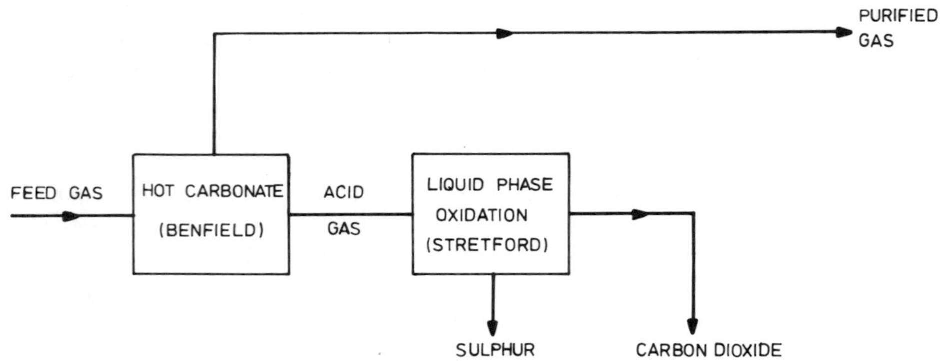

(1) HOT CARBONATE/LIQUID PHASE OXIDATION

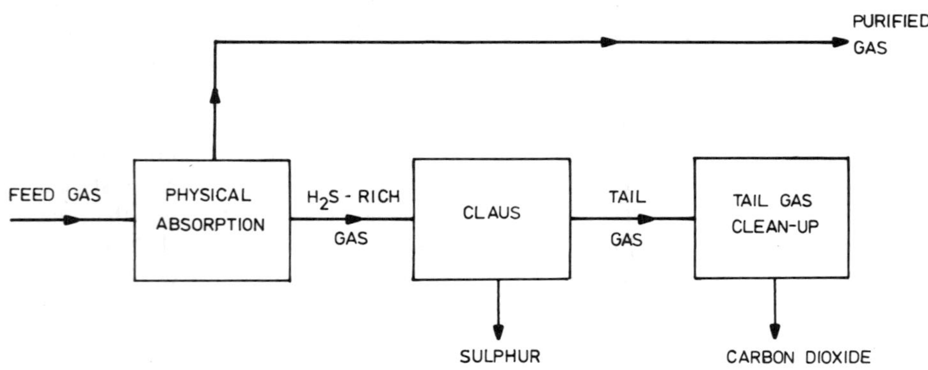

(2) PHYSICAL ABSORPTION/CLAUS

Figure 4.7 **Sulphur removal and recovery systems**

grounds and further treatment may be required. Several processes have been developed for Claus tail gas treatment. The main types are as follows.

Extensions of the Claus process

Further sulphur recovery can be obtained by employing additional stages of catalytic conversion as in the Sulfreen and IFP processes. However, the sulphur removal efficiency that can be achieved by this approach is not generally as high as with the other methods (see below).

Reduction to hydrogen sulphide

The sulphur dioxide can be converted to hydrogen sulphide by mixing the tail gases with a reducing gas and passing the mixture over a cobalt–molybdenum catalyst. The hydrogen sulphide can then be removed by one of the standard methods, for example, amine absorption and recycling of the recovered hydrogen sulphide to the Claus unit, as in the SCOT (Shell Claus Off-gas Treating) process, or liquid-phase oxidation to elemental sulphur (Stretford), as in the Beavon process.

Incineration to sulphur dioxide

The tail gases can be incinerated to convert the hydrogen sulphide to sulphur dioxide which is then removed by a regenerable flue gas desulphurisation process such as Wellman–Lord (see chapter 6). The sulphur dioxide is released when the sorbent is regenerated and is recycled to the Claus unit.

Shift reaction

The shift reaction involves the conversion of carbon monoxide to hydrogen by reaction with steam as follows

$$CO + H_2O = CO_2 + H_2 \tag{4.14}$$

A shift reactor would be employed either where hydrogen is the desired product or if it is necessary to increase the ratio of hydrogen to carbon monoxide for subsequent synthesis reactions (see below).

As discussed earlier (and indicated in table 4.1), the equilibrium conditions for the shift reaction favour hydrogen (and carbon dioxide) at low temperatures and carbon monoxide (and steam) at high temperatures. The conversion is therefore achieved by encouraging reaction 4.14 to approach equilibrium at a sufficiently low temperature.

Three types of catalyst are available: iron–chromium, copper-zinc and cobalt–molybdenum.

The catalysts are normally used in conventional fixed bed reactors. Since the shift reaction is mildly exothermic, several reactors may be required with intercooling between the stages. The main features of the catalysts are compared in table 4.5.

The iron–chromium and cobalt–molybdenum catalysts are tolerant to the presence of sulphur compounds in the gas and may be used directly after gas cleaning (and therefore before acid gas removal). This arrangement, which is illustrated in the upper part of figure 4.8, has the following two advantages.

(1) Since the gas leaves the scrubber at saturation, it can contain a considerable amount of water vapour, particularly with pressurised systems.

This water vapour contributes to the steam requirements of the shift reactor and reduces the amount of steam to be supplied separately.
(2) In general, the carbon dioxide produced by the shift reaction has to be removed before the gas passes to further processing stages. If the hydrogen sulphide can also be removed after the shift reactor, the need to cool the gas for two separate acid gas removal stages is eliminated and the gas-purification system is simplified.

Table 4.5 Comparison of shift catalysts

Catalyst	Temperature range (°C)	Sulphur tolerance
Iron-chromium	350–550	Moderate[a]
Copper-zinc	200–250	Intolerant
Cobalt-molybdenum	260–450	Tolerant[b]

[a]The tolerance of iron–chromium catalysts to sulphur compounds increases with the operating temperature.
[b]Cobalt-molybdenum catalysts require a minimum sulphur content of 20 ppm in the gas for satisfactory operation.

For most applications of synthesis gas, a high conversion of carbon monoxide to hydrogen is not required. Indeed, for substitute natural gas, methanol or liquid fuels production, the gas stream is split (as shown in the upper part of figure 4.8) so that only a proportion of the gas passes through the shift reactor, the recombined stream having the required ratio of hydrogen to carbon monoxide. In such cases, the degree of carbon monoxide conversion achieved by the high and medium temperature catalysts (iron–chromium and cobalt-molybdenum) is sufficient and these types would normally be preferred.

If, however, a high conversion of carbon monoxide to hydrogen is required, equilibrium considerations indicate that the reaction should be carried out at a low temperature. The approach usually adopted is shown in the lower part of figure 4.8. For the initial bulk conversion of carbon monoxide, an iron–chromium catalyst is employed. The gas is then cooled and a copper–zinc catalyst is used for the final conversion to high-purity hydrogen. Since copper–zinc catalysts are poisoned by sulphur compounds, the sulphur removal unit must precede shift conversion and is usually supplemented by a zinc oxide guard reactor.

The main application of the low-temperature shift system is in hydrogen production for ammonia manufacture. The catalyst used in the ammonia synthesis stage (see below) is poisoned by carbon monoxide and a high conversion is therefore essential. For coal liquefaction, however, the hydrogen produced for hydrogenating coal digests or solutions need not be of a high purity and high-temperature or medium-temperature catalysts would be used.

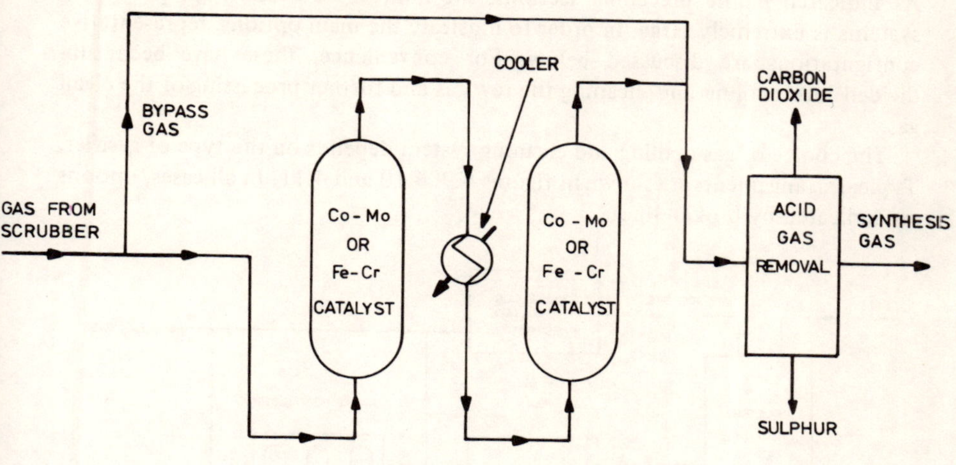

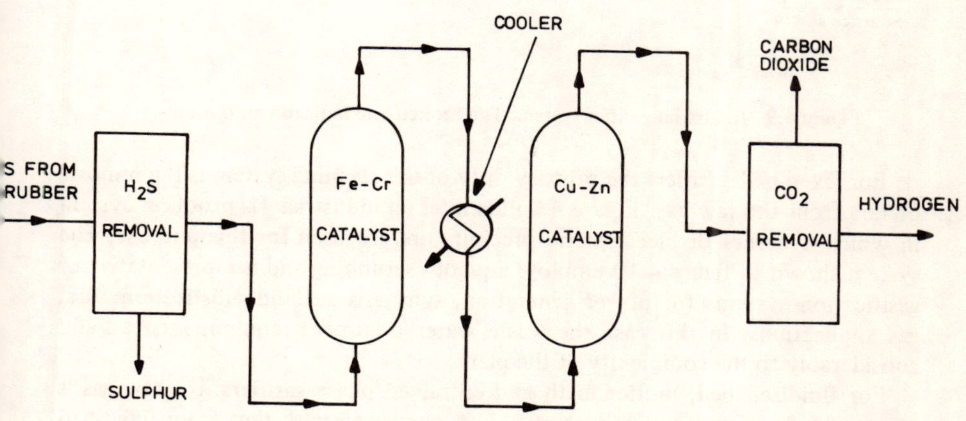

Figure 4.8 Shift reactor systems

Representative configurations

As indicated in the preceding sections, the number of possible gas-processing systems is extremely large. In order to illustrate the main options, representative configurations are discussed below. For convenience, these have been subdivided into cooling and cleaning the raw gas and further processing of the clean gas.

The choice of gas-cooling and cleaning system depends on the type of gasifier. Typical arrangements are shown in figures 4.9, 4.10 and 4.11. In all cases, options are indicated by broken lines.

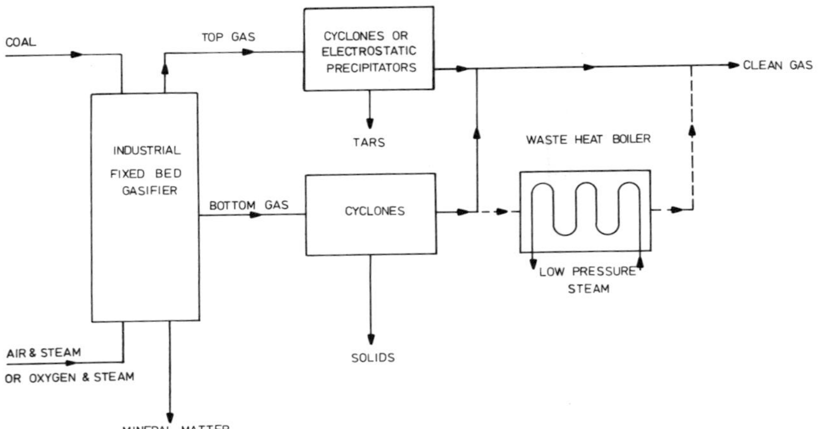

Figure 4.9 **Industrial gasifier system. The broken line indicates an option**

For fixed-bed gasifiers the primary duty of the cleaning system is the removal of tars from the raw gas. Figure 4.9 illustrates an industrial gas producer system in which cyclones or electrostatic precipitators are used for this purpose. The system shown in figure 4.10 employs aqueous scrubbing and is representative of gasification systems for power generation, synthesis gas and substitute natural gas applications. In this case the waste water treatment requirements can add considerably to the complexity of the plant.

For fluidised bed, molten bath and entrained phase gasifiers, the raw gas is essentially free from tars and phenols but, compared with that from fixed-bed gasifiers, has a high dust loading and is at a high temperature. A gas-cleaning system based on aqueous scrubbing is illustrated in figure 4.11 together with various arrangements for gas cooling and heat recovery.

The principal options for further processing of the clean gas are summarised in figure 4.12 which also defines the terminology adopted here for the various types of gas. The clean gas may be burned directly or, where sulphur emission controls apply, desulphurised prior to use as a fuel. The gas may be processed

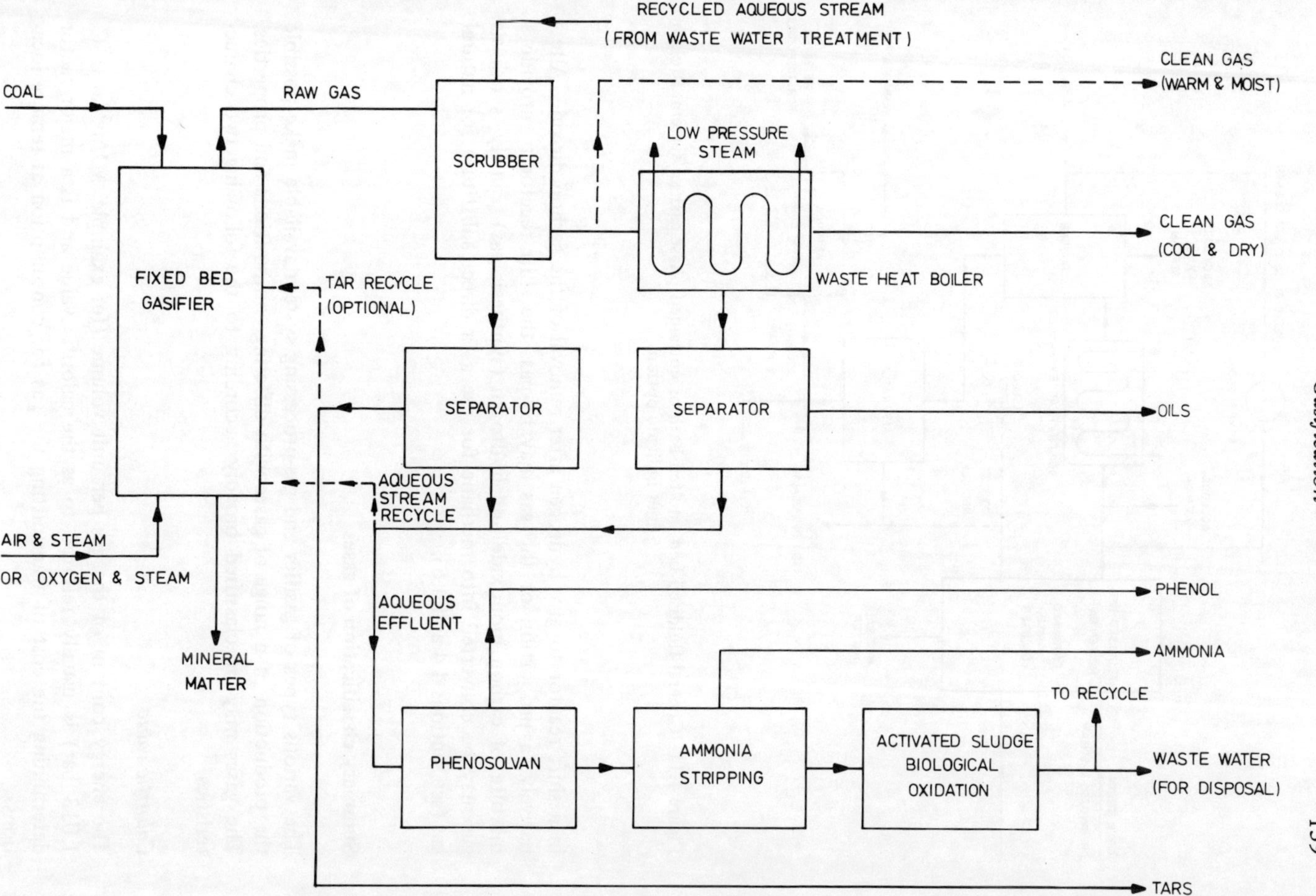

Figure 4.10 General fixed-bed gasifier system. Broken lines indicate options

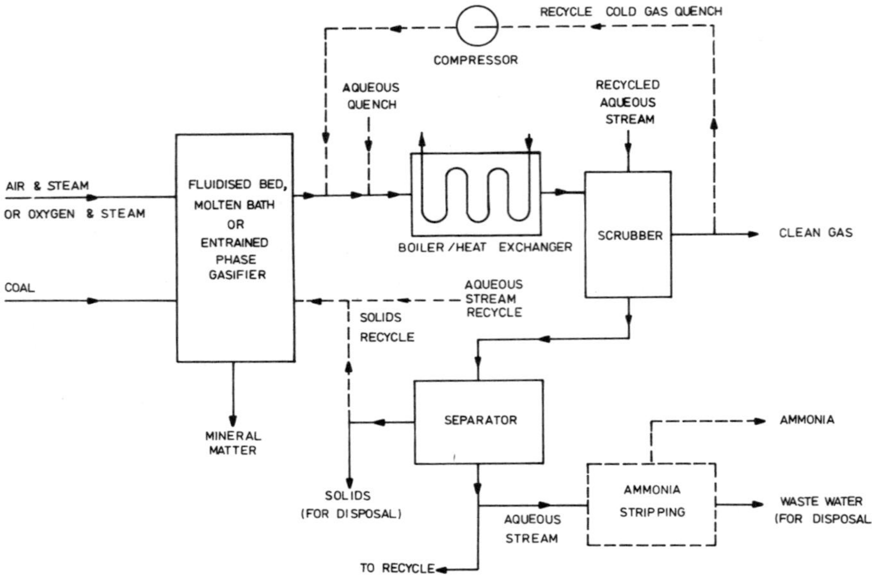

Figure 4.11 General fluidised bed, molten bath or entrained phase gasifier system. Broken lines indicate options

in a shift reactor to give hydrogen after removal of the carbon dioxide. Alternatively, a proportion of the gas may by-pass the shift reactor to provide a mixture of carbon monoxide and hydrogen (synthesis gas). Finally, synthesis gas may be converted into methane for use as a direct substitute for natural gas (see sections 4.4 and 4.6).

Summary classification of gases

The various types of gasifier and gas-processing system available make possible the production of a range of gases with differing compositions and properties. The gases may be classified broadly according to the following two characteristics.

Calorific value

The energy content of the gas per unit volume (for example, MJ/m^3 at $0\,°C$, 1.013 bar) is usually referred to as the calorific value and is a major factor determining the cost of transmitting the gas for consumption at remote locations.

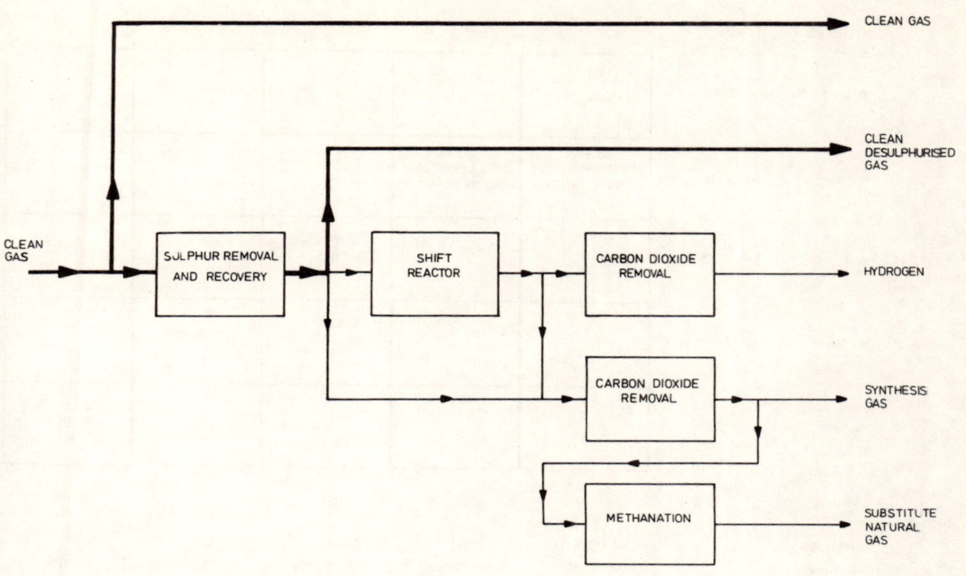

Figure 4.12 Gas processing options. The thick lines indicate options available for both air/steam and oxygen/steam gasifiers; the thin lines are options for oxygen/steam gasifiers only

The main factors affecting the calorific value are the contents of nitrogen and methane. A low calorific value gas has a substantial nitrogen content and results from air/steam gasification; a medium calorific value gas contains negligible nitrogen and is obtained by oxygen/steam gasification. A high calorific value gas (or substitute natural gas) comprises mainly methane and may be produced by upgrading a medium calorific value gas (see section 4.4).

Acid gas content

The clean gas obtained after cooling the raw gas and removing dust and tars contains minor amounts of sulphur compounds—principally hydrogen sulphide. These may be removed to give a desulphurised gas either for combustion as a low-sulphur fuel or to meet feedstock specifications for subsequent gas-processing stages. Carbon dioxide may also be removed where the product gas is required for synthesis, hydrogen or substitute natural gas.

A classification of coal-derived gases is summarised in figure 4.13 together with typical ranges for the contents of the main constituents.

	HYDROGEN SULPHIDE	CARBON DIOXIDE					
	0.1 - 1.0%	5 - 25%	CLEAN GAS (LOW CV)	CLEAN GAS (MEDIUM CV)			
	c. 1ppm	5 - 25%	CLEAN DESULPHURISED GAS (LOW CV)	CLEAN DESULPHURISED GAS (MEDIUM CV)			
ACID GAS CONTENT	c. 1ppm	UP TO 4%			SYNTHESIS GAS		
	c. 1ppm	c. 10 ppm				HYDROGEN	
	c. 1ppm	UP TO 2%					SUBSTITUTE NATURAL GAS
CALORIFIC VALUE (MJ/m^3)			5 - 7	9 - 13	12 - 15	12	c. 35
NITROGEN			40 - 50%	c. 1%	c. 1%	c. 1%	c. 3%
METHANE			UP TO 5%	UP TO 10%	UP TO 15%	c. 100ppm	93-95%
CARBON MONOXIDE			15 - 30%	20 - 60%	20-30%	c. 10ppm	BELOW 0.1%
HYDROGEN			15 - 25%	25 - 40%	60-75%	c. 99%	2-4%

Figure 4.13 Summary classification of coal-derived gases (sources: *British Gas Corporation submission to the Commission on Energy and the Environment*, 1979; Martin A. Elliot (ed.). *Chemistry of Coal Utilisation*, Second Supplementary Volume, Wiley, New York, 1981)

4.4 Synthesis technologies

Synthesis gas may be used in a variety of further processes to manufacture substitute natural gas, liquid fuels or chemicals. The processes involve chemical reactions of the two principal constituents of the gas, carbon monoxide and hydrogen, and require a catalyst. The types of catalyst used are summarised in table 4.6 together with typical operating conditions.

Table 4.6 Synthesis gas processes

Process	Catalyst	Typical operating conditions	
		Temperature (°C)	Pressure (bar)
Methanation	Nickel	290–480	20–30
Fischer–Tropsch	Iron	220–320	20–30
Methanol	Copper–zinc	250	50–100
Mobil MTG	Zeolite	350–400	20
Oxo-synthesis	Cobalt	100–200	200–300
Ammonia	Iron	400–600	150–300

Methanation

The initial studies on the synthesis of methane from carbon monoxide and hydrogen were carried out by Sabatier and Senderens as long ago as 1902. Although methanation is now a commercially available technology, it is usually employed for the removal of carbon monoxide from a gas stream and so far has been applied to the bulk production of substitute natural gas only on a development basis.

The principal chemical reaction is

$$CO + 3H_2 = CH_4 + H_2O \tag{4.15}$$

Other reactions that can take place include the shift reaction and carbon deposition reactions (the reverse hydrogasification and Boudouard reactions).

As discussed earlier, equilibrium conditions for the methanation reaction 4.15 favour methane formation at high pressures and low temperatures. In order to obtain a high conversion to methane it is therefore necessary to carry out the reactions at a low temperature using a catalyst. Nearly all methanation reactions, both existing and under development, use nickel catalysts–generally with an alumina or kieselguhr base. These catalysts are highly susceptible to poisoning by sulphur compounds and sulphur removal before methanation is therefore necessary. In addition, a zinc oxide guard reactor would normally be used.

The methanation reaction is strongly exothermic. Since excessively high gas temperatures can cause deterioration of the catalyst, a major consideration in methanator design is heat removal.

Existing methanation systems employ a series of fixed-bed reactors. A number of different arrangements for controlling the gas temperature are used and normally involve some or all of the following three design features.

Intercooling between reactor stages

As many as six reactors, each with intercooling, may be required. The last reactor may operate at a lower temperature than the others in order to increase the degree of conversion.

Dilution with recycle gas

Some of the product gas may be recycled to reduce the carbon monoxide content of the feed stream to the first methanator thereby limiting the temperature rise. The gas may be recycled either after removal of the water vapour (cold gas recycle) or before (hot gas recycle). The latter option avoids the cooling and subsequent reheating of the recycle stream but this advantage is offset by the additional compression energy requirements for recycling a hot gas.

Staged introduction of the feed gas

The carbon monoxide concentration can also be limited by splitting the introduction of the feed gas stream between several reactors.

A typical configuration employing cold gas recycle is illustrated in figure 4.14. The process usually operates at a pressure of 20 to 30 bar. The temperature of the gas at the inlet to each reactor is about 290 °C rising to about 480 °C at

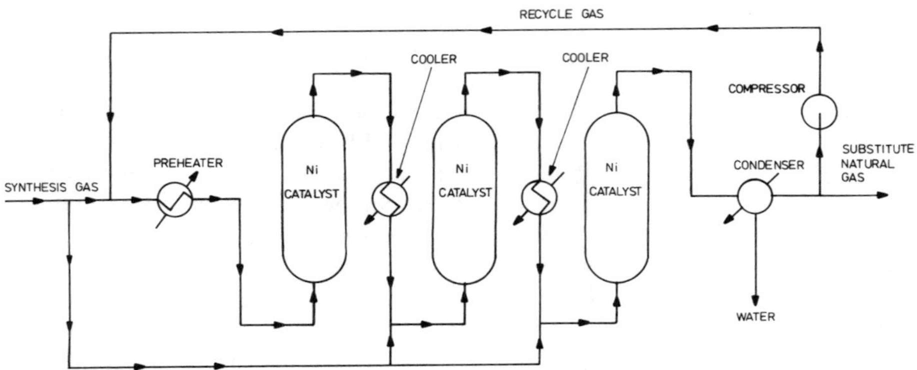

Figure 4.14 Methane synthesis (methanation)

the exit. The corresponding change in carbon monoxide concentration is from about 4 per cent (volume basis) at the inlet to about 0.05 per cent (volume basis) at the exit.

An advanced concept with particular promise for application to coal-based SNG manufacture is combined shift and methanation. This approach is under development by a number of organisations including British Gas, Parsons and Koppers and is sometimes referred to as the 'high carbon monoxide' (HCM or HICOM) route. The overall effect of the reactions may be represented by the equation

$$2CO + 2H_2 = CH_4 + CO_2 \qquad (4.16)$$

As with conventional methanation, the reactions take place in a series of fixed-bed reactors and configurations similar to those discussed above are employed to limit the temperature rise. In order to promote the shift reaction, steam is added to the feed for the first reactor and a modified nickel-based catalyst is used.

An advantage of the combined shift and methanation route is that the steam requirements are reduced. The steam released by the methanation reaction 4.15 may contribute to the steam required by the shift reaction as indicated by the form of equation 4.16. A significant improvement in the efficiency of SNG manufacture results.

Conventional methanation and the combined shift and methanation route are compared in figure 4.15. Carbon dioxide removal follows methanation in the combined shift and methanation route but precedes methanation in the conventional route.

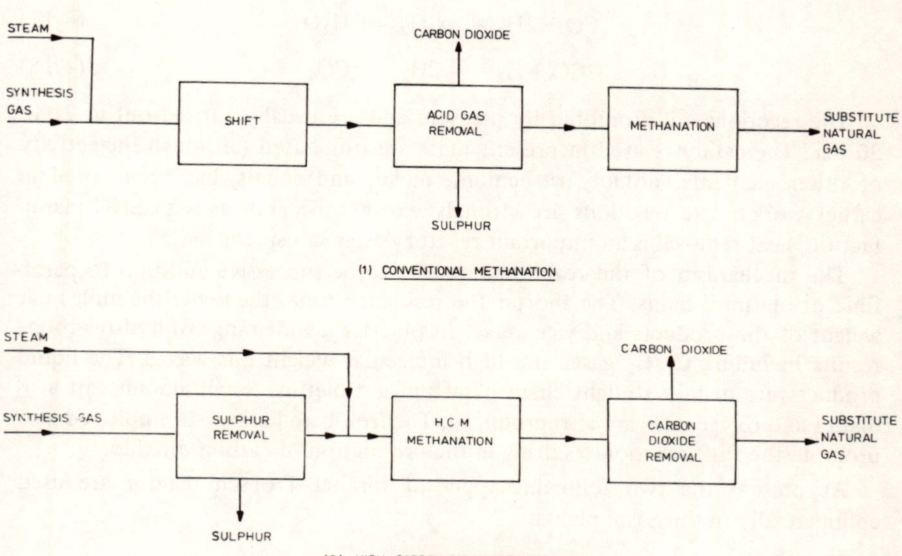

Figure 4.15 Comparison of the conventional methanation and the combined shift and methanation routes

At present, catalysts for the combined shift and methanation route are generally nickel-based and therefore require a sulphur-free gas. However, development work (for example by Bituminous Coal Research and Texaco) is being carried out on cobalt–molybdenum combined shift and methanation catalysts that are sulphur-tolerant. This would enable sulphur removal to take place after methanation at the same time as carbon dioxide removal and is therefore expected to result in a further increase in conversion efficiency.

A number of novel methanation reactor designs have been investigated because of the importance of efficient heat removal. These include fluidised bed systems with heat transfer surface immersed in the bed (Institute of Gas Technology and Bituminous Coal Research), reactors in which the catalyst is supported on heat transfer surfaces (the United States Bureau of Mines and Department of Energy 'Tube wall reactor' concept) and liquid-phase systems in which the catalyst is suspended in an inert paraffinic oil (Chem Systems). Practical problems have been encountered with each of these approaches and such systems are regarded as being at an early stage of development.

Fischer–Tropsch

The only commercially available process for manufacturing liquid fuels from coal is via the Fischer–Tropsch synthesis reactions and is employed at the Sasol plant in South Africa. The reactions may be represented as follows

$$CO + 2H_2 = -CH_2- + H_2O \tag{4.17}$$

$$2CO + H_2 = -CH_2- + CO_2 \tag{4.18}$$

The reactions are promoted by pressure and are usually carried out at 20 to 30 bar. The catalysts used in present units are iron-based (although the activity of other elements, notably ruthenium, nickel and cobalt, has been noted in earlier work). The reactions are strongly exothermic and, as with SNG manufacture, heat removal is an important reactor design consideration.

The mechanism of the reactions appears to be successive addition to paraffinic or olefinic chains. The shorter the residence time, the lower the molecular weight of the products and vice versa. In practice a wide range of hydrocarbons results including C_1/C_2 gases and high molecular weight oils/waxes. The liquid products are mainly straight chain paraffins and olefins. Small amounts of aromatics and oxygenates are also produced. The iron-based catalysts employed also promote the shift reaction resulting in the production of carbon dioxide.

At present the two following types of Fischer–Tropsch reactor are used commercially in the Sasol plants.

The fixed bed Arge reactor

This is a conventional fixed bed reactor operating at about 220 °C. The Arge process is used in the Sasol 1 plant but not in Sasols 2 and 3.

The fluidised bed Synthol reactor

This reactor is essentially a fast fluidised bed (see chapter 3) with the catalyst circulating rapidly between a cooled, turbulent transport reactor and a cyclone disengagement unit. The reactor operates at 320 °C and is used in all the existing Sasol plants.

The main difference in the product spectra produced by the two reactors is that more high molecular weight material including waxes is produced by Arge. In each case the conversion of 60 to 75 per cent of the carbon monoxide is achieved per pass.

An alternative to the Sasol Fischer–Tropsch technologies is the Koelbel liquid phase reactor. In this system the iron catalyst is suspended in a high boiling-range, process-derived oil through which the synthesis gas is passed. The reactor temperature is maintained at about 250 °C, heat being removed by internal cooling coils that raise medium-pressure steam. A 12 t/d pilot plant by Rheinpreussen–Koppers was completed in 1953. Advantages claimed for this approach include the ability to process gases rich in carbon monoxide and low methane production.

The Fischer–Tropsch process is illustrated in figure 4.16 which shows a simplified flow-sheet for the Synthol reactor system. The desulphurised feed gas

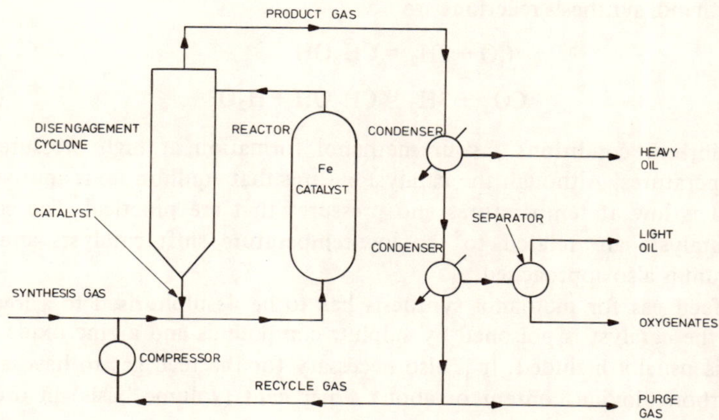

Figure 4.16 Fischer–Tropsch (Synthol) synthesis

is mixed with the recycle gas typically in a ratio of 1:2. The combined gas stream then meets the hot recirculating catalyst and passes into the reactor. The product gas is cooled in two stages, the first of which recovers a heavy oil fraction. The liquid condensed in the second stage is separated into an aqueous fraction containing the oxygenates, and a light oil fraction.

The overhead gas from the cold separator contains light hydrocarbon gases together with unconverted synthesis gas. This stream is split with part being recycled to the Fischer–Tropsch reactor to increase the conversion to liquid fuels and part proceeding to further processing. The proportion recycled is limited to avoid the methane and carbon dioxide formed by the Fischer–Tropsch reactions accumulating in the recycle gas stream.

Methanol

The manufacture of methanol from synthesis gas is a well-established technology. A patent for it was granted in 1913 to BASF who built the first commercial plant in 1923. The BASF process employed a zinc–chromium catalyst and required a pressure of 350 bar and a temperature of 350 °C. This approach has now been superseded by the new low-pressure processes the first of which was introduced by ICI in 1966. A number of similar processes have since become available, for example, Lurgi and Mitsubishi. All these processes are based on copper catalysts. The reaction pressure is 50 to 100 bar and the temperature about 250°C. These relatively mild operating conditions result in significant cost savings. Although existing methanol production is based mainly on natural gas, methanol synthesis technology is equally applicable to a coal-derived gas. The main methanol synthesis reactions are

$$CO + 2H_2 = CH_3OH \tag{4.19}$$

$$CO_2 + 3H_2 = CH_3OH + H_2O \tag{4.20}$$

Equilibrium conditions favour methanol formation at high pressures and low temperatures. Although the catalyst ensures that equilibrium is approached, the yield is low at temperatures and pressures that are practical. The copper-based catalysts are related to the low-temperature shift catalysts and shift equilibrium is also approached.

The feed gas for methanol synthesis has to be desulphurised to a low level because the catalyst is poisoned by sulphur compounds and a zinc oxide guard reactor is usually included. It is also necessary for the feed gas to have a minimum carbon dioxide content of about 4 per cent (volume basis) in order to maintain the activity of the catalyst.

A typical methanol process is shown in simplified form in figure 4.17. The synthesis gas is mixed with a recycle stream, pre-heated to the required tem-

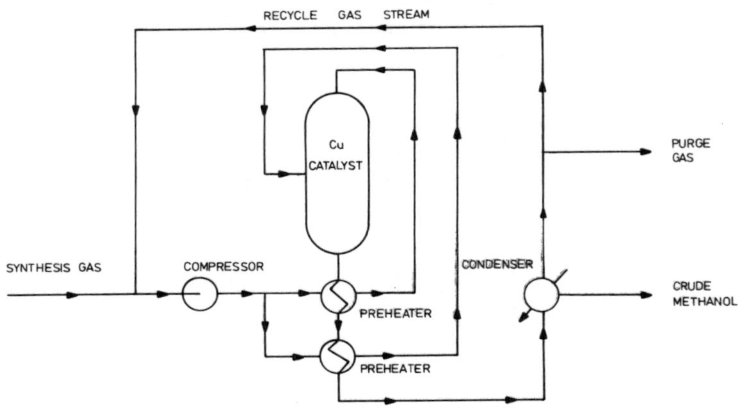

Figure 4.17 Methanol synthesis

perature and fed to the fixed bed reactor. Since the reactions are strongly exothermic, the feed stream is split with part being used as a quench to moderate the gas temperature. The product gas is cooled so that the methanol vapour, which comprises about 5 per cent (volume basis), condenses and is recovered. The crude methanol contains water and higher alcohols and may be purified to fuel-grade or chemical-grade methanol by distillation. Some of the unconverted synthesis gas is recycled to the reactor but a purge stream is necessary to avoid a build-up of inerts present in the feed gas. These may include nitrogen, argon and, if a fixed-bed gasifier is used, methane (in contrast to the Fischer–Tropsch process, only trace quantities of methane are formed during methanol synthesis).

Methanol-to-gasoline (MTG)

Development work by Mobil during the 1970s has led to a new class of catalysts which offer the prospect of more efficient and selective routes from synthesis gas to liquid hydrocarbon fuels.The conversion uses methanol as the feedstock and initial work has been directed towards the production of gasoline. The reactions may be represented as follows

$$CH_3OH = -CH_2- + H_2O \tag{4.21}$$

The initial conversion is from methanol to dimethyl ether which subsequently further dehydrates to give light and then heavier olefins. These then rearrange to form paraffins, cyclo-paraffins and aromatics.

The catalysts used are shape-selective zeolites. These have a three-dimensional pore structure comprising channels of approximately 6 Ångströms diameter. The product contains no hydrocarbons with a higher molecular weight than that of C_{10} compounds (for example, durene). Although high-molecular weight compounds are probably formed, they are too large to escape from the zeolite structure and remain until further reactions produce smaller, lower-molecular weight material that can escape. The size of the channels in the zeolite therefore determines the size of the molecules in the product.

The most advanced design concept for the Mobil MTG process involves two fixed-bed reactors as shown in figure 4.18. The system operates at about 20 bar

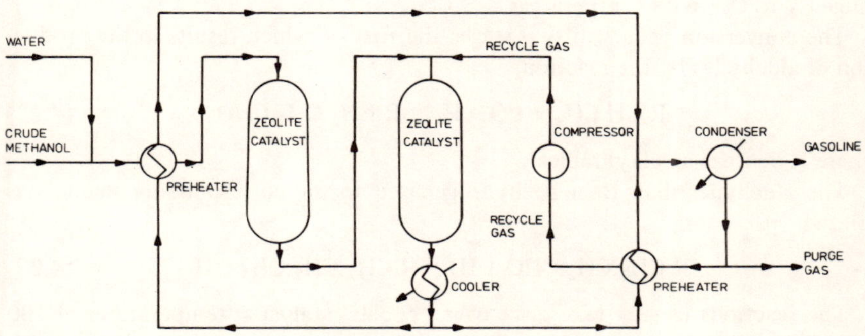

Figure 4.18 The Mobil MTG process

and for each reactor the gas temperature rises from about 350 °C at the inlet to about 400 °C at the exit.

Crude methanol from a synthesis unit (without distillation or further treatment) is mixed with water and pre-heated. The resulting vapour enters the first reactor in which it is dehydrated to give an equilibrium mixture of dimethyl ether, methanol and water. This mixture is then combined with the recycle gas stream in a ratio of about 1:8 and passes to the second reactor where the dehydration to hydrocarbon products is completed. Although the reactions during this second stage are strongly exothermic, the temperature rise is limited by the high recycle ratio. The product gas is cooled to recover liquid hydrocarbon products and water of reaction. The uncondensed light gases are recycled to the second reactor after removal of a purge stream.

An alternative process involving the use of a fast fluidised bed (similar to Synthol) is also being investigated by Mobil. This type of reactor has the advantage that the two stages required in the fixed-bed system may be combined into a single unit.

Both the fixed-bed and fluidised bed reactor systems have been demonstrated on a 0.5 t/d scale. Development prospects centre on a 1500 t/d fixed-bed plant planned in New Zealand and a 12 t/d fluidised bed pilot plant in Wesseling, West Germany.

The MTG approach is widely regarded as having considerable potential for further development and a number of active research programmes are in progress. Possible variants of the existing system include shortening the reactor residence time to maximise the production of short-chain olefins as chemical feedstocks and incorporating conventional catalysts (for example, Fischer–Tropsch catalysts) into the zeolite structure so that the methanol synthesis stage can be eliminated.

Oxo-synthesis

Oxo-synthesis involves the manufacture of alcohols by reacting olefins in the range C_3 to C_{20} with synthesis gas.

The conversion occurs in two stages the first of which results in the production of aldehydes by the reaction

$$R.CH.CH_2 + CO + H_2 = R.CH_2.CH_2.CHO \tag{4.22}$$

where R denotes an alkyl radical.

The aldehydes may then be hydrogenated to the corresponding alcohols as follows

$$R.CH_2.CH_2.CHO + H_2 = R.CH_2.CH_2.CH_2.OH \tag{4.23}$$

The reactions usually take place over a cobalt catalyst at temperatures of 100 to 200 °C and pressures of 200 to 300 bar.

The oxo-alcohols are used as feedstocks in the manufacture of a variety of products including solvents (C_3 to C_6), plasticizers (C_7 to C_{12}) and detergents (C_{12} to C_{20}).

Oxo-synthesis technology originated as a development of the Fischer–Tropsch process and has been available commercially since the 1950s.

Ammonia

The commercial manufacture of ammonia dates from 1913 when a plant was constructed by BASF in Germany. The plant employed the Haber process and, although detailed improvements have been made, this remains the basis for present installations. Ammonia synthesis may therefore be regarded as a mature technology.

Ammonia is produced by reacting nitrogen and hydrogen as follows

$$N_2 + 3H_2 = 2NH_3 \tag{4.24}$$

The reaction takes place over an iron catalyst at a pressure of 150 to 300 bar and a temperature of 400 to 600 °C. A fixed bed reactor is normally employed and a large number of designs are commercially available. Since the reaction is exothermic, heat removal is necessary to limit the increase in gas temperature. This is usually achieved either by inter-stage heat exchangers or by staged introduction of the feed gas.

A typical ammonia synthesis system is illustrated in figure 4.19. The feed gas is mixed with a recycle gas stream and pre-heated to the required temperature. The gas mixture leaving the reactor is cooled by a series of heat exchangers and,

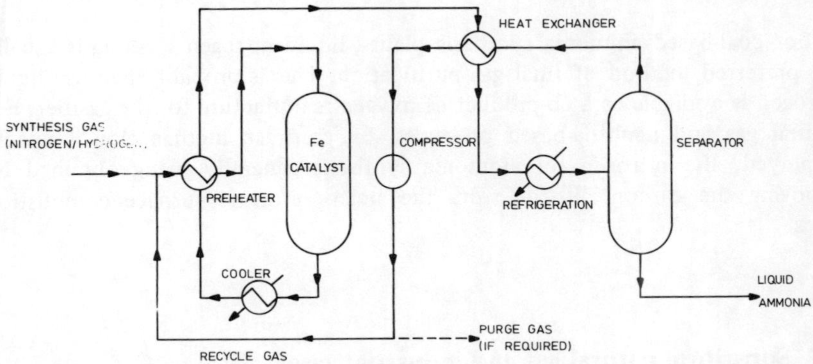

Figure 4.19 Ammonia synthesis

finally, by a refrigeration unit. The ammonia is condensed out and separated from the unconverted synthesis gas which is recycled to the reactor. A purge stream from the recycle loop may be required depending on the inerts content of the feed gas (see below).

The iron catalyst is poisoned both by sulphur compounds and oxygen compounds (including water vapour, carbon dioxide and carbon monoxide) and it is therefore important that these species are removed from the nitrogen/hydrogen synthesis gas mixture. The water vapour and carbon dioxide may be removed to the required level of purity by condensation and conventional acid gas removal respectively. Two main approaches to carbon monoxide removal exist.

Methanation

Catalytic methanation (see above) may be used to convert residual carbon monoxide in the synthesis gas to methane which is an inert in the ammonia synthesis reaction and does not affect the catalyst. In this case, however, a purge stream from the ammonia synthesis recycle loop is necessary to prevent accumulation of the methane.

Liquid nitrogen wash

Alternatively, liquid nitrogen washing may be employed to remove the carbon monoxide from the hydrogen feed. Impure hydrogen is passed upwards through a plate column counter-current to a flow of liquid nitrogen. The carbon monoxide condenses and is removed with liquid nitrogen at the bottom of the column. Purified hydrogen is obtained mixed with nitrogen as the overhead gas stream. This hydrogen/nitrogen mixture has an extremely low inerts content and no voluntary purge is required from the ammonia synthesis recycle loop since sufficient inerts are removed in solution with the ammonia to make the loop self-purging.

For coal-based ammonia synthesis plants, liquid nitrogen washing is usually the preferred method of final gas purification. This is probably because liquid nitrogen is available as a co-product of oxygen manufacture for the gasifier. For natural gas and naphtha-based processes, by contrast, methanation is widely employed, the nitrogen for ammonia synthesis generally being obtained by removing the carbon dioxide from the steam reformer furnace combustion gases.

4.5 Substitute natural gas and industrial gases

At present natural gas accounts for about 20 per cent of world energy supplies and is generally thought to have a lifetime of about 50 years at current production rates. In the longer term, the existence of gas transmission and distribution pipeline systems in industrialised countries and the advantages to energy consumers of the cleanliness and convenience of gas combine to provide powerful incentives to supplement natural gas supplies by the manufacture of a coal-derived gas.

In order that the present infrastructure and appliances can be used without modification it is likely that the main demand will be for a 'substitute natural gas' (for convenience referred to as SNG) manufactured from coal. To be compatible with natural gas, substitute natural gas will have to match existing specifications closely and therefore have to comprise mainly methane. The current size of the market for gas indicates that the manufacture of substitute natural gas could become a major use of coal in the future.

For some industrial applications, however, it may be preferable to build small, locally situated gasifiers to supply clean low or medium calorific value gases as a replacement for natural gas.

Substitute natural gas manufacture; first-generation processes

It is generally expected that the first generation of substitute natural gas plants will be based on the synthesis gas/methanation approach. A typical system is illustrated in figure 4.20. In the example shown, the raw gas from the gasifier

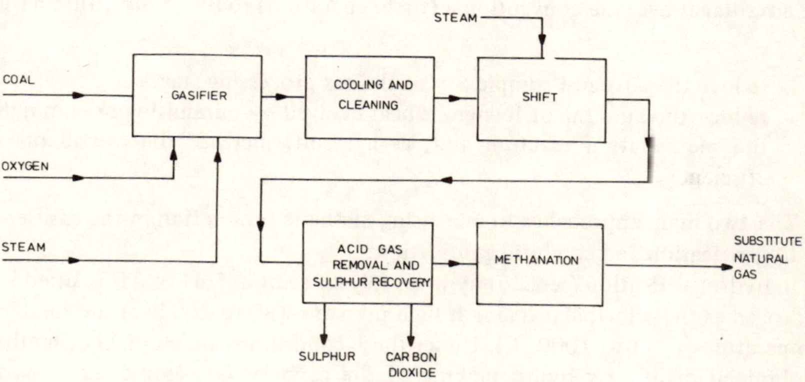

Figure 4.20 Substitute natural gas manufacture

is first cooled and cleaned and then passes to a shift reactor where some of the carbon monoxide is converted to hydrogen and carbon dioxide by reaction with steam. The gas is then processed to remove sulphur compounds and the carbon dioxide and the resulting synthesis gas passes to a catalytic methanation unit where it is converted into substitute natural gas. As discussed earlier, a number of variants of this approach exist including sulphur removal before the shift reactor and combined shift and methanation.

For such plants, elevated-pressure fixed-bed gasifiers are particularly advantageous. The raw gas produced by fixed-bed gasifiers contains a high proportion of methane compared with that from other types of gasifier so that the downstream processing required for upgrading this gas to substitute natural gas is correspondingly reduced. Pressurised operation of the gasifier is desirable in order to obtain a high throughput and to match the requirements of the gas processing units which generally operate at elevated pressure. The Lurgi gasifier and its derivatives (including the British Gas–Lurgi slagging gasifier and the Ruhr 100 gasifier) are therefore well-suited for this application.

Numerous design studies for demonstration plants have been carried out particularly in the United States. The usual plant size assumed is 7 Mm^3/d of substitute natural gas (at 0 °C, 1.013 bar) and would consume approximately 12500 t/d of coal. Between 6 and 20 gasifiers would be required depending on their size and design.

Advanced substitute natural gas processes

Several advanced processes for converting coal into substitute natural gas are under development and have the common object of increasing the conversion of coal to methane in the gasifier. It is expected that this will lead to the following advantages over the conventional (first-generation) route to substitute natural gas

- reduce the cost and complexity of the gas processing stages
- reduce the amount of low-grade heat evolved by minimising or eliminating the methanation reaction and, as a result, increase the overall process efficiency

The two main approaches to increasing methane production in the gasifier are hydrogasification and catalytic gasification.

In hydrogasification (see also hydropyrolysis, section 5.4) coal is gasified with hydrogen as the principal reactant at high pressures (80 to 200 bar) and moderate temperatures (700 to 1000 °C). Under these conditions, a high yield of methane is obtained mainly by hydrocracking of the primary tars (some direct hydrogasification of the coal/char also occurs). Hydrogasifier developments include the Hygas and Cities Service–Rockwell processes.

A typical flow-sheet for a hydrogasification process is given in the upper part of figure 4.21. The conversion of the coal in the hydrogasifier is only partial with the char residue being passed to a conventional oxygen/steam gasifier for hydrogen production. The methane product is recovered from the hydrogasifier product gas by cryogenic separation after the gas has been processed by shift and acid gas removal units. The residual hydrogen is recycled to the hydrogasifier. An alternative configuration to that shown in the figure employs only the oxygen/steam gasifier train for hydrogen supply with all of the hydrogasifier gas being upgraded to substitute natural gas by methanation.

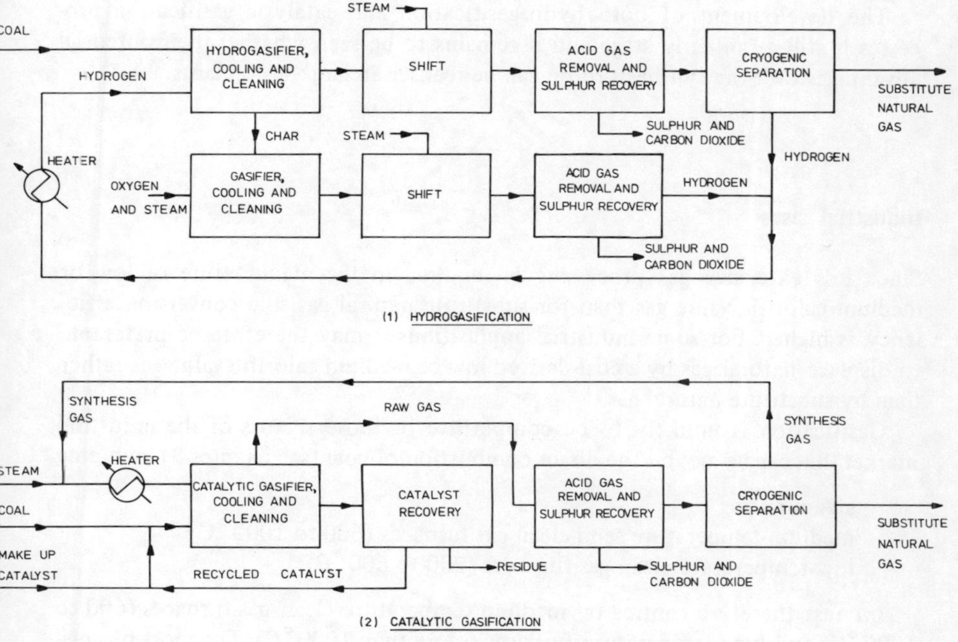

Figure 4.21 Advanced substitute natural gas processes

The catalytic gasification approach uses a catalyst in the gasifier to promote the methanation reaction. It has been known for some time that alkali metal salts of weak acids (potassium carbonate, for example) promote the gasification reactions taking place in conventional oxygen/steam systems. In the mid-1970s, however, Exxon demonstrated that the same catalysts could be used to promote methane formation in the gasifier. The Exxon process employs a fluidised bed gasifier with typical operating conditions of 700°C and 30 bar. Steam is added to the gasifier so that a heat balance is obtained between the exothermic methanation reaction and the endothermic water-gas reaction.

The Exxon catalytic gasification system is illustrated in the lower part of figure 4.21. Coal is mixed with the catalyst (at a concentration of 10 to 20 per cent of catalyst in the coal) and fed to the gasifier. A solids stream containing catalyst, unreacted carbon and mineral matter is withdrawn continuously and processed to recover the catalyst which is recycled. The raw gas leaving the gasifier is approximately at equilibrium for the shift and methanation reactions. The gas passes to a conventional acid gas removal unit which eliminates the hydrogen sulphide and carbon dioxide and then to a cryogenic separation unit where the methane product is recovered. The carbon monoxide and hydrogen remaining are mixed with steam and recycled to the gasifier.

The development of both hydrogasification and catalytic gasification processes is still at an early stage and it remains to be seen whether their potential advantages over the synthesis route can be realised in large-scale plants.

Industrial gases

Since less extensive gas processing is required in the manufacture of low or medium calorific value gas than for substitute natural gas, the conversion efficiency is higher. For some industrial applications it may therefore be preferable to displace natural gas by a coal-derived low or medium calorific value gas rather than by substitute natural gas.

Gasification is unlikely to be competitive in those sectors of the industrial market that can be met by the direct combustion of coal (see chapter 3) including

steam raising
medium-temperature semi-clean gas furnaces (600 to 1000 °C)
low-temperature clean gas furnaces (200 to 600 °C)

Interest therefore centres on medium-temperature clean gas furnaces (600 to 1000 °C) and high-temperature furnaces (more than 1000 °C). The most promising industrial sectors include

glass manufacture
pottery and ceramics manufacture
metal melting and reheating

It is widely expected that premium gas consumers (including the domestic and commercial sectors) will continue to prefer a high calorific value gas, either natural gas or coal-derived substitute natural gas.

There are the following two main approaches to the supply of industrial gases.

Industrial gas producers

As noted earlier, a number of industrial gas producers are available commercially. These are all fixed-bed designs and are normally air-blown giving a low calorific value gas (although some can operate on oxygen to give a medium calorific value gas if required). The gasifier size is usually in the range 1 to 5 t/h of coal and is of the same order as the size of the gas demand at many industrial sites. Industrial gas producers are therefore likely to be used as dedicated plants and would be situated near the point of gas consumption. For some applications it may be possible to use the gas hot and thereby to take advantage of its sensible heat content.

Local medium calorific value gas manufacture

An alternative method of meeting industrial gas demands would be to install large gasifiers (Lurgi, Winkler or Koppers-Totzek, for example) in an industrial area and to distribute the gas through a special local pipe network to a number of industrial consumers. Such gasifiers are generally oxygen-blown and therefore produce a medium calorific value gas. Throughputs lie in the range 10 to 40 t/h of coal per gasifier.

The comparative economics of substitute natural gas, industrial gas producers and local medium calorific value gas manufacture are discussed in chapter 8.

4.6 Power generation

A potentially important use of coal gasification is the manufacture of a clean gas for power generation. The main incentives are to reduce emissions of sulphur dioxide cheaply and effectively and to obtain an improvement in the efficiency of power generation compared with conventional stations by employing advanced combined cycle systems.

The relative importance of these advantages will clearly depend on the coal price and quality, the environmental legislation in existence and the utility plant stock.

Sulphur removal

As discussed in section 6.7, flue gas desulphurisation systems for conventional pulverised fuel power stations add significantly to capital costs and reduce electricity generation efficiencies. If sulphur emissions have to be controlled, gasification offers a means of producing a clean, desulphurised fuel thereby eliminating the need for sulphur removal after combustion. However, gasification also introduces additional capital costs and efficiency losses and is unlikely to be competitive with flue gas desulphurisation for retrofitting existing coal-fired power stations to reduce sulphur emissions (although it may be competitive for oil-fired stations because fuel costs would also be reduced).

For new power stations, gasification can be used in combined cycle systems and, even with present gas turbine technology, these are expected to have efficiencies and costs which are competitive with those of pulverised fuel power generation with flue gas desulphurisation.

Combined cycle systems

Combined cycles differ from the steam cycles employed in conventional pulverised fuel power stations in employing both gas and steam turbines. These two types of turbine are complementary, the former having relatively high inlet and outlet temperatures whereas the latter operates over a lower temperature range. The use of both gas and steam turbines in a combined cycle leads to a higher overall power generation efficiency (see below). Pulverised fuel steam cycles and the various categories of combined cycle are described in more detail in section 3.7.

A typical combined cycle system for use with coal gasification is illustrated in figure 4.22. Air is compressed (usually to 8 to 16 bar) and is used to burn a

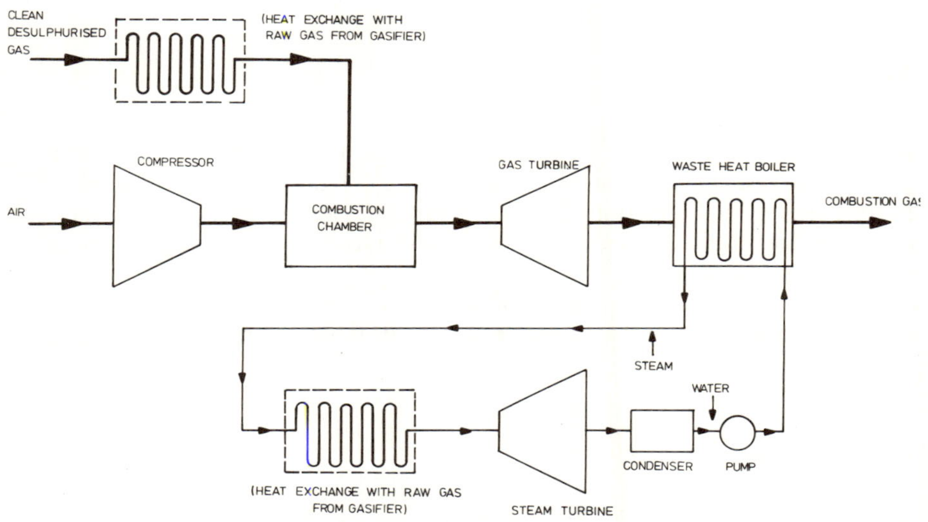

Figure 4.22 Gasification combined cycle power generation system

clean, desulphurised coal-derived gas in the combustion chamber. The hot combustion gases are expanded through the gas turbine which both drives the compressor and generates power. The exhaust gases from the turbine pass to a waste heat boiler to raise steam and to generate further power in a conventional steam cycle. If the raw gas from the gasifier is cooled in a heat exchanger before cleaning (as in fluidised bed, molten bath and entrained phase gasifiers but not as in fixed bed gasifiers) then heat may be recovered for reheating the gas and for use in the steam cycle. These are indicated as options in the figure.

The main factors affecting the efficiency of such a system are

the gas turbine inlet temperature
the proportion of the heat recovered directly in the steam cycle
the type of steam cycle used

The higher the inlet temperature to a gas turbine, the greater the amount of work that can be obtained from the gas during its expansion to atmospheric pressure. For this reason a high gas turbine inlet temperature is an important factor in obtaining a high efficiency in all gas turbine applications.

Heat that is recovered directly to the steam cycle, for example, in cooling the raw gas from the gasifier, is not used as effectively as that which contributes to both the gas and steam cycles. In order to obtain a high efficiency it is therefore desirable for the proportion of the heat recovered directly to the steam cycle to be low (and the participation of the gas turbine correspondingly high).

In order to operate an efficient steam cycle, sufficient heat must be recovered at a high temperature. An important factor affecting the amount of high-temperature heat available in a gasification combined cycle is the gas turbine inlet temperature as this in turn affects the temperature of the turbine exhaust gases. For turbine inlet temperatures above about 1300 °C there is usually enough high-temperature heat to permit an efficient reheat cycle (160 bar/565 °C/565 °C) to be employed. At lower turbine inlet temperatures, however, inferior steam conditions may have to be used (possibly with a dual pressure cycle) unless a substantial amount of high-temperature heat is recovered from the raw gas leaving the gasifier. The quantity of low-grade heat available is such that it may be unnecessary to use bleed steam for feed water heating.

The effect of turbine inlet temperature on efficiency is illustrated in figure 4.23 for two representative systems; a British Gas–Lurgi slagging gasifier cycle and a Texaco gasifier cycle.

Gas turbine developments

At present industrial gas turbines operating with an inlet temperature of 1100 °C are available. Using such turbines, however, it is unlikely that the power generation efficiencies achievable with gasification combined cycle systems would greatly exceed that of the best existing pulverised fuel power stations without flue gas desulphurisation (although it would probably be significantly better than stations with flue gas desulphurisation). This is because the advantages of combined cycle operation are offset by the losses in the gasification and gas processing stages. In order to obtain an increase in power generation efficiency with gasification combined cycles, it is therefore essential that turbines with higher inlet temperatures are developed.

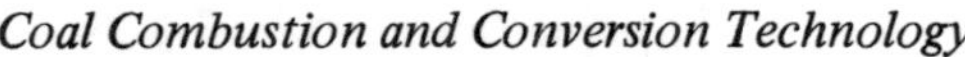

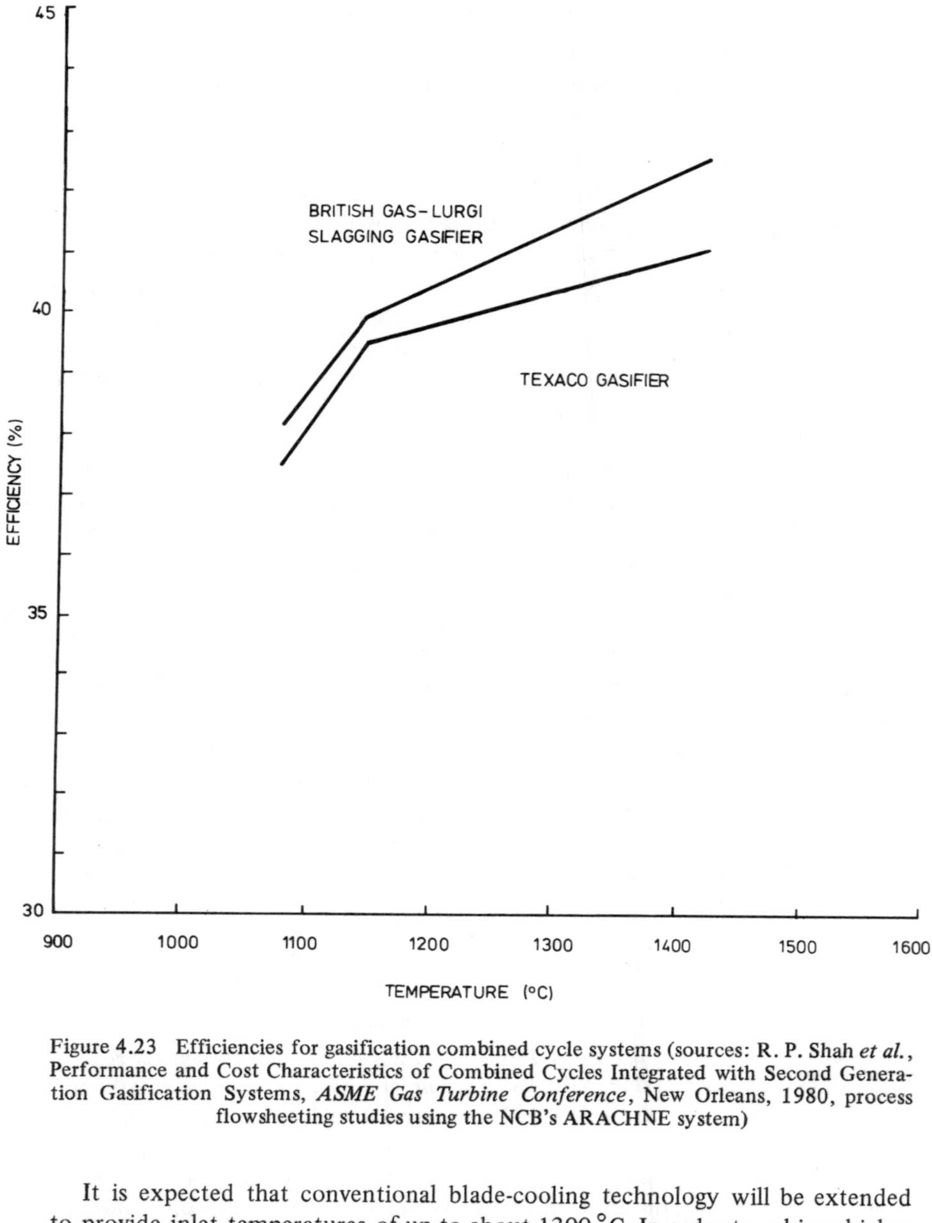

Figure 4.23 Efficiencies for gasification combined cycle systems (sources: R. P. Shah *et al.*, Performance and Cost Characteristics of Combined Cycles Integrated with Second Generation Gasification Systems, *ASME Gas Turbine Conference*, New Orleans, 1980, process flowsheeting studies using the NCB's ARACHNE system)

It is expected that conventional blade-cooling technology will be extended to provide inlet temperatures of up to about 1300 °C. In order to achieve higher inlet temperatures of up to 1600 °C, however, the development of new technologies is required and has been encouraged particularly by the 'high-temperature turbine technology' (HTTT) programme in the United States. The approaches include transpiration air cooling, water cooling and uncooled ceramic blades.

Major development efforts are being directed towards transpiration air cooling and water cooling by Curtiss–Wright and General Electric respectively and it is possible that industrial engines employing these techniques could be available by the end of the 1980s. Work on uncooled ceramic blades is less advanced but offers the prospect of high-temperature operation while eliminating the penalty of cooling losses.

For coal gasification systems generally, the clean gas obtained by cooling and aqueous scrubbing has low concentrations of particulates and alkalis and is generally regarded as of acceptable quality both for existing commercial turbines and for the advanced designs under development.

Choice of gasifier

Many types of gasifier have been proposed for use in power generation applications. The main choices are as follows.

Atmospheric or elevated pressure

Although atmospheric-pressure gasifiers have the advantage of engineering simplicity, elevated-pressure gasifiers generally give a higher system efficiency. This is because less compression energy is required for the reactants fed to a gasifier than for the resulting product gas. The costs of gas cooling and cleaning may also be reduced by the smaller gas volumes at elevated pressure.

Air-blown or oxygen-blown

During the 1970s it was widely thought that problems of lack of flexibility, capital costs and energy requirements militated against the use of oxygen plants in combined cycle power generation systems. Effort was accordingly directed mainly towards air-blown systems (for example, the air-blown Lurgi plant at Lünen, West Germany). More recently, however, it has been recognised that the disadvantages of introducing oxygen plants are not as great as previously thought and, indeed, are offset by some advantages. These arise mainly from the increased calorific value of the product gas and include less sensible heat to be removed from the raw gas directly to the steam cycle, reduced gas volumes for the gas processing stages and easier gas combustion technology.

Type of gasifier

In principle, any of the main types of gasifier (fixed bed, fluidised bed, molten bath or entrained phase) could be used for combined cycle applications. Fixed bed gasifiers have an advantage in producing a relatively cool gas thereby minimising the amount of sensible heat recovered for use directly in the steam cycle. However, the other types of gasifier are generally more tolerant of the properties of the feed coal.

Although there are widely differing views on the relative merits of the various gasification systems for combined cycle applications, it seems likely that the next generation of such plants to be built will be based on elevated-pressure, oxygen-blown gasifiers (such as the Texaco and the Brtish Gas–Lurgi slagging gasifier).

Comparison with pressurised fluidised bed combustion

Both pressurised fluidised bed combustion and gasification combined cycle systems offer the prospect of more efficient power generation together with reduced sulphur emissions and may therefore be regarded as competing technologies.

Gasification has the following three possible advantages when compared with pressurised fluidised bed combustion.

Combustion gas quality

Cooling and scrubbing a combustible gas is established as an effective means of removing constituents that may otherwise have a damaging effect on gas turbine blades. The gases from a pressurised fluidised bed combustor, in contrast, have to be cleaned hot and it is at present uncertain whether an adequate gas turbine blade lifetime can be obtained.

Gas turbine inlet temperature

Since the temperature of a fluidised bed combustor is limited to 775 to 950 °C in order to avoid ash sintering, the turbine entry temperature for a pressurised fluidised bed combustion cycle is similarly limited. In fact, the turbine entry temperature may have to be restricted to below 700 °C to ensure an adequate blade life. It follows that pressurised fluidised bed combustion is unable to take advantage of the increases in cycle efficiency that are expected to arise from the development of advanced, high entry-temperature gas turbines.

Sulphur retention

For pressurised fluidised bed combustion, the maximum sulphur retention efficiency that can be achieved is about 90 per cent; above this, the amount of sulphur acceptor required becomes impractically large. In gasification systems, however, sulphur can be removed from the product gas with an efficiency of 99.99 per cent (see above). If, therefore, stringent sulphur-emission regulations were introduced requiring the removal of significantly more than 90 per cent of the sulphur, gasification could be used but pressurised fluidised bed combustion would be impracticable.

These possible advantages of gasification combined cycle systems are offset by the efficiency losses in the gasifier and gas processing stages and, as noted earlier, with present gasifier and turbine technology, the power generation efficiency is likely to be less than that of pulverised fuel power stations. Pressurised fluidised bed combustion, however, offers the possibility of sulphur retention together with a modest improvement in power generation efficiency using existing gas turbines. Future gasification combined cycle systems may provide higher efficiencies than can be achieved with pressurised fluidised bed combustion but will require high inlet-temperature gas turbines and may need a longer development time-scale.

Stand-alone configurations

The previous discussion has concentrated on combined cycle systems in which the gasifier and gas processing stages are integrated with the power generation system. There is, however, interest in systems that operate in a 'stand-alone' mode (that is, systems in which the only mass or energy transfer between the gasification system and the power generation cycle is the clean desulphurised gas). Such systems would offer greater flexibility than integrated systems; when gas is not required for power generation, a stand-alone facility could produce gas for storage or for other applications (for example, SNG manufacture). Stand-alone plant may also be attractive to utilities in relieving them of some of the capital expenditure required for integrated combined cycle plant.

Some efficiency loss for stand-alone operation is inevitable and arises mainly in the recovery of the sensible heat from the raw gas. The loss is therefore greatest for air-blown gasifiers and gasifiers having a high raw gas exit temperature. The preferred type of gasifier for stand-alone operation is therefore the oxygen-blown fixed bed gasifier for which the sensible heat content of the raw gas is small and may be utilised to provide gasifier steam and air-separation power requirements. The efficiency penalty for stand-alone operation in this case can be as little as 1.5 percentage points.

Retrofitting

Besides combined cycle power generation, gasification may also be used to convert (retrofit) existing stations to firing with a clean desulphurised coal-derived gas. The main opportunities for this are likely to occur with oil-fired and natural gas-fired stations for which the fuel has become too expensive or unavailable and possibly with oil-fired stations where sulphur emissions have to be reduced. A wide range of gasifier types, including atmospheric and elevated pressure, air and oxygen-blown systems have been proposed for retrofitting schemes.

4.7 Liquid fuels and chemicals

The manufacture of liquid fuels and chemicals from coal by gasification and synthesis is one of the two main routes to these products. The other route (direct liquefaction) is described in chapter 5.

Most of the synthesis technologies relevant to coal conversion are already commercial and cover a wide range of current markets as shown in figure 4.24.

Oxygen/steam gasifiers are used in order to provide a nitrogen-free gas. As synthesis processes operate at elevated pressure, pressurised gasifiers are likely to be preferred when the second-generation processes are commercial. The choice between fixed-bed, fluidised bed, molten bath and entrained phase gasifiers is discussed later.

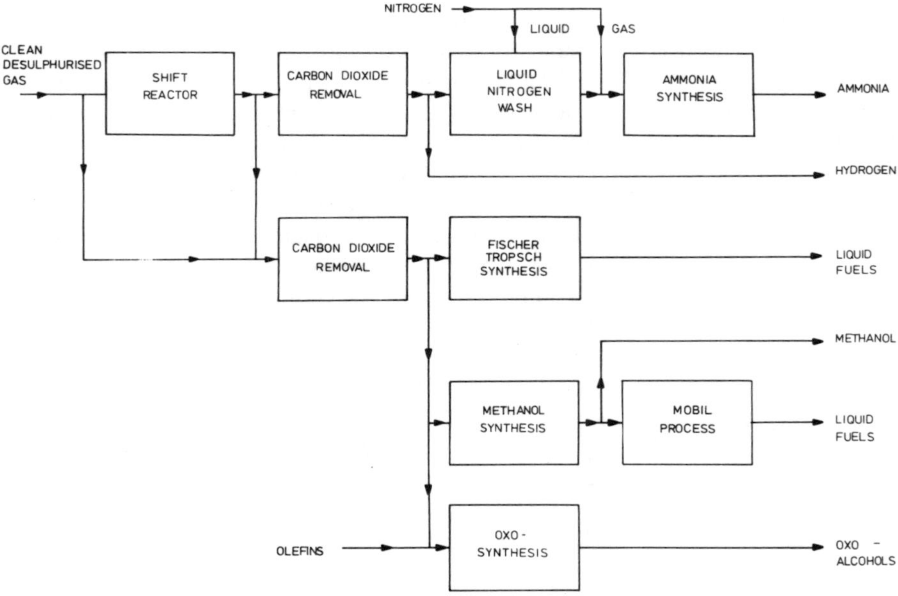

Figure 4.24 Liquid fuels and chemicals from coal

Liquid fuels

The main process options for the manufacture of liquid fuels from coal are Fischer–Tropsch, methanol and methanol-to-gasoline (MTG).

Although methanol is at present used as a chemical feedstock rather than as a fuel, it is included as an option because it is generally cheaper to manufacture from synthesis gas than conventional liquid fuels. It has been shown that methanol can be used as a motor engine fuel and is less polluting than conventional fuels. However, its use in existing engines is probably impractical because the modifications required are extensive. Storage and handling of pure methanol for road vehicles may also pose problems. These disadvantages may be avoided by using methanol as a blend stock; up to 15 per cent of methanol can be introduced into gasoline without requiring major changes to engines or to the supply infrastructure. If widely adopted, this would represent a large potential market for coal.

Both the Fischer-Tropsch and Mobil MTG processes produce conventional hydrocarbon fuels for which the product spectra are given in table 4.7.

Table 4.7 Typical product spectra for synthesis processes[a]

Component	Yield (mass %)		
	Arge	Synthol	Mobil MTG
Methane (CH_4)	7.8	13.1	1.4 (methane, ethane and ethylene combined)
Ethane (C_2H_6)	2.5	5.8	
Ethylene (C_2H_4)	0.7	4.4	
Propane (C_3H_8)	2.2	3.4	5.5
Propylene (C_3H_6)	3.9	12.8	0.2
Butanes (C_4H_{10})	2.4	3.2	11.9
Butenes (C_4H_8)	2.5	10.0	1.1
Gasoline (C5–C11)	24.8	33.4	79.9
Diesel (C12–C20)	14.7	5.1	–
Heavy oil etc. (C21+)	36.2	–	–
Oxygenates	2.3	8.8	–

[a] Sources: C. Y. Wen and E. Stanley Lee (eds.), *Coal Conversion Technology*, Addison Wesley, Reading, Massachusetts, 1979; Martin A. Elliot (ed.), *Chemistry of Coal Utilisation*, Second Supplementary Volume, Wiley, New York, 1981.

The Fischer-Tropsch processes produce a wide spectrum of products comprising mainly straight-chain alkanes and olefins. The principal refining stages required (other than fractionation and blending) are:

(1) The C3-C4 fraction contains 60 to 80 per cent of olefins and may be passed to a catalytic polymerisation unit to produce a high octane C6+ gasoline blend stock.

(2) The C5+ fraction contains 50 to 70 per cent of olefins together with some oxygenates and is usually hydrotreated to reduce the contents of these components.

The Mobil MTG process differs from the Fischer-Tropsch processes in being highly specific; nearly 80 per cent of the product occurs in the C5-C11 gasoline range. In particular, no diesel boiling-range material is produced and the yield of C1-C2 gases is low. Apart from fractionation and blending, the main refining operation carried out on the product is alkylation of the C4 fraction. In this process, butene and isobutane combine in the presence of a catalyst to produce high octane iso-paraffins as gasoline blending components.

Chemicals

At present, world production of synthesis chemicals is dominated by natural gas and petroleum feedstocks. The size of the market is indicated in table 4.8 which gives production data for ammonia, methanol, hydrogen (for oil refining) and oxo-alcohols.

Table 4.8 Production of chemicals[a]

Chemical	World production[b]	Typical large plant (product output) (t/d)	Equivalent coal input[c] (Mt/a)
Ammonia	42.5 Mt/a	1000	83
Methanol	8.4 Mt/a	1500	12
Hydrogen (for oil-refining)	46.5 Gm^3/a		32
Oxo-alcohols	3.5 Mt/a		3

[a]Source: *Yearbook of Industrial Statistics*, United Nations, 1980; Martin A. Elliot (ed.). *Chemistry of Coal Utilisation*, Second Supplementary Volume, Wiley, New York, 1981.
[b]World production figures refer to 1980 (ammonia and methanol) and 1974 (hydrogen and oxo-alcohols).
[c]'Equivalent coal input' refers to the coal required if all the world production were coal-based.

The contribution of coal to these industries is small. About 2 Mt/a of ammonia are manufactured by coal gasification and a further 7.5 Mt/a from coke oven gas; only minor quantities of the other chemicals are made from coal.

As indicated in the table, the potential size of the market for coal if it were to replace existing feedstocks for synthesis chemicals is significant, totalling over 130 Mt/a. This may be seen in the context of present world coal gasification capacity of about 4 Mt/a (coal) for ammonia manufacture and a further 17 Mt/a (coal) for liquid fuels manufacture at Sasols 1, 2 and 3.

Furthermore, growth in the world market for synthesis gas chemicals may be expected to continue, linked to world economic growth, and may even accelerate if diversification occurs (for example, the production of methanol as a fuel or hydrogen for coal liquefaction).

As synthesis gas technology evolves, other markets may also become attractive. For example, there is particular interest in the development of more specific synthesis routes to the lower olefins possibly by dehydration of the corresponding alcohols. Current world markets for ethylene and propylene are approximately 30 Mt/a and 15 Mt/a respectively.

The co-production of substitute natural gas

A common feature of many synthesis processes for the manufacture of liquid fuels and chemicals is that methane is obtained as a co-product. The methane can be disposed of either by steam reforming to produce a synthesis gas which is then recycled to the synthesis reactor or by cryogenic separation and sale as a substitute natural gas.

Although steam reforming has the advantage of simplifying the product slate, there is a severe efficiency penalty associated with this option, typically of 12 percentage points. For this reason, where there exists a demand for a gaseous fuel, it will generally be preferable to sell the methane as a substitute natural gas co-product. Viewed in economic terms, methane is valued more highly than synthesis gas as a fuel but less than synthesis gas as a chemical feedstock.

Methane arises as a co-product of synthesis processes either as a product of gasification, if a fixed-bed gasifier is used, or as a by-product of the synthesis stage (this applies mainly to the Fischer–Tropsch processes).

The technical and economic implications of methane co-production can provide an incentive to minimise the amount of methane generated. In designs of future coal-based synthesis plants, it may therefore be expected that:

(1) Second-generation pressurised entrained phase, fluidised bed or molten bath gasifiers will be preferred to pressurised fixed-bed gasifiers because of their low product gas methane contents.

(2) Specific synthesis processes that minimise methane production (Mobil MTG, for example) will be preferred to less specific processes that produce significant quantities of methane (such as Fischer–Tropsch).

CHAPTER 5

LIQUEFACTION

5.1 Chemistry of liquid fuels

Coal and oil belong to the family of fossil fuels which also includes natural gas, oil shales, tar sands and peat. Although all fossil fuels may be broadly regarded as hydrocarbon compounds, there are significant chemical differences which, in particular, determine the possible approaches to coal liquefaction.

Crude oil is a complex mixture of organic compounds containing principally paraffins, olefins, naphthenes and aromatics in varying amounts. Crude oils may be characterised by the proportions of these classes of compound, usually referred to as the 'PONA' analysis. The median molecular weight typically lies in the range 150 to 250 corresponding approximately to between 12 and 20 carbon atoms per molecule.

In terms of chemical structure, crude oil therefore differs significantly from coal. The differences are reflected in the hydrogen contents of the two fuels. The comparatively low hydrogen content of coal (generally less than 6 per cent) corresponds to the condensed polyaromatic structure whereas the higher hydrogen content of crude oil (generally greater than 10 per cent) corresponds to the lower molecular weight and more paraffinic and naphthenic nature of crude oil.

A further difference lies in the elemental compositions of crude oil and coal. The heteroatom content of crude oil is generally much lower than that of coal. Also, sulphur is usually the main heteroatom in crude oil whereas oxygen predominates in coal. Sulphur occurs principally in the higher molecular weight species as heterocyclic aromatics and naphthenes although some sulphur is also present in the lower molecular weight species, mostly as mercaptans. Nitrogen and oxygen are present in smaller quantities; nitrogen occurs mainly as heterocyclic aromatics and oxygen as paraffinic and naphthenic acids.

An important characteristic of a liquid fuel is the boiling-range. The distribution of boiling-points is one of the main methods of classifying crude oils and of defining product fractions. The boiling-range of components is closely related to their molecular structure. The main factor is the molecular weight (or, equivalently, the number of carbon atoms per molecule). As shown in figure 5.1, the boiling-point increases with the size of the molecule; the smallest molecules correspond to the gases and gasoline fraction and the largest molecules correspond to the residual fuel oil.

For many crude oils there is a preponderance of paraffinic compounds in the lower-boiling fractions with the naphthenic and aromatic species being concen-

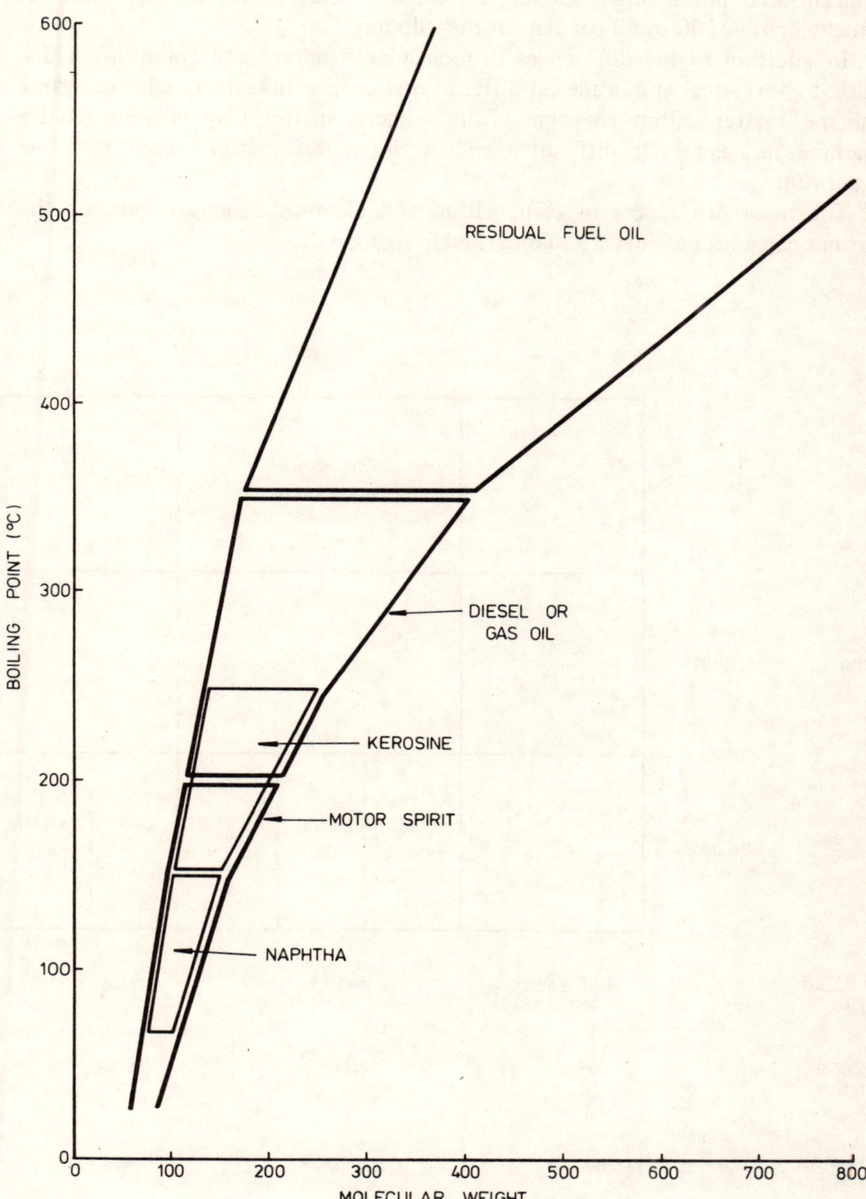

Figure 5.1 Relationship between boiling-point and molecular weight (source: *Our Industry Petroleum*, The British Petroleum Company, London, 1977)

trated in the higher boiling-range fractions. The heteroatoms are almost always concentrated in the higher-boiling fractions; residual fuel oil, for example, usually contains 80 to 90 per cent of the sulphur.

In addition to the differences in molecular structure and composition discussed above, coal and crude oil differ in that coal, as mined, contains inorganic mineral matter. Although some of the mineral matter may be removed by washing, in general it is difficult to reduce the mineral matter content to below 5 per cent.

The main differences in composition and chemical structure between the various categories of fuel are summarised in figure 5.2.

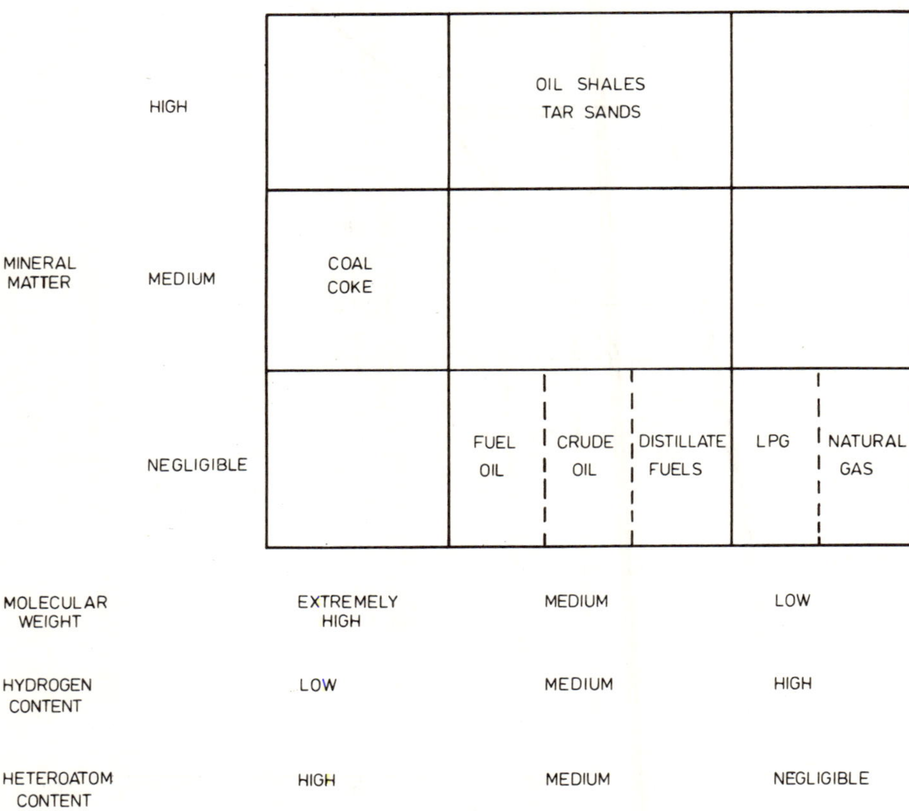

Figure 5.2 Types of fuels

5.2 Classification of processes

Comparison of the physical and chemical characteristics of coal and crude oil (for example figure 5.2) indicates that, in order to convert coal into synthetic liquid fuels, four interrelated changes must be accomplished.

(1) The complex, high-molecular weight structure of coal must be broken down into relatively small molecules. This may be achieved by three chemical processes: pyrolysis (thermal decomposition); hydrogenation; partial or total oxidation.
(2) The hydrogen content must be increased. Two approaches are possible: adding hydrogen or rejecting carbon.
(3) The heteroatom content must be reduced. This is achieved mainly by hydrogenation, the oxygen, nitrogen and sulphur being removed as water, ammonia and hydrogen sulphide.
(4) The mineral matter must be removed. This implies the conversion of the coal substance into a fluid: a liquid (or liquid solution) or a gas (or condensible vapour).

Three basic conversion routes capable of effecting the above changes have been developed.

Pyrolysis

Coal is decomposed by the action of heat into hydrogen-rich species (gases and tars) and a low hydrogen content char residue which retains the mineral matter or ash. Further refining of the tars is necessary, including hydrotreatment to reduce the heteroatom content, and a market must exist for the char residue. A simplified diagram illustrating the pyrolysis route is given in figure 5.3.

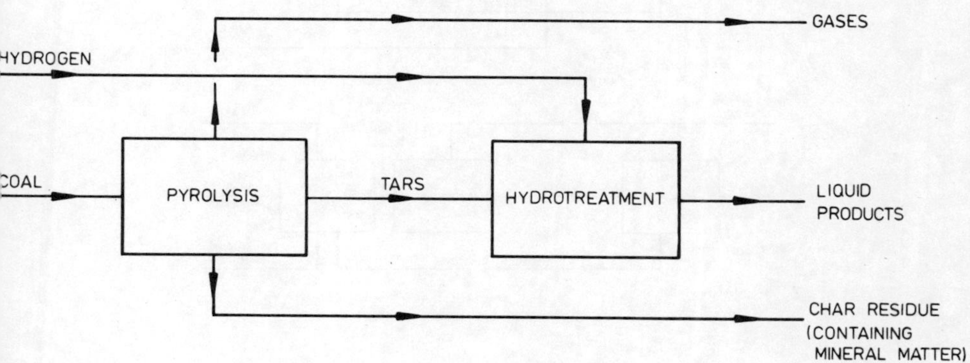

Figure 5.3 The pyrolysis route

Direct liquefaction

The main stages in this route are as follows.

(1) Dissolving the coal in a solvent (extraction).
(2) Reaction with hydrogen (hydrogenation).
(3) Separation of the mineral matter.
(4) Recovery of the products and recycle of the solvent.

The various approaches to direct liquefaction can be classified according to the order in which the above processes take place and the extent to which they are combined. In particular, the processes may be sub-divided into single-stage processes (in which extraction and hydrogenation take place in the same reactor) and two-stage processes (in which separate reactors are employed). Representative configurations are illustrated, in simplified form, in figure 5.4. The solvent extraction, hydrogenation, mineral matter separation and product refining processes are discussed in later sections.

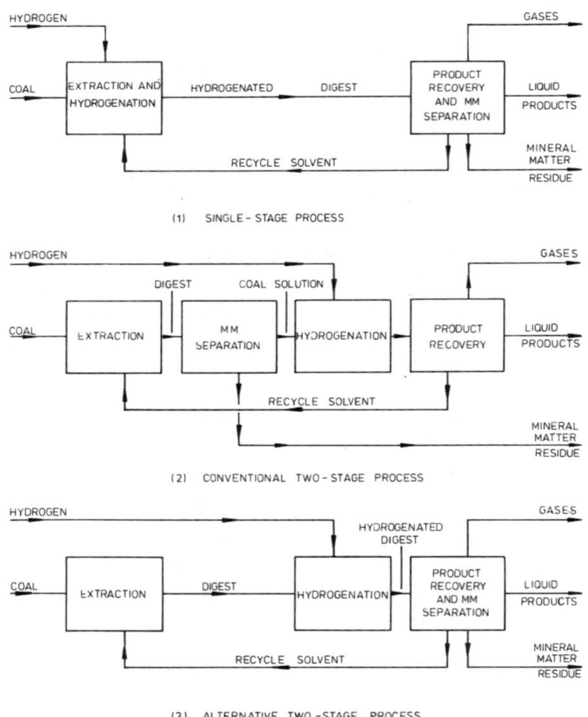

Figure 5.4 The direct liquefaction route

Indirect liquefaction

As discussed in chapter 4, coal may be converted into a synthesis gas which, after purification, may in turn be converted in the presence of a catalyst into a range of liquid fuels and chemicals.

In the indirect route, the complex molecular structure of coal is broken down completely by oxidation, producing mainly carbon monoxide and hydrogen. This route contrasts with the pyrolysis and hydrogenation routes in which a comparatively mild degradation of the molecular structure is achieved preserving those components that are required in the liquid fuel products.

The carbon monoxide is hydrogenated using hydrogen produced during the gasification stage and possibly supplemented by hydrogen from a shift conversion stage (see chapter 4). Some of the mineral matter is removed during gasification with the remainder being removed by the gas purification stage together with the heteroatoms other than oxygen. Finally, the oxygen is eliminated by the synthesis reaction. The indirect route is illustrated schematically in figure 5.5.

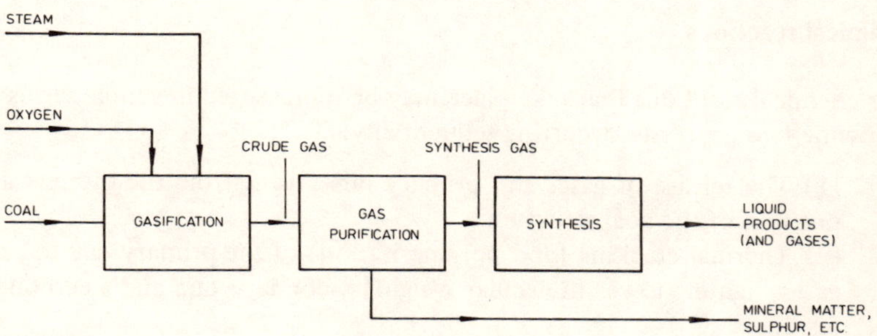

Figure 5.5 The indirect liquefaction route

In the remainder of this section the component technologies for the conversion of coal into synthetic liquids by the pyrolysis and direct liquefaction routes are considered together with the options for their integration into complete systems. The main features of these routes are also compared with those of the indirect route, the technology of which has already been discussed in chapter 4.

5.3 Pyrolysis

Pyrolysis is defined as the thermal decomposition of coal in a substantially oxygen-free environment.

To a greater or lesser extent, pyrolysis is a component of most coal processes including combustion, gasification and direct liquefaction systems. Pyrolysis alone is also the traditional process for the commercial manufacture of metallurgical coke and towns gas. In these cases, the processes employed produce a small yield of by-product liquid hydrocarbons (tar and light oil). Although typically less than 5 per cent of the input coal is evolved as liquids, in the past these processes have provided important sources of fuels and chemicals (see section 2.3).

The operation of these traditional pyrolysis processes was optimised, however, for the production of coke or gas rather than for the liquid hydrocarbon by-products. Recently, attention has been concentrated on the development of pyrolysis processes designed to maximise the yield of liquid fuels and hydrocarbon gases. It is with such applications of pyrolysis (that is, for coal liquefaction) that the present section is mainly concerned.

Chemical reactions

The chemical reactions that take place may be represented in simple terms as the following two processes occurring sequentially.

(1) The release of gases and primary tars arising from the thermal decomposition of the coal structure.
(2) Thermal cracking (and polymerisation) of the primary tars to produce gases, lighter (lower molecular weight) secondary tars and a carbonaceous residue.

The reaction scheme is illustrated in figure 5.6.

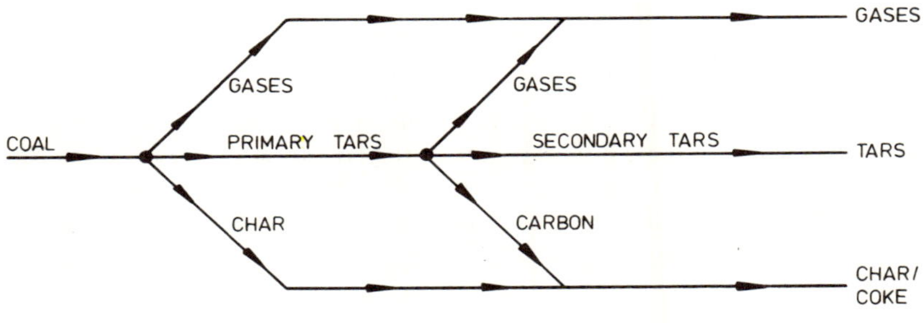

Figure 5.6 Stages in pyrolysis

As coal is heated, the first stage of decomposition is associated with the breaking of the links between the aromatic cluster units. In a coking coal, this results in the coal particles softening and eventually forming a coherent plastic mass. The aliphatic and oxygen-containing side chains (for example, methyl and hydroxyl) and bridges (for example, methylene and ether) break away from the cluster units together with low molecular weight fragments of the coal structure. These changes account for the rapid evolution of primary tars, hydrocarbon gases and water during this stage.

Some free radicals are created and these encourage polymerisation of the cluster units leading to the formation of a coke or char. As heating progresses, volatile matter continues to be released although mainly as gases. At high temperatures, hydrogen gas is the predominant species and its evolution may be attributed mainly to the loss of aromatic hydrogen. The final product is a carbon that has a 'layer-plane' structure but with the layers lacking the mutual alignment of graphite.

Thermal cracking of the primary tars is thought mainly to occur either on their passage through the coal particles or at the external surfaces of the particles. A possible reason is that the reaction may be catalysed by the free radicals created during the thermal decomposition.

Thermal cracking increases substantially with temperature. In some systems the volatile matter is heated with the char to a high temperature before quenching. Under such conditions, more thermal cracking may be expected than if the volatiles are removed and quenched at their temperature of release.

Rate-controlling mechanisms

The pyrolysis reactions may be constrained by three possible rate-controlling mechanisms depending on the design of the reactor:

heat transfer to the coal particles
heat transfer across the coal particles (that is, from the surface to the centre)
chemical reaction kinetics

For indirectly heated, fixed-bed reactors (including conventional coke ovens and gas retorts), the rate-controlling stage is heat transfer from the reactor walls to the particles at the interior of the charge. This mechanism limits the particle heating rates to values of the order of 0.1 K/s.

With fluidised bed and entrained phase systems, heat transfer to the coal particles is rapid and substantially higher heating rates can be achieved (see below). In such systems, the rate-controlling mechanism depends on the particle size and the heating rate; with large particles or high heating rates, heat transfer across the coal particles is limiting whereas for small particles or low heating rates, chemical reaction kinetics are limiting. The particle size at which chemical reaction kinetics replace heat transfer as the rate-controlling mechanism decreases from 2 mm at 100 K/s to 0.2 mm at 10 000 K/s. In practice, the design and operation of most of the pyrolysis processes being developed as a route to liquid fuels is such that chemical reaction kinetics are rate controlling.

Kinetics

As coal is heated above the temperature at which thermal decomposition begins, the rate of evolution of volatile matter increases rapidly to a maximum value. The tars are released during this initial stage, the incremental yield at high temperatures comprising gases.

The maximum rate of mass loss is generally of the order of 10^{-3} K^{-1} and is approximately independent of heating rate for values from 0.1 to 100 000 K/s.

At the low heating rates in conventional coke ovens and gas retorts (about 0.1 K/s), thermal decomposition begins at 350 to 400 °C and the maximum release rate usually occurs at 450 to 500 °C. The effect of increasing the heating rate is to delay the release of the volatile matter typically by 100 K for each order of magnitude increase in the heating rate above 0.1 K/s. For example, at 10 000 K/s the maximum release rate (and mean temperature of release) does not occur until about 1000 °C.

The above description relates generally to bituminous coals, sub-bituminous coals and lignites; anthracites differ in not producing tars and in releasing the volatile matter at somewhat higher temperatures than for the lower-rank coals.

Pyrolysis reactions appear to be too complex to be modelled in detail although several simple descriptions of the kinetics have been proposed. A widely adopted approach is to regard the net effect of the volatile release and primary tar-cracking reactions as different parts of the molecular structure decomposing at different temperatures. This leads to a system of parallel first-order reactions corresponding to a distribution of activation energies.

Rapid pyrolysis

Since the early 1960s, interest has been directed mainly towards 'rapid' pyrolysis techniques employing high heating rates. Two methods of achieving high heating rates in commercial-scale plant are under development.

Fluidised bed systems

The high heat transfer coefficients obtainable in a fluidised bed enable high particle-heating rates for crushed coal to be achieved. This approach was adopted in the multiple fluidised bed pyrolysis unit used in the COED and COGAS processes.

Entrained phase systems

Still higher heating rates can be obtained with entrained phase systems in which pulverised coal is used. The coal may be heated by the gas stream alone or, as in the Occidental (formerly Garrett) process, by hot recycled char mixed with gas.

By using the above techniques, coal particle-heating rates of 100 to 50 000 K/s can be achieved compared with values of about 0.1 K/s for the conventional, fixed-bed systems.

The most important feature of rapid pyrolysis is that high yields of volatiles and, particularly, tars can be obtained. Compared with the conventional, low heating-rate processes, an increase of up to 50 per cent in the total volatile matter yield is possible together with an increase of up to a factor of four in the yield of tars.

It is mainly because of these high yields that recent development work has concentrated on rapid pyrolysis systems.

Operating conditions

Initially it was assumed that the high heating rate was directly responsible for the increased yield of volatile matter obtained during rapid pyrolysis, the underlying mechanism being a change in the way in which the molecular structure decomposed. However, the conditions (particularly concentration and residence time) necessary to obtain rapid pyrolysis differ significantly from those for low heating rates. Over recent years it has become widely accepted that the heating rate itself is relatively unimportant, the main factor being the effect of the pyrolysis conditions on the extent of the thermal-cracking reactions.

For these reasons, the preservation of the primary tars can be regarded as the main criterion in the choice of operating conditions for pyrolysis processes when a high yield of liquid products is required. The main parameters are as follows.

Temperature

If the reaction temperature is too low, the release of primary tars will be incomplete and the yield of liquid products low. A high reaction temperature, however, will lead to thermal cracking of the primary tars and will also result in a low yield of liquid products. There is, therefore, an optimum temperature corresponding to the maximum tar yield.

Residence times

A short solids residence time can limit the release of primary tars because the thermal decomposition of the coal will be unable to proceed to completion. A short gas residence time, however, can reduce thermal cracking.

In a fluidised bed pyrolysis system, the gas residence time is considerably shorter than that of the solids so that thermal decomposition can be completed but thermal cracking is limited by the rapid removal of the volatiles from the bed in the gas stream.

In an entrained phase system, the solids and gas residence times are generally similar and there therefore exists an optimum residence time for the maximum tar yield. For this reason, it is common in such processes to quench the products rapidly after the pyrolysis reactor to preserve the high initial tar yield.

Pressure

Theoretically, a decrease in pressure would be expected to encourage both the thermal-decomposition and thermal-cracking reactions (by Le Chatelier's principle). It is found that pyrolysis under reduced pressures (vacuum pyrolysis) typically increases the total volatile matter yield by 10 to 20 per cent and also results in a substantial increase in the tar yield. It is possible that the main effect is the increased volatility of the primary tars. This enables them to escape more readily from the coal particles and thereby reduces the opportunity for thermal cracking (see also 'Gas–solids contacting' below).

Vacuum pyrolysis does not at present appear to be a favoured option for commercial process development.

Gas–solid contacting

Since thermal cracking of primary tars appears mainly to occur on particle surfaces, the tar yield can also be improved by encouraging the primary tars to escape more quickly from the vicinity of the particle and to become diluted in the gas stream (thereby reducing the opportunity for thermal cracking). This can be achieved by either reducing the particle concentration or using finer particles.

Entrained phase pyrolysis reactors offer the advantage of permitting both low particle concentrations and the use of fine particles.

Hydrogen atmosphere

An important method of minimising thermal cracking is to stabilise the primary tars by carrying out the pyrolysis in a hydrogen atmosphere. This technique, which has a number of advantages over pyrolysis in an inert atmosphere and has received considerable attention in recent years, is discussed separately in the next section (hydropyrolysis).

The main effects of the operating conditions on the pyrolysis reactions (thermal decomposition of the coal and thermal cracking of the primary tars) and on the resulting product slate are summarised in table 5.1. If liquids are the required product, thermal decomposition must be promoted but thermal cracking must be inhibited. However, if gases are required, both the thermal-decomposition and thermal-cracking reactions must be promoted.

Table 5.1 Effects of operating conditions on pyrolysis products

Parameter	Reactions[a]		Products[a]		
	Decomposition	Cracking	Gases	Liquids	Total
Temperature[b]	+	+	+	Optimum	+
Solids residence time	+	0	+	+	+
Gas residence time[b]	0	+	+	–	–
Pressure[b]	–	+	+	–	–

[a] Notation: + indicates an increase, 0 indicates no change, – indicates a decrease, when a parameter is increased.
[b] Range of conditions: temperature 400 to 900 °C, gas residence time 0.01 to 20 s, pressure 0.01 to 100 bar.

General features of pyrolysis processes

Pyrolysis processes can be applied to the majority of bituminous and sub-bituminous coals. Lignites may also be used for some applications but generally give a comparatively low tar yield. Anthracites are unsuitable because of their low total yield of volatile matter. As bituminous and sub-bituminous coals form the major part of world reserves, considerations of coal type are not a limiting factor in potential applications of pyrolysis.

One consequence of the increased yield of liquid products obtained in rapid pyrolysis processes is that the quality of the liquid products is lower. In particular, the liquids have a relatively high content of high molecular weight material (pitch, for example) and heteroatoms. Under low heating rate conditions the increased extent of thermal cracking of the primary tars results in a lower molecular weight tar product (although at a reduced yield) with more of the heteroatom content of the coal reporting in the gaseous products. More extensive refining of the liquids produced by rapid pyrolysis is necessary before marketable fuels are obtained.

Process configurations

A feature of all pyrolysis processes is the co-production of a carbonaceous residue (char or coke).

In the conventional slow pyrolysis processes, the coke was either the main product (as in metallurgical coke manufacture) or could find attractive applications as a smokeless fuel or gasifier feedstock (for producer gas or water-gas manufacture).

Design configurations for the new, rapid pyrolysis processes, however, are generally based on the assumption that the use of the char has to be integrated with the overall process. Some of the char can be burned to meet process heat requirements and some gasified to produce the hydrogen required to upgrade the crude liquid product to saleable fuels. For char remaining in excess of these requirements, a number of process options have been considered for commercial application and may be classified in terms of their final products.

SNG and liquids

In this option, the char is fed to a total gasification process. Some of the gas is used to manufacture hydrogen to upgrade the crude liquid product and the remainder is used for SNG manufacture (see the upper part of figure 5.7).

SNG and electricity

Alternatively the char can be used for power generation as illustrated in the middle part of figure 5.7. In this case, the pyrolysis unit is operated so as not to produce liquids.

Liquids and electricity

If the pyrolysis unit is operated so as to maximise the yield of liquid products and the char is used for power generation (either directly as boiler fuel, or indirectly via a gasification system) then a configuration of the type illustrated in the lower part of figure 5.7 is obtained.

Recent developments

In general, the development of new pyrolysis processes for liquid or gaseous fuels does not have as high a priority in research and development programmes as other approaches to coal conversion. The two main processes for which pilot plants have been built are COED and Occidental.

The COED process employs a multiple fluidised bed pyrolysis unit and has been operated on a 25 t/d scale. Initially, work concentrated on configurations for the co-production of liquid products and electricity. However, the sulphur content of the char was high, making this concept appear unattractive. Effort was then directed towards the COGAS process in which the char from the multiple fluidised bed pyrolysis unit is gasified using a novel gasification system (see chapter 4). The final products are SNG and liquid fuels. Work on both of these options has now stopped.

The Occidental pyrolysis process is based on an entrained phase reactor and has been operated on a 3 t/d scale. Configurations for this process involve using the char for electricity generation (with either SNG or liquids as the co-product).

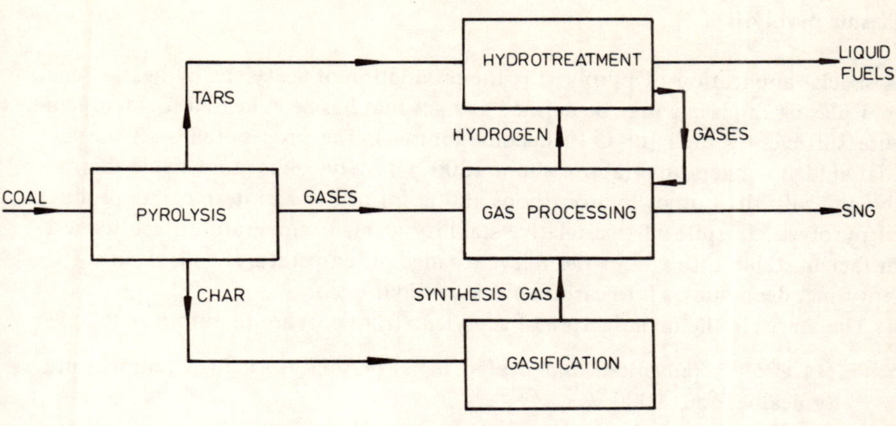

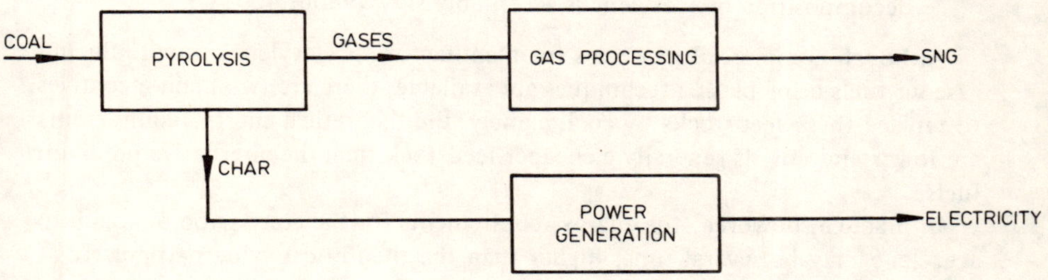

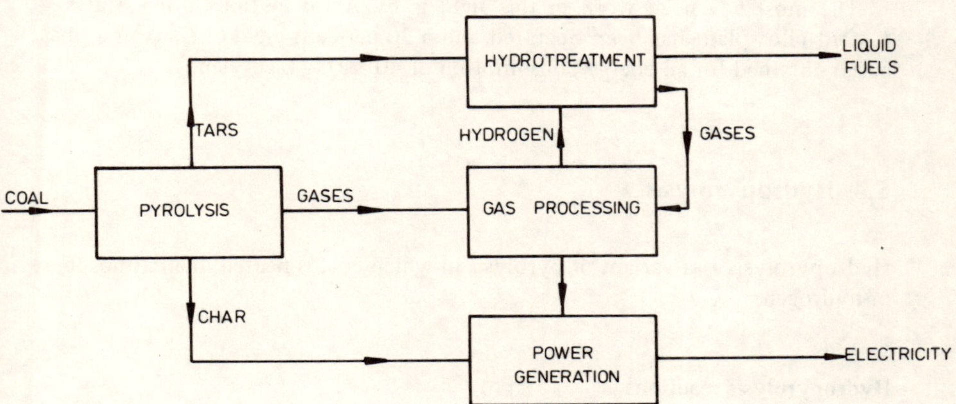

Figure 5.7 Pyrolysis process configurations

Plasma pyrolysis

A special application of pyrolysis is the production of acetylene by heating coal in a plasma (a plasma may be defined as a gas that has been heated to a temperature sufficiently high for it to become ionised). The process relies on the fact that at high temperatures (above about 1300 °C) acetylene is more stable thermodynamically than other hydrocarbons and is formed as an intermediate product of pyrolysis. In spite of this relative stability at high temperatures, acetylene is in fact unstable throughout the relevant range of temperatures and, given sufficient time, decomposes into carbon black and hydrogen.

The stages in the manufacture of acetylene from coal are therefore as follows.

(1) Coal is pulverised and heated in a plasma to a high temperature, typically about 2000 °C.
(2) The coal remains in the plasma for a short residence time, usually a few milliseconds, chosen to optimise the yield of acetylene.
(3) The products are quenched rapidly to a temperature below which the decomposition of acetylene is acceptably slow (about 400 °C).

Although commercial processes for manufacturing acetylene from liquid and gaseous fuels using plasma techniques are available, there are two main incentives to replace these feedstocks by coal, namely, the theoretical energy requirements are lower and coal is generally a cheaper feedstock than the alternative premium fuels.

In practice, however, the energy requirements for the conversion of coal into acetylene may be several times higher than the theoretical value particularly at the throughputs and conversion yields required for economic reasons. Also, the use of coal introduces additional material problems, notably erosion and corrosion.

The most advanced work in this field is by AvCo in the United States. A 1.3 t/d pilot plant has been operated and a 30 per cent yield of acetylene (mass basis) obtained for an energy consumption of 40 MJ/kg (acetylene).

5.4 Hydropyrolysis

Hydropyrolysis is a variant of pyrolysis in which coal is heated in an atmosphere of hydrogen.

Hydropyrolysis reactions

The hydrogen inhibits the thermal cracking of the primary tars by stabilising the reactive species initially formed by the decomposition of the coal. Some hydrocracking of the primary tars to light hydrocarbon liquids (predominantly benzene, toluene and xylenes) and/or hydrocarbon gases (predominantly methane and ethane) occurs together with some hydrogasification of the char residue to methane. The extent of these reactions depends on the residence time, temperature and pressure. Under the conditions usually adopted for hydropyrolysis, the contribution of hydrogasification is, however, small. The possible reaction paths are summarised in figure 5.8.

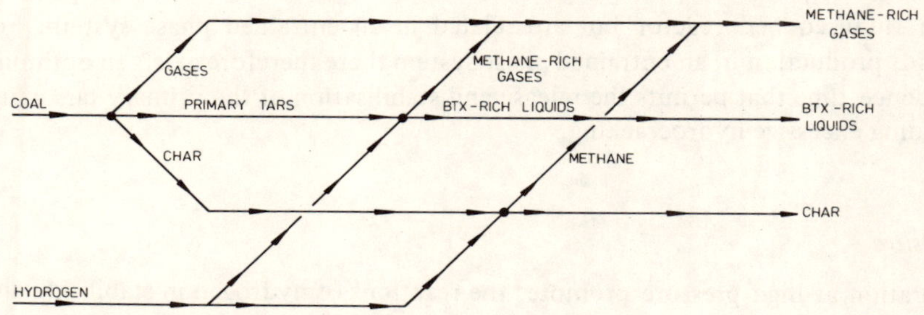

Figure 5.8 Stages in hydropyrolysis

The stabilisation of the primary tars occurs quickly and is followed by hydrocracking although this is a slower process. Even so, complete hydrocracking of the tars to methane can be achieved in less than one second under favourable conditions (high pressures, high temperatures and fine particles). Hydrogasification of the char is extremely slow by comparison.

Operating conditions

The development of hydropyrolysis systems has been linked to the rapid pyrolysis technology discussed in the previous section and effort has concentrated on entrained phase and fluidised bed designs. The most advanced development is the Cities Service-Rockwell process which uses rocket engine technology to achieve efficient entrained phase gas-solid contacting. A pilot plant of 25 t/d capacity has been operated. Fluidised bed hydropyrolysis was proposed for the Coalcon project (now suspended).

The main effects of the choice of operating conditions on the yields obtained from hydropyrolysis are as follows.

Temperature

An increase in temperature increases both thermal decomposition of the coal and hydrocracking of the tars. As a result, the total yield of volatiles increases with temperature but the yield of liquids passes through a maximum value. The pattern is therefore similar to that determined by pyrolysis reactions.

Residence times

A short solids residence time can limit the release of primary tars and a short gas-phase residence time can reduce hydrocracking of the tars and hydrogasification of the char. The solids residence time must always be adequate to permit the thermal decomposition of the coal. The choice of gas-phase residence time, however, depends on the required product with long residence times favouring gas production and short residence times favouring tar production.

As noted previously, the gas-phase and solids residence times are independent in a fluidised bed reactor but are related in an entrained phase system. For liquids production in an entrained phase system there therefore exists an optimum residence time that permits the release and stabilisation of the primary tars while avoiding excessive hydrocracking.

Pressure

Operation at high pressure promotes the reactions of hydrogen in stabilising and hydrocracking the primary tars and hydrogasifying the char. Since stabilisation of the primary tars proceeds rapidly, the effects of increasing the hydrogen pressure are to increase the yield of gas and the total yield of volatile matter.

Gas–solid contacting

Good gas–solid contacting, for example by using fine particles, also assists in promoting the reactions of the hydrogen with effects similar to those of increasing the pressure.

The main effects of the operating conditions on the reactions and product slate are summarised in table 5.2. If the required product is liquid fuels, the thermal decomposition and primary tar stabilisation reactions must be promoted but the tar-hydrocracking and char-hydrogasification reactions inhibited. For gas production, however, all the reactions must be promoted.

Comparison with rapid pyrolysis

The main advantages of hydropyrolysis over rapid pyrolysis are as follows.

(1) A higher yield of volatiles (liquids or gas, depending on the conditions) is obtained. Typically, 60 to 70 per cent of the coal input can be converted into liquids and gases in a hydropyrolysis reactor as compared with 40 to 50 per cent with rapid pyrolysis.

(2) The liquids produced by hydropyrolysis are of a higher quality than those obtained directly by rapid pyrolysis. In particular, the molecular weight and heteroatom content are comparatively low and the yield of BTX is high. However, liquids of similar quality can be obtained by hydro-

Table 5.2 Effects of operating conditions on hydropyrolysis products

Parameter	Reactions[a]			Products[a]		
	Decomposition and stabilisation	Hydrocracking	Hydrogasification	Gases	Liquids	Total
Temperature[b]	+	+	+	+	Optimum	+
Solids residence time	+	0	+	+	+	+
Gas residence time[b]	0	+	+	+	–	+
Pressure[b]	0	+	+	+	–	+

[a] Notation: + indicates an increase, 0 indicates no change, – indicates a decrease, when a parameter is increased.
[b] Range of conditions: temperature 400 to 900 °C, gas residence time 0.01 to 20 s, pressure 0.01 to 100 bar.

treating the crude tar product of rapid pyrolysis. In this respect, an advantage of hydropyrolysis may be the combination of the pyrolysis and hydrotreatment processes into a single stage.

The demand for hydrogen for the hydropyrolysis reactor, together with the high conversion of the coal to liquid and gaseous products, can lead to the situation in which all of the char produced is required for hydrogen manufacture. A balanced process can therefore be obtained, in contrast with most rapid pyrolysis systems.

The Cities Service-Rockwell process is typical of the hydropyrolysis configurations under consideration and is illustrated in figure 5.9.

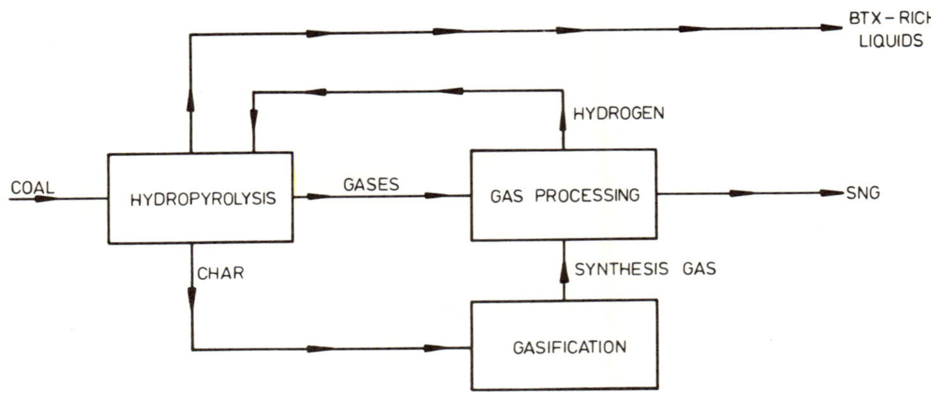

Figure 5.9 Hydropyrolysis process configuration

5.5 Solvent extraction

The direct hydrogenation route for converting coal into liquid fuels is characterised by the thermal decomposition of coal in the presence of a liquid solvent under conditions that ensure that most of the carbonaceous material in the coal dissolves. Such a form is convenient for the reaction of the coal with hydrogen and for the separation of the mineral matter. The main approaches to direct liquefaction are illustrated in figure 5.4 (see section 5.2), which also introduces some of the terminology. The initial product obtained by dissolving the coal in a solvent is termed the 'digest'. Following the removal of the insoluble residue from the digest, the resulting liquid is referred to as the 'coal solution'. The coal-derived material dissolved in the solvent is the 'coal extract'.

The present section discusses the conditions necessary for solvent extraction; hydrogenation and solids separation form the subjects of subsequent sections.

Mechanisms of solvent extraction

Four mechanisms may contribute to the extraction of coal by a solvent, depending on the operating conditions and the nature of the solvent. These may be considered in relation to their effects on the coal structure, as follows.

Solution

Poor solvents are able to dissolve only a small proportion of the coal (typically 3 per cent for a bituminous coal). The extract comprises predominantly aliphatic waxes and resins captive in the coal structure.

Dispersion

More effective solvents are absorbed by the coal and cause swelling to occur. This releases some of the smaller, more weakly bonded structures (possibly those bonded only by Van der Waals' forces) from the coal into the solvent. The chemical structure and composition of the extracted material appear similar in most respects to those of the undissolved residue and the original coal except that the average molecular weight of the extracted material (in solution) is of the order of 1000.

The degree of extraction depends on the chemical structure of the solvent but, for all solvents, increases substantially with temperature even before decomposition of the coal begins.

Thermal decomposition

If the temperature at which the extraction takes place is such that the coal is beginning to decompose, a high degree of extraction is generally obtained. This is thought to be because thermal rupture of the linkages joining the aromatic nuclei releases more of the structure for dispersion in the solvent.

Hydrogenation

If extraction takes place above the decomposition temperature of the coal, the free radicals formed may repolymerise to produce insoluble high molecular weight material. In order to prevent this, the reactive species must be stabilised, as formed, by hydrogenation (as with hydropyrolysis, discussed earlier). This can be achieved in three ways:

- by transfer of hydrogen from the solvent to the coal
- by reaction with elemental hydrogen
- by internal rearrangement of the hydrogen in the coal

Solvent properties

Solvents that possess the ability to decompose partially and to release hydrogen to the coal are termed 'hydrogen donors' and play an important role in hydrogen transfer to the coal in all direct liquefaction processes. In particular, reaction with elemental hydrogen and rearrangement of internal hydrogen also appear to be promoted if the solvent possesses hydrogen donor properties (that is, either is a hydrogen donor or is capable of being hydrogenated to a hydrogen donor state). This may be because the solvent acts as a hydrogen carrier in these cases.

Solvents for coal have been classified according to which of the above mechanisms is dominant. The usual terminology is, however, rather confusing. Solvents that extract only the waxes and resins are termed non-selective whereas solvents that extract a significant proportion of the coal (and for which the extract is similar to the coal) are referred to as selective. Solvents that act as hydrogen donors are termed reactive. Commonly quoted examples of non-selective, selective and reactive solvents are benzene, pyridine and tetralin respectively. These categories can, however, be misleading because the mechanism by which a solvent operates may depend on temperature. Benzene, for example, is a poor solvent at ambient temperature but at high temperatures (and elevated pressure) can extract a significant proportion of the coal.

The most effective solvents for coal are capable of extracting no more than 40 per cent of the coal substance (dry, mineral-matter-free) at 'near' ambient temperatures. It has been found that the only conditions under which a high extract yield of around 90 per cent of the coal substance can be obtained involve carrying out the extraction within a range of temperatures defined by the following limits.

(1) The temperature has to be above that at which the coal begins to decompose (generally about 375 °C). This is probably because solvents alone are unable to break the covalent cross-linkages between the aromatic nuclei; thermal decomposition is required to assist this process.

(2) The temperature has to be below that at which thermal cracking of the extract to give gas and an insoluble char begins (generally about 475 °C).

The direct liquefaction processes under development generally operate within the comparatively narrow range of temperatures specified above. The remainder of this chapter is devoted to extraction under these conditions because of their commercial importance.

The rate of extraction at the operating temperatures for liquefaction processes is high, the time constant typically being of the order of one minute. However, longer residence times, sometimes of up to one hour, may be specified for the extraction stage in order to allow slow changes to occur to the extract in solution. These arise from progressive hydrogenation and reduction in the molecular weight of the dissolved material.

A wide range of partially hydrogenated polynuclear aromatic compounds are effective solvents for coal under the conditions of interest. Such compounds predominate in the heavy distillate fraction of hydrogenated coal extract and similar materials can be obtained by hydrogenating the anthracene oil fraction of coal tar. Three factors contribute to the suitability of these compounds as coal solvents.

Chemical structure

On the principle of 'like dissolving like', polynuclear aromatic compounds may be expected to be effective solvents for coal (compared with aliphatic or highly polar compounds) because their chemical structure is closely related to that of coal.

Hydrogen donor properties

As discussed earlier, the reactive species formed while coal decomposes can be stabilised by hydrogenation, inhibiting repolymerisation to form an insoluble residue. The extraction yield increases with the degree of hydrogen transfer up to about one per cent (daf coal, mass basis) above which little further improvement is obtained. Hydrogen transfer in excess of one per cent is, however, employed in single-stage processes to produce a high direct yield of light products (see below).

Boiling-range

In order to carry out the extraction at temperatures above the decomposition temperature of the coal, the boiling-range of the solvent must be sufficiently high at the operating pressure. Processes that employ high pressures in the extraction stage (to promote hydrogenation) can operate using a low boiling-range solvent fraction, which in turn implies a low mean molecular weight for the solvent.

For the above reasons, all the direct liquefaction processes under development use a recycle stream from the crude liquid product as solvent. Liquefaction processes generally have to maintain self-sufficiency in solvent so that losses arising, for example, by hydrocracking to lighter fractions, must be compensated for by the production of material within the selected boiling-range from hydrogenation of the extract.

Since certain components of the solvent will be hydrogenated and cracked preferentially, the recycling solvent will progressively accumulate those species that are most resistant to hydrogenation and cracking. The least reactive species are extremely high molecular weight, high boiling-point, condensed aromatic compounds. A purge stream (allied with excess solvent production) may therefore be necessary to prevent the accumulation of stable, unreactive species in the recycle solvent, particularly when a high boiling-range fraction is used.

Effect of coal type

Suitability for use in direct liquefaction processes varies considerably with coal type and the problem of coal selection and characterisation for this application has received much attention. A number of coal properties appear to be important and these can be considered in the following three main categories.

Structure and composition

The traditional coal classification systems devised for combustion and coking applications cannot be used alone to assess coals for liquefaction. However, because the classification systems reflect trends in molecular structure, a broad indication of suitability is possible.

The high molecular weight of the aromatic nuclei and extensive cross-linking in anthracites effectively prevents the breakdown of the structure and these coals are almost completely resistant to solvent extraction. Anthracites cannot therefore be used for direct liquefaction.

In contrast, the smaller aromatic nuclei and less extensive cross-linking in bituminous, sub-bituminous and lignitic coals enable thermal degradation and solvent action to breakdown the coal structure and these coals can all be extracted readily.

The maximum extraction yields are obtained with bituminous coals, somewhat lower yields are obtained with sub-bituminous coals and lignites. A further disadvantage of sub-bituminous coals and lignites is that more hydrogen is required for heteroatom removal because of the high oxygen content of these coals.

There is, however, evidence that sub-bituminous coals and lignites are more reactive than bituminous coals which implies shorter residence times or lower hydrogen pressures for the same degree of hydrogenation.

Maceral analysis

As noted in chapter 1, coal is not homogeneous but consists of a microscopic mixture of components termed 'macerals', arising from the decay of different parts of the original plant material. Each maceral component behaves differently under extraction conditions so that it is possible for two coals to have the same analysis (in terms of carbon, hydrogen, oxygen, nitrogen and sulphur) but to give significantly different conversion yields under extraction.

A number of maceral classification systems have been devised. A simple and widely used terminology is as follows.

Exinite This is derived from resins, spores, leaf cuticles and algae and has a relatively high hydrogen content.
Vitrinite This is derived from woody tissue and is the principal component of bright coals. The hydrogen content is, as a result, similar to that of the aggregated coal.
Inertinite This includes fossil charcoal, and has a relatively low hydrogen content.

It is generally found that exinite and vitrinite can be extracted readily but that little of the inertinite is taken into solution. As a result, the extraction yield is found to correlate broadly with the content of the reactive macerals exinite and vitrinite. It follows that coals with a high inertinite content are not favoured for direct liquefaction.

Mineral matter

It is known that some of the constituents of the mineral matter contained in coal, principally iron sulphides and oxides (for example, pyrites, haematite), can act as catalysts for processes in which the extraction takes place in the presence of hydrogen (see next section). In liquefaction processes of this type, a high iron content in the mineral matter is therefore advantageous particularly if no other catalysts are employed. It is relevant to note that an iron ore called 'red mud' obtained as a by-product of bauxite-processing was one of the catalysts added to the coal in the Bergius process.

It is generally a disadvantage for the coal to have a high mineral matter content (particularly inherent mineral matter, see section 1.1) as this imposes an additional burden on the solids separation system and may (depending on the design) involve increased capital costs or more of the input carbon reporting in the mineral matter residue stream.

Supercritical gas extraction

A novel approach to coal liquefaction that has similarities both with conventional solvent extraction and pyrolysis is supercritical gas extraction.

It has been found that a proportion of the coal substance can be extracted by a gas; the extent of the extraction depends mainly on the gas density. However, the proportion extracted remains low unless a gas density approaching that of liquid solvents is achieved. In practice, this means employing high molecular weight gases and operating at high pressures in order to extract a substantial fraction of the coal substance.

A liquefaction process based on these principles has been investigated by the National Coal Board. As with extraction using liquid solvents, the preferred extraction temperatures lie in the range where the coal is beginning to decompose thermally. Typically, the extraction is carried out at 350 to 450 °C and at pressures of 100 to 200 bar. A process-derived fraction having a critical temperature of 300 to 400 °C is used as the solvent and the extraction can therefore take place at supercritical temperatures.

Under these conditions, up to half of the mass of the coal can be extracted into the gas phase. The extracted material can be precipitated and the solvent recovered simply by reducing the temperature or pressure. As in pyrolysis processes, the char residue remaining after extraction can be gasified to provide the hydrogen required for subsequent hydrogenation of the liquid product.

At present, work has been confined to operation on the kg/h scale. Although the technique has revealed information about coal structure and the mechanisms of thermal breakdown, larger-scale work is necessary before this approach can be regarded as the basis for a commercial coal liquefaction technology. A number of areas for further development, including the disengagement of the solids from the gas stream, remain and coal liquefaction work by the National Coal Board is now concentrating on the more conventional liquid solvent extraction process.

5.6 Hydrogenation

The term 'hydrogenation' applies generally to any reaction in which hydrogen is added and is used here to refer to the primary liquefaction of the coal. The more specific terms 'hydrocracking' (hydrogenation under severe conditions so that cracking occurs, with an associated reduction in molecular weight) and 'hydrotreatment' (hydrogenation under mild conditions to remove heteroatoms and saturate aromatics) are used to describe secondary refining operations (see section 5.9). However, both of these processes generally occur during primary liquefaction so that the use of the terminology is, to some extent, arbitrary.

The role of hydrogenation

In order to convert coal into marketable fuels, it is necessary to add the equivalent of 5 to 7 per cent of hydrogen (daf coal, mass basis). Usually, 3 to 4 per cent of hydrogen is added in the initial hydrogenation stage to produce distillate fuels, the remaining hydrogen being added in the subsequent refining processes (see section 5.9). If more hydrogen is added initially, then less is required during refining and vice versa. The severity of the refining stage therefore depends on the amount of hydrogenation achieved initially.

As discussed in section 5.2, the liquefaction processes being developed differ in the conditions under which the initial hydrogenation takes place. Two main categories exist.

Single-stage processes

In single-stage processes (see figure 5.4(1)), extraction and hydrogenation take place in the same reactor. The mineral matter is then removed from the hydrogenated digest and the products recovered for subsequent refining. This approach formed the basis of the Bergius process used extensively in Germany during the 1930s and 1940s (see section 2.3).

Two-stage processes

In two-stage processes, the extraction and hydrogenation take place in separate reactors. Two variants are possible. Processes that follow the Pott–Broche approach (also see section 2.3) incorporate a mineral matter removal stage immediately after extraction. This produces a clean coal solution for hydrogenation (see figure 5.4(2)). Alternatively, the digest can be hydrogenated immediately after extraction, with mineral matter separation occurring subsequently (see figure 5.4(3)). In both cases, the products are recovered for downstream refining.

The comparative simplicity of single-stage processes has resulted in most of the large-scale research and development effort over recent years being directed towards this approach. Recently, however, increasing interest has been shown in two-stage processes because of the prospect of improved performance and flexibility.

With single-stage processes, the reactor conditions have to be chosen as a compromise between the requirements for extraction and hydrogenation. In practice, these are in conflict; in particular, a high temperature is necessary for a good yield during extraction but results in excessive gas production during hydrogenation. Carrying out extraction and hydrogenation separately in a two-stage process enables the conditions to be optimised for each stage. The main advantages claimed for two-stage operation are as follows.

(1) High coal conversion by the appropriate choice of extraction conditions.
(2) High liquids yield and low gas yield by the appropriate choice of hydrogenation conditions.
(3) Good flexibility to process a wide range of coal types.
(4) High hydrogen utilisation efficiency.

Although a number of development programmes have been initiated, work on two-stage processes is still at a comparatively small scale, and it has yet to be established whether these benefits can be realised economically on a large scale.

For both single-stage and two-stage processes, hydrogen transfer to the coal digest or solution can be promoted in the following three ways.

By using a hydrogenated solvent

If the solvent is hydrogenated catalytically before the extraction stage, this hydrogen can be transferred to the coal during extraction. Hydrogen transfer by this means alone is limited to one or two per cent (daf coal, mass basis).

By using hydrogen under pressure

Pressures of more than 100 bar are necessary for hydrogenation to be effective, and pressures of up to 700 bar have been employed. For single-stage processes, the extraction must therefore take place at high pressures. For two-stage processes, the extraction pressure can be as low as 15 bar, sufficient only to prevent the solvent boiling at the extraction temperature, although pressures similar to those for single-stage processes are often used instead. The low-pressure approach offers easier coal–solvent slurry feeding (centrifugal pumps rather than reciprocating pumps may be employed), less wear on pressure control valves and a lower extraction vessel cost. However, extraction at high pressures may be expected to have advantages of lower solvent to coal ratios (and therefore lower thermal losses), shorter residence times for extraction and a choice of a wider range of boiling points for the solvent.

By using a catalyst

A catalyst may be used to promote the hydrogenation of a coal digest or solution. The effect of a catalyst is to assist the reaction of gaseous hydrogen with both the coal and the solvent; hydrogen transfer from a donor solvent to the coal appears to be unaffected by a catalyst. It follows that a catalyst is used only where hydrogen is present under pressure and that, for such conditions, the use of a hydrogenated solvent is unnecessary because the catalyst will promote hydrogenation of the solvent. Single-stage processes that employ a catalyst may be particularly susceptible to catalyst poisoning (see later).

The configurations and conditions for the main direct liquefaction processes of current research and development interest are summarised in section 5.8.

Characterisation of hydrogenation products

The nature of the products of hydrogenation changes progressively as the amount of hydrogen addition increases. This may be illustrated by reference to the following three widely used parameters.

Boiling-range

One of the most important effects of hydrogenation is a reduction in the molecular weight of the products. In particular, this is reflected in their boiling-range.

As hydrogen addition progresses, the yield of distillates increases and the yield of distillation residue is correspondingly reduced. Furthermore, the distillates become 'lighter' in character so that a greater proportion of the product slate occurs in the lower boiling fractions. In itself, this is generally desirable as the lower boiling fractions are required for premium applications such as transport and chemicals. However, a further consequence of a high degree of hydrogenation is a high gas yield and, depending on the plant location, this may be an economic disadvantage.

Solubility

The species formed as coal is hydrogenated to produce distillate fuels may also be classified by their solubility in various solvents.

Materials soluble in hexane are generally referred to as 'oils' and materials insoluble in hexane but soluble in benzene are generally referred to as 'asphaltenes'. The terminology for materials insoluble in benzene is less well established although the term 'preasphaltenes' is widespread and is adopted here. An effective solvent, often quinoline or pyridine, is used to distinguish between 'extract' and 'insolubles' in laboratory determinations.

The classification of components is illustrated in figure 5.10.

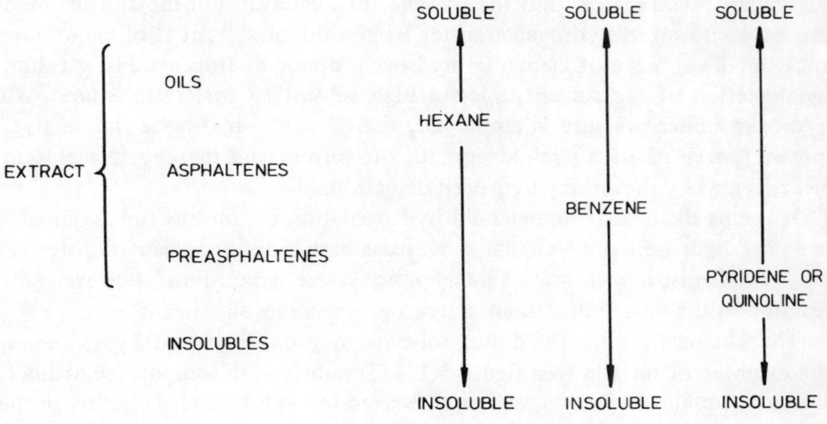

Figure 5.10 Classification of extracted material by solubility

The increasing solubility of the products is a further reflection of the reduction in molecular weight with hydrogen addition. Initially, the main product of extraction is preasphaltenes but, as hydrogenation progresses, these are broken down into asphaltenes and oils.

Heteroatom removal

As noted earlier, one of the main effects of hydrogenation is the removal of the heteroatoms oxygen, nitrogen and sulphur from the coal.

Initially, the removal of heteroatoms proceeds rapidly, the addition of about two per cent of hydrogen being sufficient for the extracts from a range of coals to attain a similar chemical composition. Thereafter, heteroatom removal becomes more difficult and the low heteroatom content required for marketable fuels is achieved only by severe hydrogenation involving the addition of up to 7 per cent of hydrogen.

The progressively greater difficulty of heteroatom removal reflects the distribution of molecular structures in which the heteroatoms occur. The most easily removed heteroatoms are those present as side groups or cross-links, for example, hydroxyl or ether compounds. The most difficult heteroatoms to remove are those in heterocyclics where breakage of the ring structure itself is necessary.

Reaction mechanisms

Reaction of hydrogen with coal in solution may be achieved in two ways: transfer from a hydrogen donor solvent or reaction with elemental hydrogen.

Which of the above mechanisms dominates depends on the conditions, particularly temperature and the presence of a catalyst. For most of the modern processes, however, hydrogen transfer from a donor solvent is of major importance. In the absence of elemental hydrogen, donor hydrogen transfer is limited by depletion of the solvent unless a high solvent to coal ratio is used. Where hydrogen under pressure is employed, the elemental hydrogen and catalyst (if present) serve partly to rehydrogenate the solvent and thereby to maintain its effectiveness by preventing hydrogen depletion.

It seems likely that, under mild hydrogenation conditions (up to about one per cent hydrogen transfer, daf coal, mass basis), donor solvent transfer is the only mechanism operating. Under more severe conditions, however, direct reaction of the coal with elemental hydrogen becomes significant.

The chemical action of donor solvents may be illustrated by reference to the example of tetralin (see figure 5.11). Tetralin can decompose, releasing four atoms of 'donor' hydrogen, and is converted to naphthalene. Tetralin can therefore act as a donor solvent; the corresponding 'spent' solvent is naphthalene. The reaction is, however, reversible and naphthalene may be rehydrogenated to tetralin. The rehydrogenation may also proceed further and produce decalin. Although decalin contains more hydrogen than tetralin, it is more stable and therefore a less effective donor solvent for coal.

It is generally the case that donor solvents can be 'over-hydrogenated'. This is because the effectiveness of a donor solvent depends not on its total hydrogen content but on the content of hydrogen that can be released readily to the coal.

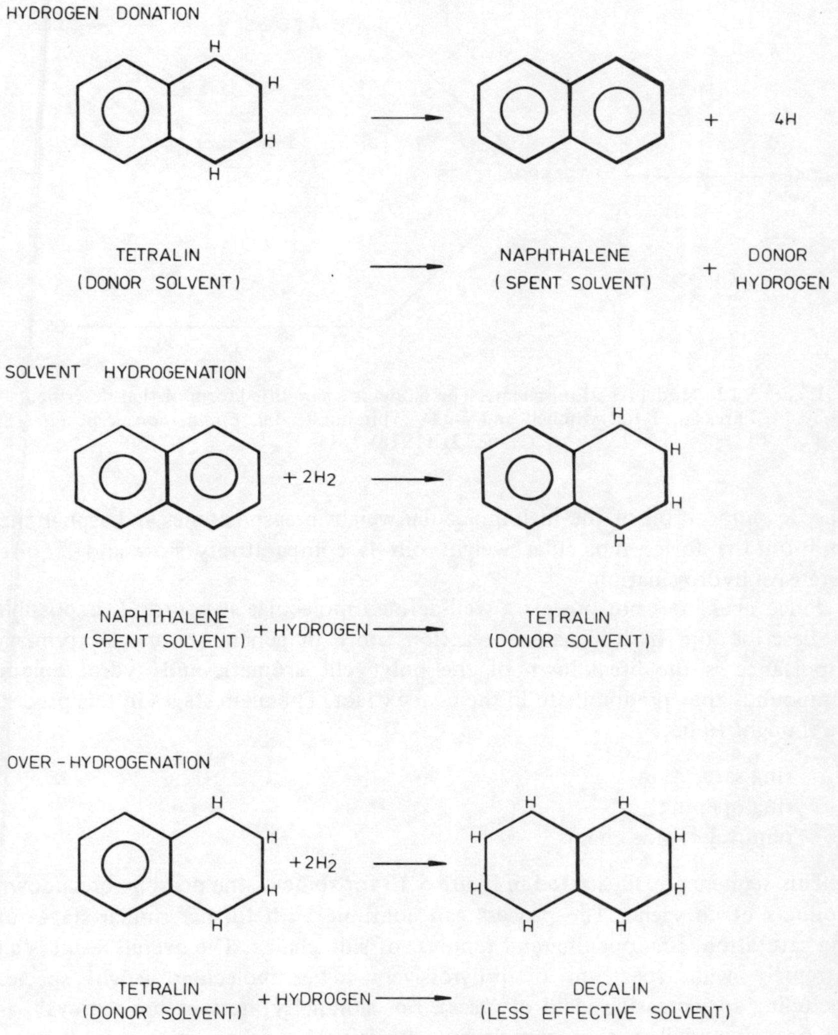

HYDROGEN DONATION

TETRALIN (DONOR SOLVENT) → NAPHTHALENE (SPENT SOLVENT) + DONOR HYDROGEN (4H)

SOLVENT HYDROGENATION

NAPHTHALENE (SPENT SOLVENT) + HYDROGEN ($+2H_2$) → TETRALIN (DONOR SOLVENT)

OVER - HYDROGENATION

TETRALIN (DONOR SOLVENT) + HYDROGEN ($+2H_2$) → DECALIN (LESS EFFECTIVE SOLVENT)

Figure 5.11 Donor solvent reactions

Excessive hydrogenation can form saturated compounds that do not transfer hydrogen readily and are relatively poor solvents.

Various models have been proposed to describe the kinetics of extraction and hydrogenation. A simple example is shown in figure 5.12. The coal breaks down first into preasphaltenes which in turn break down into asphaltenes and oils. The first stage—the decomposition of the coal to form preasphaltenes—is a comparatively rapid process and is responsible for the high rate at which coal is taken into solution. It is thought that this reaction may be a purely thermal process.

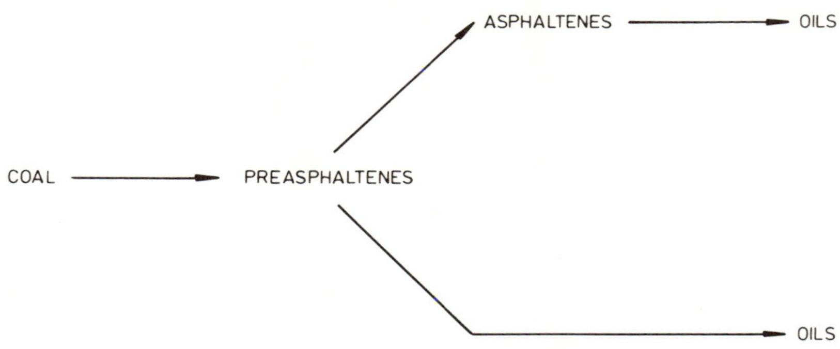

Figure 5.12 Model reaction scheme. The model is a simplified form of that described by M. Farcasiu, T. O. Mitchell and D. D. Whitehurst, *Am. Chem. Soc., Div. Fuel Chem.*, **21** (1976) 7, 11

The decomposition of the high molecular weight preasphaltenes and asphaltenes to form the lower molecular weight oils is comparatively slow and requires extensive hydrogenation.

Since coal does not possess a well-defined molecular structure, it is possible to describe the hydrogenation reactions only in general terms. Of primary importance is the breakdown of the polycyclic aromatic and hydroaromatic compounds that predominate in the coal extract. The main stages in this process are thought to be:

ring saturation
ring opening
removal of side chains

This sequence is illustrated in figure 5.13 for some of the possible breakdown products of chrysene. The process can continue with further similar stages of ring saturation, ring opening and removal of side chains. The overall result is an extremely wide spectrum of progressively lower molecular weight species (including an increasing yield of gases). For simplicity, heteroatom removal has not been included in the example but, as discussed earlier, this process occurs concurrently with the changes described above.

Catalysts

All coal-liquefaction processes use a catalyst (or the mineral matter in the coal) to promote hydrogenation at some stage. The main categories of catalysts that have been used in coal-liquefaction processes are as follows.

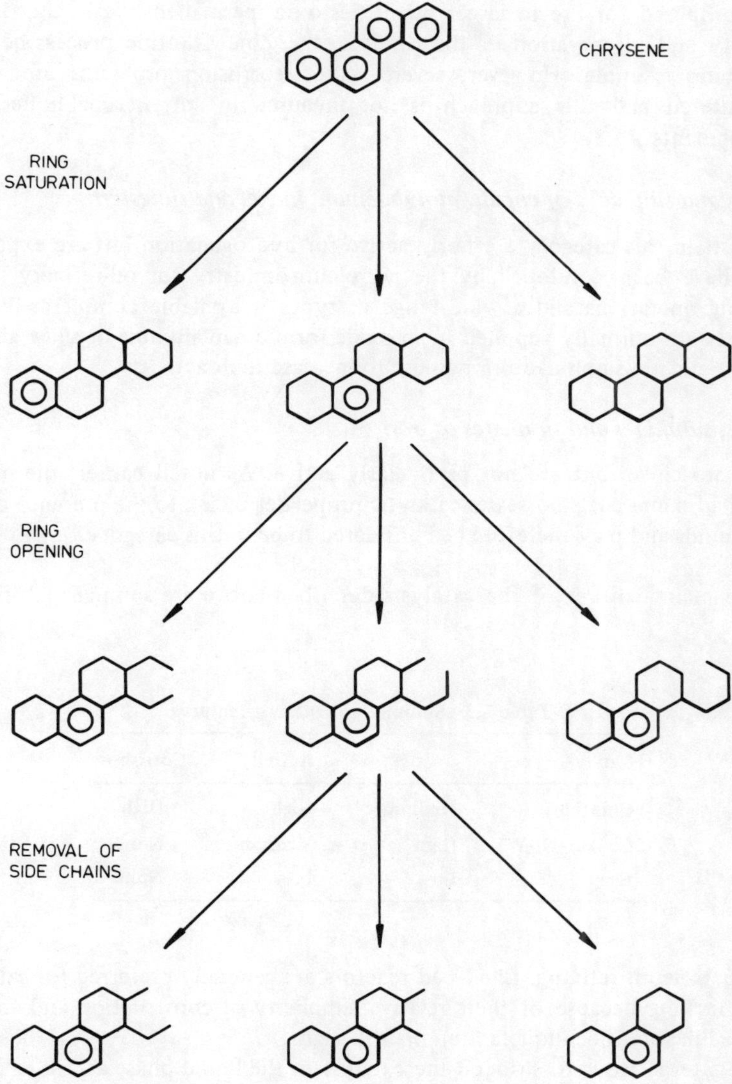

Figure 5.13 Some of the possible hydrogenation products of chrysene

Chloride (and other halides) of zinc, tin and ammonia

These are the most active of the hydrogenation catalysts and may be used on a 'throw away' basis in small quantities. For example, chlorides of ammonia and tin were employed as additives in the Bergius process in this way. Chlorides have also been proposed for use in larger quantities on a 'permanent' basis, that is with recovery and regeneration of the catalyst, the Zinc Chloride process being an important example. However, severe halide corrosion problems are always encountered and this approach is not favoured in current coal-liquefaction developments.

Oxides and sulphides of cobalt, molybdenum, nickel and tungsten

Catalysts in this category are fairly active for hydrogenation but are expensive. They have been developed by the petroleum industry for oil-refinery hydrocracking operations and a wide range of types is available commercially. The catalysts are normally supplied in an oxide form on an alumina or silica/alumina support and are sulphided prior to use to increase their activity.

Oxides, sulphides and sulphates of iron

These are cheap but are not particularly active. As noted earlier, the mineral matter of some coals possesses catalytic properties owing to the presence of iron compounds and may therefore be considered to be in this category.

The main features of the catalysts described above are summarised in table 5.3.

Table 5.3 Summary of catalyst features

Catalyst type	Cost	Activity	Corrosion
Chlorides	Medium	High	High
Co–Mo–Ni–W	High	Medium	None
Fe	Low	Low	None

In petroleum refining, fixed bed reactors are generally preferred for catalytic hydrocracking because of their relative simplicity of construction and are also favoured in some coal-liquefaction plant designs.

The alternative is to suspend the catalyst in the liquid phase as, for example, in a fluidised bed. The main disadvantages of this approach are that the reactor is considerably more complex than fixed catalyst bed designs and good fluidisation must be maintained at all times in order to avoid hot spots. However, suspension systems have the following three principal advantages.

Temperature control

Hydrogenation reactions are significantly exothermic and good temperature control is essential to avoid excursions which may adversely affect the product composition and damage the catalyst. In fixed-bed systems, it is therefore necessary to use a number of reactors in series separated by intercoolers and to introduce cold gas at intermediate points in the reactors in order to cool the reactants. In a well-mixed suspension system, however, these measures are unnecessary and a single reactor is generally sufficient.

Deposition

The abrasion in a suspension system tends to keep the catalyst particle surfaces more free from deposits although degradation of the catalyst by attrition may also occur.

Catalyst removal

As a result of deactivation, the catalyst has to be removed periodically either for regeneration or replacement. With a fixed catalyst bed, this requires the unit to be shut down whereas with a suspension system continuous on-stream catalyst removal and replacement is possible. Indeed, a suspension system is the only feasible option when the catalyst is used on a once-through basis.

The main technical problem associated with the use of catalysts in coal liquefaction is deactivation. Liquefaction catalysts depend for their effectiveness on having a high active surface area. This is provided by a micropore structure with much of the porosity being in the size range 0.001 to 0.01 microns. Deactivation results from a reduction in the effective surface area, and can occur for a number of reasons.

Organic deposits

Coke deposition can occur during the hydrogenation of coal digests and solutions and may lead to blocking of the catalyst pores and to consequent deactivation. It has also been suggested that the deposition of heavy liquids (preasphaltenes and asphaltenes) within catalyst pores may be responsible for some deactivation. Such deposits may be removed by carefully controlled combustion so that the activity of the catalyst is restored.

Inorganic deposits

It has been found that catalyst deactivation also occurs as a result of deposition in the pore structure of minerals and metals.

The mineral deposits include clays (for example, aluminosilicates), sulphides of iron, silica and alumina and are derived directly from the mineral matter in the coal. Metals deposited include titanium, calcium and sodium and are thought to be present in the coal as organometallic compounds such as porphyrins. The metals are deposited when the compounds containing them decompose on the catalyst.

The mineral matter is present in an extremely wide range of particle sizes with a significant amount in the submicron size range. The organometallic compounds are soluble in the solvent. For these reasons, catalyst poisons are difficult to remove and, even with an efficient filtration system, the ash content of the coal extract cannot be reduced to much below 0.1 per cent.

Sintering

Sintering, particularly of the catalyst support material, can be responsible for a loss of active surface area. The factors important in catalyst sintering are not fully understood although deactivation due to sintering appears to increase with the presence of steam in the reactor (for example, from a high moisture or oxygen content coal).

There are three main design options for minimising the effects of catalyst deactivation.

Operating on a once-through basis

A catalyst can be used on a once-through basis (that is, discarded when deactivated) only if it is sufficiently cheap. In practice, this method of operation is limited to iron-based and chloride catalysts.

Iron-based catalysts were employed in the Bergius process on a once-through basis and are also proposed for the Ruhrkohle process. The SRC 2 process, although not using a catalyst as such, relies on the catalytic properties of coal mineral matter and may, therefore, also be considered to fall into this category.

Reducing exposure to catalyst poisons

Catalysts in coal liquefaction may be required to hydrogenate coal digests (before mineral matter separation), coal solutions (after mineral matter separation) and the recycle solvent.

The greatest exposure to catalyst poisons occurs when hydrogenating coal digests as in this case both the mineral matter and the high-molecular weight material responsible for carbon deposition are present.

A lower rate of deposition may be expected when hydrogenating coal solutions because the mineral matter has been removed. As noted earlier, however, neither the organometallic compounds in solution nor some of the finely divided (submicron) mineral particles are eliminated by filtration and, as a result, the ash content of coal extract is usually of the order of 0.1 per cent. The presence of these constituents will, therefore, continue to give rise to some catalyst deactivation. Carbon deposition from the preasphaltenes and asphaltenes present in the solution may also occur but may be reduced in extent by the hydrogenation achieved during the extraction stage.

Hydrogenation of the recycle solvent performed in a separate unit is expected to be relatively free from catalyst-deactivation problems. The fraction used as the solvent (usually boiling in the range 180 to 450 °C) avoids the high molecular weight materials that are largely responsible for carbon deposition and contain most of the organometallic compounds. Furthermore, because the solvent is obtained by distillation, it has an extremely low mineral-matter content.

Relying on regeneration

Deactivation arising from carbonaceous deposits may be reversed by periodic regeneration. In general, however, there is a need to develop catalysts that are resistant to the poisons present in coal liquids.

The main design options adopted in the liquefaction processes under development are summarised in section 5.8.

Disposable iron catalysts are normally used only for hydrogenating a coal digest, as the greater activity of cobalt–molybdenum or nickel–molybdenum catalysts is preferred for the relatively protected conditions of hydrogenating a coal solution or recycle solvent. Usually, a fixed-bed system is chosen when hydrogenating a coal solution or recycle donor solvent whereas a suspension system is used for hydrogenating a coal digest. This is partly because of the practical difficulties in feeding a digest containing a high concentration of mineral matter particles through a fixed bed. Also, with a digest, the greater exposure to catalyst poisons necessitates a comparatively high rate of catalyst removal for which a fixed-bed system would be inconvenient.

5.7 Solids separation

The choice of solids-separation technique depends on the type of liquefaction process and, in particular, on the amount of hydrogenation occurring at the extraction stage. The four main categories of solids-separation techniques—distillation, assisted settling, filtration and density separation devices—are discussed below.

Distillation

The term distillation refers generally to processes for vapour–liquid separation. If solids are present, these will remain almost entirely with the liquid phase so that the vapour-phase material is recovered with an extremely low mineral-matter content. Distillation can therefore be used as a simple and effective method of solids separation.

The main disadvantage of distillation is that only distillate materials are separated from the mineral matter. As a result, the approach is restricted to conditions where the amount of undistillable material is small, as such material remains with the mineral matter and is unrecovered. For distillation to be feasible, it is therefore necessary that a substantial amount of hydrogenation has taken place and, in this case, distillation is usually the preferred solids separation technique.

The crude product from hydrogenation typically leaves the reactor at a temperature of 350 to 475°C and a pressure of 100 to 300 bar. Under these conditions, both a vapour phase and a liquid phase are present (although, in some reactors, separate offtakes for the two phases may be provided). Normally, a number of stages are used for product separation, including the processes of flash distillation, atmospheric distillation and vacuum distillation.

Many configurations are possible, a representative example being given in figure 5.14. This shows the two-phase product leaving the hydrogenation unit at 450°C, 200 bar, and passing to a separator. The vapour carried overhead is cooled to 40°C and a hydrogen-rich gas suitable for use in hydrogenation processes is recovered. Water is removed and the liquid hydrocarbon product is let down to ambient pressure for fractionation by atmospheric distillation. The heavy oil liquid-phase product (possibly containing insoluble carbonaceous material and mineral matter) from the primary separator is flashed to atmospheric pressure to recover gas and light oil and then vacuum distilled. Recycle solvent is usually obtained from the atmospheric distillation bottoms and the heavy distillate fraction from the vacuum distillation column. The mineral matter is removed with the vacuum bottoms. In order that the vacuum bottoms have a sufficiently low viscosity for removal from the column, it may be necessary to limit the amount of distillate material obtained in the vacuum distillation stage.

Assisted settling

Where distillation is not applicable, assisted settling is an option. Processes based on this technique are under development by Lummus and Kerr–McGee. In both cases, distillation is used as an initial separation stage with the distillation residue then being mixed with a solvent chosen to promote separation of the solids.

In the Lummus process, the low boiling-point fractions are removed by distillation and the remaining fraction, which contains the process solvent, is

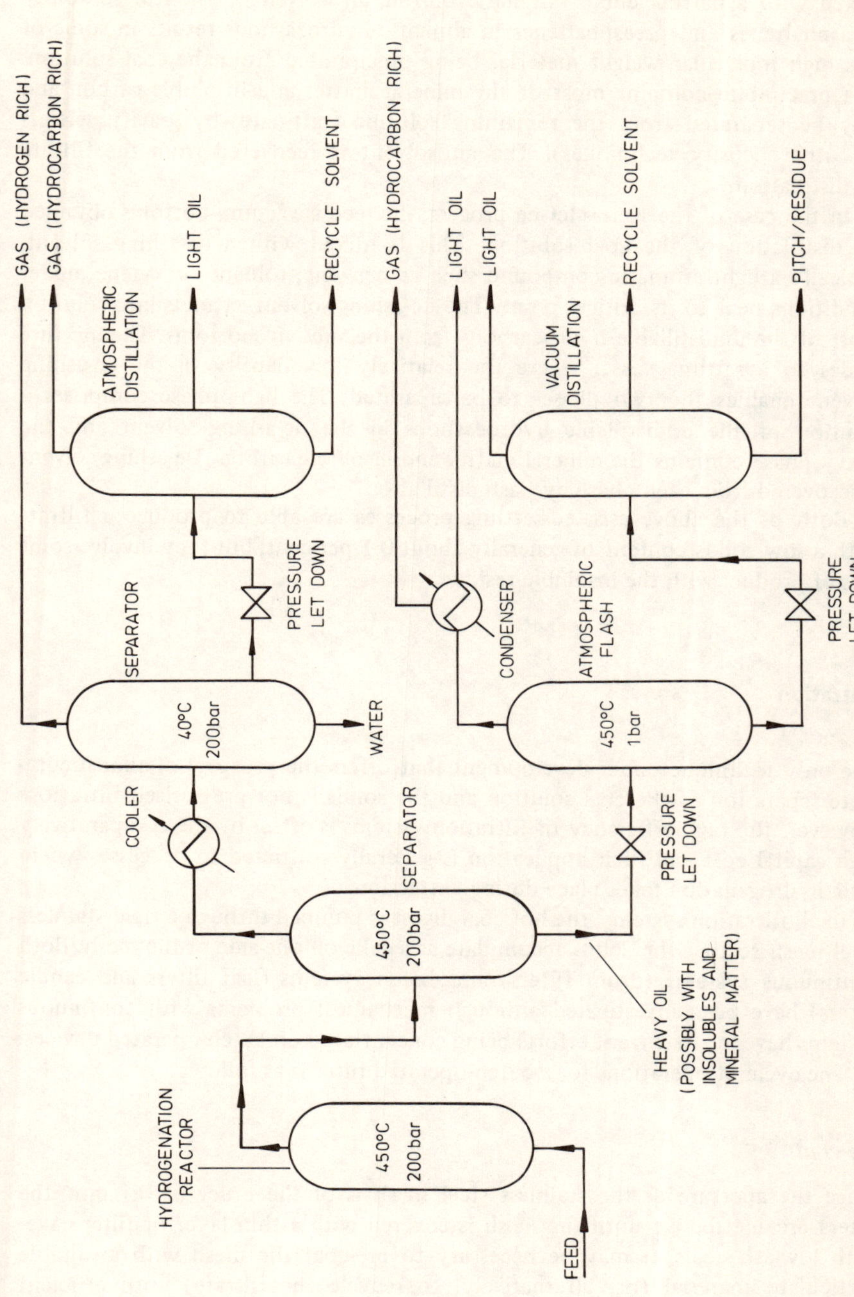

Figure 5.14 Solids separation by distillation

mixed with a narrow cut of aliphatic oil (an 'antisolvent'). The low solubility of asphaltenes and preasphaltenes in aliphatic hydrocarbons results in some of this high molecular weight material being precipitated from the coal solution. The precipitate contains most of the mineral matter and insoluble carbon and may be separated from the remaining solution (raffinate) by gravity settlers (or other density techniques). The antisolvent is recovered from the filtrate by distillation.

In the case of the Kerr–McGee process, the feed is vacuum bottoms obtained by distillation of the coal solution. This is mixed with a de-ashing solvent, typically a light aromatic compound such as benzene, toluene or xylene, under conditions near to its critical point. The de-ashing solvent extracts into solution most of the undistillable hydrocarbons from the vacuum bottoms. The mixture passes to a settling vessel where the relatively low density of the de-ashing solvent enables the two phases to be separated. The light phase comprises a solution of the undistillable hydrocarbons in the de-ashing solvent and the heavy phase contains the mineral matter and insoluble carbon. De-ashing solvent is recovered from each phase by flash distillation.

Both of the above assisted-settling processes are able to produce a filtrate with a low solids content of generally about 0.1 per cent, but they involve some loss of product with the insoluble residue.

Filtration

The only technique under development that offers the prospect of almost complete separation of the coal solution and the solids is hot pressurised filtration. However, the high efficiency of filtration systems is offset by the comparatively high capital cost and their application is generally restricted to processes where little hydrogenation takes place during extraction.

In all filtration systems, the hot coal digest is pumped through a rigid stainless steel mesh so that the solids accumulate as a cake on one side of the mesh. Both continuous systems (drum filters) and batch systems (leaf filters and candle filters) have been investigated although mechanical problems with continuous systems have led to current efforts being concentrated on batch-operated devices.

The cycle of operations for a batch-operated filter is as follows.

Pre-coat

Since the aperture of the stainless steel mesh is of the order of 0.1 mm, the filters are ineffective until the mesh is covered with a thin layer of filter cake. With low-ash coals, it may be necessary to pre-coat the mesh with a suitable particulate material (or, alternatively, to recycle the filtrate) until efficient filtration begins.

Filtration

The efficiency of filtration depends on a number of factors including the extraction conditions and the solids concentration. Adding a 'body feed' such as crushed coke to the digest, although increasing the amount of solids to be handled, can reduce the filter cake resistance substantially.

Washing

During filtration, some coal extract is retained within the filter cake. The amount is typically 10 to 15 per cent of the total extract processed and is recovered by pumping clean solvent through the filter cake. The resulting solution is dilute and is usually returned to the extraction stage as a solvent for the feed coal.

Drying

After washing, the filter cake retains solvent which may be recovered by depressurising and vacuum flash drying.

Discharge

The cake is then discharged from the filter mesh mechanically (for example, by using a knife or by rotating/vibrating/impacting the filter) and the system is repressurised for the next cycle.

On the basis of pilot-scale trials, filtration can achieve a solids concentration in the filtrate of 0.1 per cent.

Density separation devices

A number of solids-separation devices based on density difference have been investigated for use in coal-liquefaction processes, the principal types being as follows.

Gravity settling vessels

Gravity settling vessels are simple and cheap and are the preferred separation system in the assisted-settling processes described earlier. Their use alone requires the liquid phase to have a low viscosity and specific gravity.

Hydroclones

Hydroclones achieve only a poor solids–liquid separation when applied to coal solutions. Typically, the overflow stream contains 75 per cent of the liquids and 25 per cent of the solids; the underflow stream contains the remaining 25 per cent of the liquids and 75 per cent of the solids. In particular, the solids content of the overflow stream is unacceptably high for marketable products and hydroclones are therefore unsuitable for use alone in coal liquefaction.

Hydroclones may, however, be used as an initial separation stage in conjunction with other solids-separation systems. The overflow stream is normally sent to a hydrogenation unit either downstream in the process or, if appropriate, by recycling to the extraction stage. It is likely that only a proportion of the hydroclone overflow can be returned to the extraction stage in order to avoid a build-up of fine solids in the recycle system. The hydroclone underflow is passed to a further solids-separation system. Distillation appears to be the preferred option for this secondary separation stage although assisted settling and filtration have also been investigated.

Centrifuges

Several different types of centrifuge have been considered for application in coal-liquefaction processes. Although solids removal to the required efficiency can be achieved, reliability has been low because of mechanical problems and costs appear to be high. For these reasons, centrifuges are not generally favoured in current development programmes.

5.8 Process configurations

Representative processes

As indicated in the preceding sections, coal liquefaction involves a number of different operations for which there exist various technical solutions and various ways in which these may be combined in an integrated plant. The following processes are representative of the principal technical options and include the recent major demonstration projects currently in progress. The main features of the processes are summarised in table 5.4 and the status of the development programmes is described in appendix 1.

Bergius

In the Bergius process, crushed coal is mixed with recycle solvent and a disposable catalyst. The catalyst is principally the 'red mud' by-product of bauxite processing (also known as 'Luxmasse' or 'Bayermasse'), the active component of which is ferric sulphate. Chlorides of ammonia and tin are also added in small quantities. The prepared slurry is then fed with hydrogen at 300 to 700 bar to the hydrogen reactors. The vapour-phase products are condensed, distilled at atmospheric pressure and then undergo further hydrocracking to produce the required fuels. The heavy oil slurry, containing unconverted coal and mineral matter, is centrifuged and recycled as solvent to the slurry preparation stage.

Table 5.4 Summary of direct liquefaction processes

Process	Extraction pressure (bar)	Hydrogenation			Main solids separation method
		Catalyst	Environment	Reactor type	
Pre-war					
Bergius	300–700	Fe, disposable	Digest	Suspension	
Pott–Broche[a]	100	–	–	–	Filtration
Single-stage					
EDS	130	Ni–Mo	Solvent	Fixed bed	Distillation
H-Coal	200	Co–Mo	Digest	Suspension	Distillation
SRC 1[b]	100	–	–	–	Filtration
SRC 2	150	Mineral matter	Digest	Suspension	Distillation
Ruhrkohle	300	Fe, disposable	Digest	Suspension	Distillation
Two-stage					
Chevron	100–170	Co–Mo, Ni–Mo or Ni–W	Digest	Fixed bed	Gravity settling
Lummus	165	Ni–Mo	Solution	Suspension	Assisted settling
NCB	15	Co–Mo	Solution	Fixed bed	Filtration

[a]The filtered coal solution obtained from the Pott–Broche process was used as feedstock to a Bergius reactor.
[b]The SRC 1 process may also be considered as a two-stage liquefaction process if the filtered coal solution is used as a feedstock for hydrogenation rather than to produce a solvent refined coal.

Pott–Broche

The Pott–Broche process involves digesting pulverised coal at a pressure of about 100 bar in a hydrogenated middle oil obtained from coal tar pitch. The resulting solution is then cooled, filtered to remove the mineral matter, and used as a feedstock to the Bergius process (or for the manufacture of carbon electrodes).

Exxon Donor Solvent (EDS)

The main distinguishing feature of the EDS process is that the solvent is prepared by catalytic hydrogenation of a selected product fraction. A conventional Ni–Mo catalyst is used in a fixed-bed reactor system. This gives an effective hydrogen donor solvent which is then slurried with crushed coal and fed with hydrogen at about 130 bar to a non-catalytic reactor. Sufficient hydrogenation is achieved under these conditions for solids separation to be by distillation.

H-Coal

The H-Coal process is based on the application of the H-Oil technology used to convert heavy petroleum residuals into lighter products. Coal is crushed, slurried with a recycle solvent and pressurised to about 200 bar. The slurry is reacted with hydrogen in an ebullated bed (that is, a turbulent gas–liquid fluidised bed) of Co–Mo catalyst which can be withdrawn and replaced continuously.

The process can be operated in a 'syncrude' mode or a 'fuel oil' mode, depending on the operating conditions in the reactor (for example, the residence time). In the syncrude mode, the product spectrum is sufficiently light for distillation to be used as the main method of solids separation. In the fuel oil mode, however, this approach is not possible because the amount of undistillable liquids is too high and assisted settling appears to be the preferred option.

Solvent Refined Coal (SRC 1)

In the SRC 1 process, coal is pulverised, mixed with a solvent and delivered with hydrogen at a pressure of 100 bar to a non-catalytic reactor. The digest is then filtered to remove the mineral matter. The amount of hydrogenation achieved in the reactor is limited and the main product is a brittle solid at ambient temperatures. This is termed 'solvent refined coal' (SRC) and is suitable for use as a clean boiler fuel.

The SRC 1 process is related to the Pott–Broche process in employing non-catalytic digestion followed by filtration but, unlike Pott–Broche, hydrogen is used in the digestor. For this reason, SRC 1 may be regarded more accurately as being derived from the technology described in the Ude and Pfirrmann patents of 1936 and 1939.

Solvent Refined Coal (SRC 2)

The SRC 2 process differs from the SCR 1 process by recycling some of the digest to the reactor after removing the gases and light liquids. This gives the heavier fractions a longer residence time in the reactor and also permits the build-up of the mildly catalytic mineral matter. The result is a lighter product spectrum and vacuum distillation is used for solids separation.

Ruhrkohle

The Ruhrkohle process employs a disposable iron catalyst which is mixed with the coal and solvent, and fed with hydrogen at 300 bar to the hydrogenation reactors. The process is a direct descendant of the Bergius technology, the main difference being that the Ruhrkohle process employs vacuum distillation to recover liquids from the heavy oil slurry, rather than centrifuges. This approach avoids recycling the relatively unreactive asphaltenes to the hydrogenation reactor with the solvent, and consequently enables better conversion to be obtained under less extreme conditions.

Chevron Coal Liquefaction Process (CCLP)

The first stage of the Chevron coal liquefaction process is a non-catalytic dissolver to which a slurry of pulverised coal and solvent is fed with hydrogen at 100 to 170 bar. The digest then passes to a conventional hydrocracker which operates at a lower temperature than the dissolver. The hydrocracker is followed by solids separation, the main method being gravity settling. After removing the distillate product, the liquid stream is cooled to near ambient temperature in order to precipitate the asphaltenes and preasphaltenes before being recycled as the solvent.

Lummus Integrated Two-Stage Liquefaction (ITSL)

In the Lummus liquefaction process, pulverised coal is dissolved in the solvent and thermally cracked using a short contact time (SCT) reactor. This operates at a pressure of 165 bar and with a hydrogen atmosphere. The coal solution passes to an assisted settling unit which removes the mineral matter and then to an ebullated bed hydrocracker reactor, based on the H-Oil approach. The Lummus process may therefore be regarded as related to H-Coal, but differing in that it is a two-stage process.

National Coal Board Liquid Solvent Extraction (NCB LSE)

In the NCB LSE process, coal is pulverised, mixed with a recycle solvent and delivered, without hydrogen, to a non-catalytic digester operating at about 15 bar. Undissolved coal and mineral matter are separated from the digest by filtration. After recovery of some of the solvent, the coal solution is pressurised to 200 bar and fed with hydrogen to a fixed-bed hydrocracking unit employing a conventional Co–Mo catalyst. The products are then fractionated and the remaining solvent recovered. In carrying out digestion without hydrogen and in employing filtration as the solids separation technique, the NCB LSE process may be regarded as a descendant of the Pott–Broche approach.

Residue disposal and hydrogen production

All direct liquefaction processes produce a stream containing a carbonaceous residue (in general, part liquid and part solid) and mineral matter. The carbon content of the residue is too high for it to be discarded but the mineral matter content is such that it is unmarketable as a fuel. The residue must, therefore, be used within the process. As the processes also have requirements for hydrogen and process-steam production, the residue can be used to contribute to these demands.

In most of the liquefaction processes, a number of approaches to these aspects of plant design are under consideration; the final decision is regarded as one of economic optimisation after the basic technology has been demonstrated.

Disposal of the residue by direct combustion is not generally favoured in conceptual liquefaction-plant designs because the sulphur content is usually high. The remaining options are as follows.

Coking

If the solids separation process is such that a significant amount of hydrocarbons remains in the residue, then additional liquids and gases may be obtained by thermal cracking in a coker. The resulting coke may be used as a feedstock for gasification (see below). Coking is not favoured if filtration is used for solids separation because of the low yields of liquids and gases obtained from the filter cake.

Oxygen–steam gasification

The residue or coke from thermal cracking may be gasified using oxygen and steam to produce a synthesis gas suitable for conversion into hydrogen (see chapter 4). Pressurised entrained phase gasifiers (Texaco or Shell, for example) appear to be preferred for this purpose. Gasification of the residue may have the advantage that it can be pumped into a pressurised system directly as a slurry (that is, without the use of water or the cost of solids-feeding equipment). This advantage is, however, offset by the additional liquids and gases that may be obtained by coking the residue and gasifying the coke.

Air–steam gasification

Alternatively, residue or coke may be gasified using air and steam to produce a fuel gas for process heating.

Most coal liquefaction design concepts envisage the process being self-sufficient in hydrogen and process steam. As implied above, the strategy adopted for residue disposal is closely linked to that of hydrogen and process steam supply. In both cases, the alternative feedstocks to residue (or coke) for meeting these requirements are coal and hydrocarbon gases produced as a by-product of hydrogenation.

The main options for hydrogen supply may therefore be summarised as:

coal gasification (oxygen/steam)
residue gasification (oxygen/steam)
coke gasification (oxygen/steam)
steam reforming of the hydrocarbon gases

Similarly, the main options for meeting process steam and heat requirements are:

direct coal combustion
residue gasification (air/steam)
coke gasification (air/steam)
direct combustion of the hydrocarbon gases

The main choices available for residue disposal, hydrogen production and process steam/heat supply are summarised in figure 5.15 with the preferred

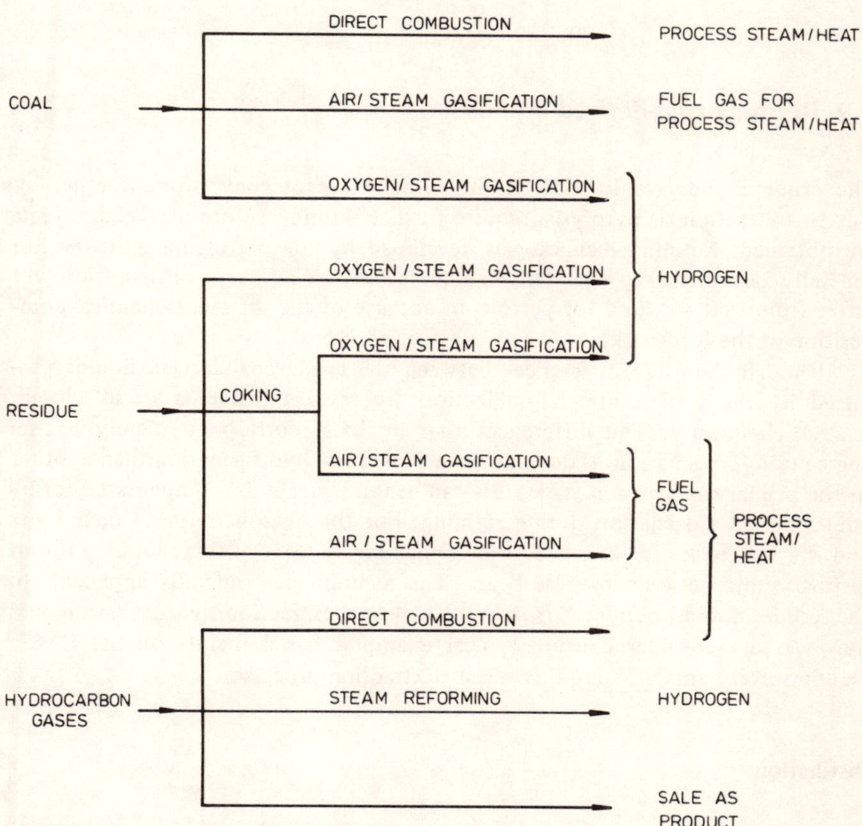

Figure 5.15 Options for residue disposal, hydrogen production and process steam/heat supplies

approach depending on site-specific factors. In particular, the plant location and the local cost of natural gas have a dominant influence as these factors will determine the relative economics of using the hydrocarbon gases within the process and of selling them as part of the product slate.

A related design area is that of power requirements. In some processes, sufficient steam is raised by heat recovery and from residue disposal to enable all internal power requirements to be met with possibly some power for export. In other cases, however, a choice exists between importing power from a grid and raising additional steam within the plant so that self-sufficiency in power supplies is attained. The efficiency and economies of scale achieved in large central power stations generally favour importing power. However, local factors are also important in this case and, for a remote plant location, self-sufficiency in power supplies may be the preferred option.

5.9 Refining coal-derived liquids

The crude liquids resulting from the initial stage of conversion, whether by solvent extraction or pyrolysis, require further refining before marketable fuels are obtained. Refining technologies developed by the petroleum industry can generally be applied to coal-derived liquids but the refinery configuration will differ from that required for petroleum because of the different chemical composition of the feedstock.

Although there are differences between the crude coal-derived liquids produced by the various direct liquefaction processes, the liquids are all closely related chemically. The differences that do exist correspond mainly to the degree of hydrogenation. If one process achieves less hydrogenation than another in the primary conversion stage, this can be (and usually is) compensated for by additional hydrogenation during refining. For this reason, refining techniques and finished fuels are described in the remaining sections of this chapter without distinguishing between process types. The technologies generally applicable to the crude liquids obtained from both solvent extraction/hydrogenation and pyrolysis are considered using typical examples based mainly on the Exxon Donor Solvent and NCB Liquid Solvent Extraction processes.

Distillation

The main purpose of distillation is the separation of the crude liquid product into fractions of different boiling-range. As discussed earlier, however, distillation also serves as a means of recovering the solvent for recycle and can be used for separating the liquid product from the insoluble residue and mineral matter.

In modern oil refineries, the crude oil may split into as many as ten fractions. For simplicity, however, the present discussion is limited mainly to two intermediate products. These are a C5–200°C stream (termed naphtha) and a 200°C + stream (termed gas oil).

Typical properties of coal-derived naphtha and gas oil are shown in table 5.5 and are compared with similar fractions obtained from crude oil.

Table 5.5 Characteristics of coal-derived and petroleum-derived liquids

Composition	Coal-derived[a]		Petroleum-derived	
	Naphtha	Gas oil[b]	Naphtha	Gas oil
H	11.8%	8.7%	15.9%	13.2%
O	13700 ppm	19300 ppm		
N	2097 ppm	7130 ppm		
S	9978 ppm	4900 ppm	500 ppm	7000 ppm
Paraffins	13%	} 14%	50%	40%
Naphthenes	43%		35%	40%
Aromatics	25%	86%	15%	20%
Olefins	10%			
Polars	9%			
Sources[c]	(1)	(2)	(3)	(3)

[a]The data for the coal-derived liquids refer to the Exxon Donor Solvent process.
[b]The PNA analysis for the coal-derived gas oil is for the 200 to 526 °C fraction.
[c](1) Gim Tan and A. J. DeRosset, *US DOE report FE-2566-25*, UOP Inc., February 1979, p. 7.
(2) F. J. Riedl and A. J. DeRosset, *US DOE report FE-2566-30*, UOP Inc., August 1979, pp. 12, 14.
(3) J. Owen, paper to *Eurochem 80 Conference*, NEC, Birmingham, NCB Coal Research Establishment, June 1980.

In general, coal-derived liquids differ chemically from petroleum liquids in two important respects: coal-derived liquids have a higher heteroatom content, particularly oxygen and nitrogen, than corresponding petroleum fractions and coal-derived liquids are mainly aromatic and naphthenic whereas petroleum fractions are mainly paraffinic and naphthenic.

With both coal-derived and petroleum-derived liquids, the contents of heteroatoms and aromatics increase with boiling-point and the hydrogen content decreases. In the case of the coal-derived liquids, most of the paraffins and naphthenes present are concentrated in the naphtha fraction.

Hydrotreatment

Hydrotreatment (also known as hydrofining) refers to the catalytic reaction of petroleum-derived or coal-derived liquids with hydrogen under conditions designed to achieve the removal of heteroatoms as H_2O, NH_3 and H_2S and the saturation of aromatics to produce naphthenes.

The preferred catalysts are the Co–Mo–Ni–W type as used in the hydrogenation of coal solutions or extracts. The conditions are, however, relatively mild; temperatures typically are 320 to 420 °C and pressures 25 to 70 bar.

The effect of hydrotreating crude coal-derived naphtha and gas oil is illustrated in table 5.6. With the exception of the high severity hydrogenation/hydrotreatment case for gas oil, the feedstocks are the naphtha and gas oil described in table 5.5. Hydrotreatment can be carried out at different severities depending on the product specification required. Results at three levels of severity are given for gas oil.

As indicated in the table, both the naphtha and gas oil fractions can be converted into a predominantly naphthenic material with the heteroatom content reduced to a few ppm. However, hydrotreatment of the gas oil fraction requires

Table 5.6 Hydrotreatment of coal-derived liquids[a]

Composition	Naphtha	Gas oil		
		Low severity	Medium severity	High severity
Mass balance				
Feed	100.0%	100.0%	100.0%	
Hydrogen	1.6%	2.4%	4.0%	
Total	101.6%	102.4%	104.0%	
C1–C4	0.6%	0.9%	3.6%	
C5–200 °C	98.3%	2.8%	25.1%	
200 °C +		95.5%	71.6%	
Heterogases	2.7%	3.2%	3.7%	
Total	101.6%	102.4%	104.0%	
C5 + product[b]				
H	13.1%	10.7%	11.9%	13.1%
O	98 ppm	1583 ppm	367 ppm	
N	0.2 ppm	1388 ppm	3 ppm	5 ppm
S	0.1 ppm	48 ppm	11 ppm	1 ppm
Paraffins	13%	} 13%	} 47%	0%
Naphthenes	65%			96%
Aromatics	22%	87%	53%	4%
Sources[c]	(1)	(2)	(2)	(3)

[a]The data refer to the Exxon Donor Solvent process, except for the high-severity case which refers to hydrogenation/hydrotreatment by the NCB Liquid Solvent Extraction process.

[b]The analyses of the C5 + product for the medium-severity case refer to the 200 °C + fraction.

[c](1) Gim Tan and A. J. DeRosset, *US DOE report FE-2566-25*, UOP Inc., February 1979, pp. 7, 13, 14.

(2) F. J. Riedl and A. J. DeRosset, *US DOE report FE-2566-30*, UOP Inc., August 1979, pp. 17, 18, 20, 24, 27.

(3) J. Owen paper to *Eurochem 80 Conference*, NEC, Birmingham, NCB Coal Research Establishment, June 1980.

considerably more hydrogen and more severe processing than hydrotreatment of the naphtha fraction. Some cracking of the gas oil to naphtha is indicated as the severity of hydrotreatment increases.

Hydrotreatment is an important preliminary step in processing coal-derived liquids for two reasons.

(1) Aromatics are resistant to thermal cracking and catalytic cracking processes and their conversion to naphthenes by hydrotreatment renders the feedstock more suitable for subsequent processing by these techniques.
(2) Nitrogen and sulphur can act as catalyst poisons in catalytic cracking and reforming and high contents of these elements in the finished fuel may also be unacceptable on environmental grounds. The results in table 5.6 indicate that the heteroatom content of coal-derived liquids may be reduced to an acceptable level by hydrotreatment.

Thermal cracking

Thermal cracking (pyrolysis) is used to produce light fractions from heavy fractions by heating the feed to a high temperature, usually in the range 500 to 850 °C. At these temperatures, large hydrocarbon molecules decompose (crack) into smaller molecules. Paraffins are most susceptible to thermal cracking, followed by naphthenes. Aromatics are comparatively resistant.

A number of thermal cracking processes exist and can handle a range of feedstocks. The processes include the following.

Delayed coking

Thermal cracking of distillation residues under severe conditions enables further distillates to be recovered leaving a coke residue. This is one of the options for disposing of the carbonaceous residue remaining after extraction and hydrogenation of coal solutions and is discussed in section 5.8.

Visbreaking

Under less severe conditions, thermal cracking of petroleum distillation residues can be used to reduce their viscosity and to produce a satisfactory fuel oil without using distillate fractions as blending agents. Visbreaking is likely to have only a limited role in refining coal-derived liquids because most of the residues contain mineral matter and are therefore unsuitable for conversion to fuel oil.

Steam cracking

Thermal cracking of distillates in a steam cracking or steam pyrolysis process can be used to produce olefins and single ring aromatics mainly as feedstocks for the manufacture of petrochemicals.

Examples of the product spectra obtained by steam cracking coal-derived liquids are shown in table 5.7. The comparisons made in the table indicate that similar yields of ethylene and substantially higher yields of BTX can be obtained by steam cracking coal-derived fractions as compared with petroleum-derived fractions.

For the coal-derived gas oil, prior hydrotreatment is essential to maximise the yields of chemical feedstocks (olefins and BTX). As discussed above, this is because the aromatics that make up an important fraction of the crude gas oil are not easily broken down by thermal cracking whereas the naphthenes formed by hydrotreatment can be cracked more easily.

Catalytic cracking

Catalytic cracking refers to thermal cracking in the presence of a catalyst. It is usually employed to convert heavy distillate fractions into the gasoline and naphtha boiling range. The role of the catalyst is to ensure a high yield of these fractions; non-catalytic thermal cracking inevitably results in a high yield of gases.

Catalytic cracking usually takes place at 450 to 550 °C. Either a fluidised bed or entrained phase reactor is used and the catalyst is mixed with the feed upstream of the reactor. Carbon is deposited on the catalyst during the cracking process and catalyst is therefore removed continuously from the reactor and fed to a fluidised bed regenerator. The catalyst is reactivated in the regenerator by burning off the carbon with air and is then returned to the reactor.

The effect of catalytic cracking on coal-derived gas oil is shown in table 5.8 together with data for a petroleum gas oil. The untreated gas oil is described in table 5.5 and the hydrotreated gas oils are described in table 5.6.

As shown in the table, the yields of gasoline and naphtha increase with the amount of hydrotreatment of the feed so that initial hydrotreatment is essential to maximise the yield of these products. As with thermal cracking, the main reason is that hydrotreatment converts the aromatics (which are resistant to cracking) to naphthenes (which are amenable to cracking). Some coal-derived gas oils contain a significant amount of asphaltene-like material which is a coke precursor and can act as a catalyst poison. The removal of these components by hydrotreatment therefore also has a beneficial effect.

Comparisons of cracking behaviour for coal-derived and petroleum-derived gas oils indicate that, taking into account differences in hydrogen content, they follow a consistent pattern.

Catalytic cracking has established an important position in modern oil-refinery practice as a means of upgrading heavy distillates to obtain a lighter product spectrum. For this purpose, it is generally favoured over the alternative technology of hydrocracking described in the next section. The need for severe hydrotreatment of coal-derived liquids before catalytic cracking, however,

Table 5.7 Thermal cracking of coal-derived liquids[a]

Product composition (mass basis)	Coal-derived[b]			Petroleum-derived	
	Naphtha (%)	Gas oil		Naphtha (%)	Kerosine (%)
		Low-severity hydrogenation/ hydrotreatment	High-severity hydrogenation/ hydrotreatment		
Methane	10	7		14	12
Ethane	2	2	13 (Methane, Ethane, C3 and C4 paraffins combined)	3	2
C3 and C4 paraffins	1			1	
Ethylene	17	9	22	23	25
C3 and C4 olefins	14	5	11	20	16
Benzene	14	8	15	8	8
Toluene	14	6	6	6	4
Xylenes	3	2	2	3	1
Balance	25	61	31	22	32

[a] Sources: G. O. Davies, Paper 2, *New Coal Upgrading Processes: Information Symposium, EEC Round Table Conference*, Vol. 1, September 1979, p. 253; G. O. Davies, Paper 2, *Coalchem 2000 Conference*, I. Chem. E., Sheffield, September 1980.

[b] The data for the coal-derived liquids refer to the NCB LSE process.

Table 5.8 Catalytic cracking of gas oil (approximately 200 °C +)[a]

Composition	Coal-derived[b]			Petroleum gas oil[c] (%)
	No hydrotreatment (%)	Low severity hydrotreatment (%)	Medium severity hydrotreatment (%)	
Mass balance				
Feed	100.0	100.0	100.0	100.0
C1–C4	7.0	15.2	18.1	20.6
C5–200 °C	15.0	38.2	49.2	
200 °C +	58.6	37.3	25.8	
Carbon	19.4	9.3	6.9	5.0
Total	100.0	100.0	100.0	
C5 + product[d]				
Paraffins			} 40.1	
Naphthenes				
Aromatics			53.4	
Olefins			6.5	
C5–200 °C (volume % feed)	18.0	43.0	54.7	68.4
200 °C + (volume % feed)	57.8	33.2	22.5	15.1

[a]Source: F. J. Riedl and A. J. DeRosset, *US DOE report FE-2566-30*, UOP Inc., August 1979, pp. 11, 31, 39, 43, 48.
[b]The data for the coal-derived liquids refer to the Exxon Donor Solvent process.
[c]The hydrogen content of the petroleum gas oil is taken to be 13.2 per cent.
[d]The analyses of the C5 + product for the medium severity case refer to the C5–200°C fraction.

detracts from this approach and, for this reason, hydrocracking appears to be the preferred approach in present studies of coal liquefaction plant refinery configurations.

Hydrocracking

The term hydrocracking is used to describe the catalytic hydrogenation of hydrocarbon liquids under conditions at which thermal cracking takes place. As noted above, hydrocracking is an alternative to catalytic cracking for upgrading heavy feedstocks to materials in the naphtha/gasoline boiling-range.

The catalysts used are of the Co–Mo–Ni–W type, similar to those employed for hydrotreatment. The conditions are, however, more severe with pressures typically being 150 to 170 bar and temperatures about 430 °C. Hydrocracking may take place in one or two stages depending on the properties of the feed and the desired product spectrum. If the feed content of nitrogen or aromatics is high, or if the maximum yield of gasoline and naphtha is required, then two-stage operation is preferred. In this case, the first stage performs the function of a hydrotreatment unit in preparing the feedstock for the second stage.

Hydrocracking can be used to obtain total conversion to light distillates by fractionating the hydrocrackate product and recycling the bottoms to extinction. In the case of a two-stage process, recycle is normally applied to the second stage only.

The effect of hydrocracking coal-derived gas oil is illustrated in table 5.9. The feedstock is as described in table 5.5.

Table 5.9 Hydrocracking of gas oil (approximately 200°C +)[a]

Composition	Coal-derived[b]		Petroleum-derived	
	Low severity	High severity	Low severity (%)	High severity (%)
Mass balance				
Feed	100.0 %	100.0 %	100.0	100.0
Hydrogen	2.5 %	6.3 %		
Total	102.5 %	106.3 %		
C1–C4	3.9 %	16.5 %		
C5–200°C	15.6 %	85.3 %	31.9	85.7
200°C +	78.6 %		63.1	
Heterogases	4.4 %	4.5 %		
Total	102.5 %	106.3 %		
C5 + product				
H	13.4 %	13.8 %		
O	226 ppm	154 ppm		
N	0.2 ppm	0.3 ppm		
S	365 ppm	0.1 ppm		
Paraffins		18 %		
Naphthenes		64 %		
Aromatics		18 %		

[a] Sources: F. J. Riedl and A. J. DeRosset, *US DOE report FE-2566-33*, UOP Inc., November 1979, pp. 67, 69, 88, 92, 93; A. J. DeRosset, Gim Tan and L. Hilfman, *US DOE report FE-2566-26*, UOP Inc., May 1979, figures 5 and 7.

[b] The data for the coal-derived liquids refer to the Exxon Donor Solvent process.

As may be expected, the products from hydrocracking have reduced heteroatom contents and are less aromatic/more naphthenic than the crude gas oil. The yield of light distillates depends on the severity of the hydrotreatment; complete conversion of the 200 °C + gas oil to a C5–200 °C naphtha is possible with 'high severity' hydrocracking.

The yield pattern obtained by hydrocracking coal-derived gas oil is similar to that obtained with petroleum gas oil under the same conditions.

Catalytic reforming

The main reaction taking place during catalytic reforming is usually the dehydrogenation of naphthenes to produce the corresponding aromatics.

Since aromatics have a high octane rating when compared with naphthenes, catalytic reforming has become an important component stage in oil refining and is standard practice for the conversion of naphtha into gasoline blend stock.

The catalysts are usually platinum-based on an alumina support and are highly susceptible to poisoning by sulphur and nitrogen compounds. The content of these elements in the feedstock therefore has to be reduced to a low level (typically 1 ppm) by hydrotreating prior to catalytic reforming. The catalysts are also deactivated by multi-ring naphthenes which cause coke formation and the amount of these compounds must therefore be controlled by careful selection of the end-point of the fraction. The reactions are usually carried out at 490 to 540 °C and 10 to 30 bar.

The dehydrogenation reactions are strongly endothermic and, for this reason, several reactors operating in series are usually necessary so that the feed can be reheated to the appropriate temperature between conversion stages. Considerable quantities of hydrogen are obtained as a by-product of catalytic reforming and can contribute to hydrogen requirements for hydrotreatment (and hydrocracking) elsewhere in the refining process.

Results obtained by catalytic reforming of coal-derived naphtha are given in table 5.10. The feedstock is the hydrotreated naphtha specified in table 5.6. The highly naphthenic nature of the coal-derived naphtha makes it an excellent reformer feedstock after hydrotreatment because of the high aromatics content, and hence octane rating, of the reformate product. As illustrated in the table, the yields of C5+ product obtained by reforming a Middle East petroleum naphtha at the same severity (that is, to produce the same reformate octane rating) are considerably lower than for the coal-derived naphtha.

Depending on the operating conditions of the reformer, some conversion of normal paraffins in the feed to aromatics (by dehydrocyclisation and dehydrogenation) and to iso-paraffins may also occur. Isomerisation of normal paraffins is encouraged by reforming at relatively low temperatures (about 150 °C) and may be employed in refining crude oil to increase the octane rating of the 'straight-run' gasoline cut (30 to 70 °C). This process is, however, unlikely to be important in refining coal-derived liquids from pyrolysis or direct liquefaction because of their relatively low content of normal paraffins.

5.10 Product applications and properties

It is generally expected that coal-derived liquids will have to meet the existing specifications for liquid fuels (based on petroleum-derived products), at least

Table 5.10 Catalytic reforming of naphtha (approximately C5–200°C)[a]

Composition	Coal-derived[b]		Petroleum-derived	
	Low severity	High severity	Low severity (%)	High severity (%)
Mass balance				
Feed	100.0 %	100.0 %		
Hydrogen	2.2 %	3.4 %		
C1–C4		0.6 %		
C5 +	97.8 %	96.0 %		
Total	100.0 %	100.0 %		
C5 + product				
H	11.6 %	10.2 %		
O	98 ppm			
N	0.1 ppm	0.1 ppm		
S	0.1 ppm			
Paraffins	16 %	16 %		
Naphthenes	25 %	6 %		
Aromatics	59 %	78 %		
C5 + product (volume % feed)	94 %	90 %	85	75

[a]Sources: Gim Tan and A. J. DeRosset, *US DOE report FE-2566-25*, UOP Inc., February 1979, pp. 16–21; A. J. DeRosset, Gim Tan and L. Hilfman, *US DOE report FE-2566-26*, UOP Inc., May 1979, figure 1.
[b]The data for the coal-derived liquids refer to the Exxon Donor Solvent process.

initially. As the availability of coal-derived liquid fuels increases and as the utilisation technologies develop, new specifications may, of course, be introduced.

Current specifications for liquid fuels may be grouped into the following three main categories.

Physical properties

Specifications usually include limits on the distillation range, volatility, density, calorific value, viscosity and low temperature properties (for example, cloud point and pour point). These properties generally affect fuel handling and storage.

Combustion properties

The combustion properties are assessed by empirical tests which depend on the application. For motor spirit, the properties are reported as an octane rating and are usually quoted as the research octane number. For heating oil and gas turbine fuel, the combustion properties are described by the smoke point and for diesel fuel the cetane number is used.

The main classes of hydrocarbons present in fuels (both coal-derived and oil-derived) each have a characteristic combustion performance as measured by the above indices. These are summarised in table 5.11.

Table 5.11 Markets for various chemical types

	Motor spirit (octane rating)	Heating oil and gas turbine fuel (smoke point)	Diesel (cetane number)
n-Paraffins	Poor	Good	Good
iso-Paraffins	Good	Good	Poor
Aromatics	Good	Poor	Poor
Naphthenes	Intermediate	Good	Intermediate
Olefins	Good	Poor	Poor

Effects of heteroatoms

A high content of oxygen, nitrogen or sulphur in the fuel is usually undesirable for a number of reasons.

The stability of the fuel is adversely affected by the presence of some oxygen and nitrogen compounds and specifications may include empirical indices of stability to ensure that the fuel does not deteriorate (for example, deposit gum) with age or at high storage temperatures.

Certain types of oxygen and sulphur compounds, such as acids and mercaptans, are corrosive and acceptable limits for corrosion rates may therefore be specified using empirical tests such as copper strip corrosion.

The sulphur content of the fuel is also important because, on combustion, this element is released as sulphur dioxide. A low sulphur content is therefore desirable both to limit corrosion of downstream materials and, from an environmental viewpoint, to reduce emissions of this pollutant to the atmosphere.

The methods by which the available refining technologies may be used to produce finished fuels to meet the specifications for the main markets are discussed below.

Gases

Hydrocarbon gases (methane, ethane, propane and butane) are produced as a by-product of cracking processes. The greater need for these processes in coal liquefaction may be expected to result in higher yields of gases than those obtained by refining crude oil. Some of the gases may be used for hydrogen production or boiler fuel within the process, with the remainder being sold as SNG (methane), LPG (propane and butane) and gasoline blend stock (butane).

Motor spirit

Motor spirit (petrol, gasoline) is a distillate fraction within the range C4–200 °C and may be prepared as follows.

(1) Direct use of the C4 and C5 fractions as blend stock.
(2) Hydrotreatment and catalytic reforming of the straight-run naphtha.
(3) Catalytic reforming of the naphtha obtained by hydrocracking the gas oil fraction. Initial hydrotreatment of this hydrocrackate naphtha is probably unnecessary.
(4) Hydrotreatment and catalytic cracking of the gas oil fraction. The resulting C5–200 °C product may be used directly as gasoline blend stock without catalytic reforming.

The above routes to gasoline are summarised, in simplified form, in figure 5.16.

The predominantly cyclic structure of coal-derived liquids results in the products obtained from catalytic cracking or reforming having a high content of aromatics. As noted in table 5.11, aromatics have good combustion properties (that is, octane rating) for motor spirit. Indeed, coal-derived motor spirit can be produced to satisfy the present octane specifications without the need for the lead additives commonly used as octane improvers in petroleum-derived motor spirit, although some engine modifications (valve seats, for example) may be necessary. There appears to be little difficulty in coal-derived motor spirit meeting the other specifications (volatility, for example) with appropriate choice of distillation range.

Jet engine fuel

At present, jet engine fuel (and industrial gas turbine fuel) is commonly a straight-run 150 to 250 °C fraction of crude oil, although fractions with a lower initial boiling point are also in use ('wide cut gasoline'). The combustion property relevant for this application is the smoke point and is affected mainly by the content of aromatics in the fuel (see table 5.11).

In order to produce jet fuel from coal-derived liquids, the appropriate middle distillate fraction must be hydrogenated or hydrotreated sufficiently to reduce the aromatics content to less than 20 per cent. With such preparation, it is expected, with the exception of the density, that the other specifications (for example, corrosion, stability, low temperature behaviour) can generally be met.

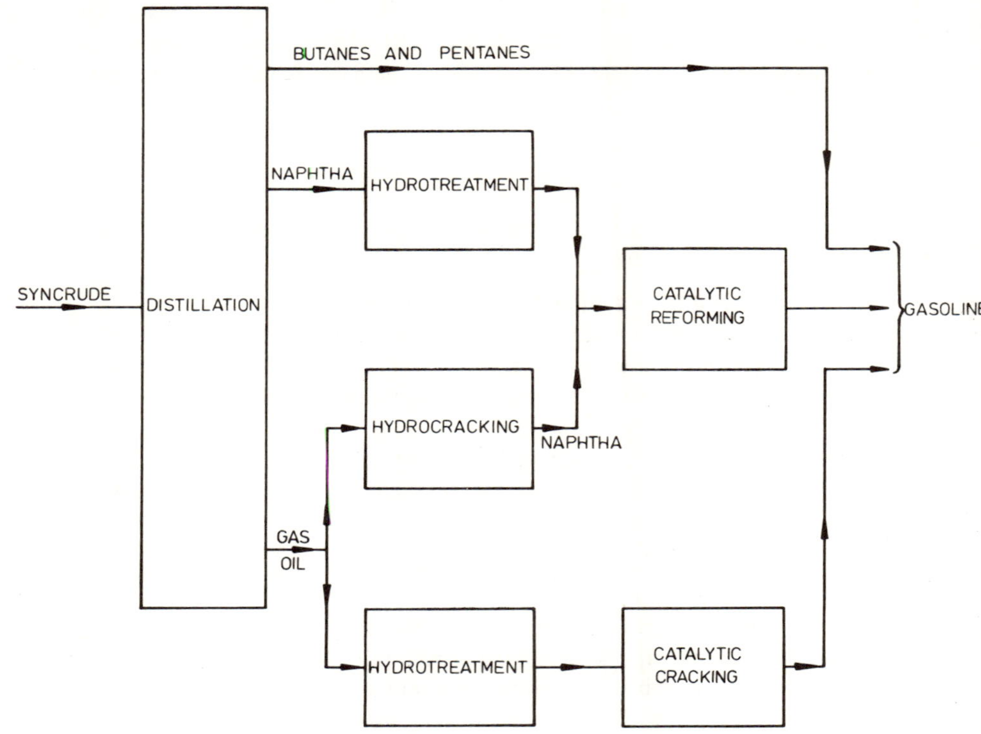

Figure 5.16 Summary of refinery options for gasoline production

Coal-derived jet fuel has a significantly higher density than present fuels but, as the calorific value per unit mass is similar, this difference is not regarded as an important disadvantage.

Kerosine heating oil

Kerosine heating oil has the same boiling-range as jet engine fuel (150 to 250 °C). The specifications are also similar except that less stringent limits are set for corrosion, stability and low temperature behaviour. It is therefore possible to produce an acceptable kerosine heating oil from coal-derived liquids provided that the comparatively high density can be tolerated.

Indeed, the severe hydrotreatment necessary to produce a satisfactory jet fuel may not be required to meet existing specifications for heating oil. However, concern has been expressed about the carcinogenic properties of the aromatic hydrocarbons in unhydrogenated heating oils and severe hydrotreatment may be employed to reduce the aromatics content to a low level.

Diesel fuel

The diesel fraction generally boils in the range 200 to 350 °C, although blends for low speed (for example, marine) engines may incorporate higher boiling-range material.

The index used to indicate the combustion properties for this application is the cetane number. As shown in table 5.11, only n-paraffins offer 'good' diesel combustion characteristics. Since conventional refining techniques cannot produce n-paraffins in the correct boiling-range from the predominantly cyclic coal-derived liquids, diesel fuel from coal has to be based on naphthenes.

It has been found that severe hydrotreatment to convert almost all of the aromatics in the appropriate coal-derived fraction to naphthenes gives a diesel fuel that is above the minimum specification although it has a lower cetane number than the best petroleum-derived fuels. It is possible that future diesel engine developments (spark-assisted engines, for example) will enable the minimum specification to be relaxed so that less severe hydrotreatment is needed.

In other respects, such as sulphur content and low-temperature properties, coal-derived diesel meets, and in some cases substantially exceeds, existing specifications. As with jet fuel and heating oil, coal-derived diesel has a higher density than petroleum diesels.

Chemical feedstocks

The main raw materials required by the petrochemicals industry are synthesis gas, olefins (principally ethylene) and BTX (principally benzene). At present, synthesis gas is manufactured mainly by steam-reforming natural gas with small amounts being obtained by steam-reforming naphtha and the partial oxidation of heavy oil. Olefins are manufactured by the steam-cracking of naphtha, gas oil and ethane extracted from natural gas (where available) with BTX also being obtained from some of these processes. Most BTX, however, is produced by the catalytic reforming of naphtha.

In substituting for oil products in these markets, it is unlikely that coal-derived naphtha will be used as a feedstock for synthesis gas as it will be more economic to manufacture synthesis gas directly from coal by gasification (see chapter 4). As shown earlier, however, steam-cracking of hydrotreated coal-derived naphtha produces satisfactory yields of chemical feedstocks although less ethylene and more benzene are obtained than with petroleum-derived naphtha.

Clean boiler fuel

For power generation and bulk steam-raising applications, it has been proposed that a solvent refined coal (SRC) is prepared. This involves solvent extraction, mild hydrogenation of the coal to reduce the sulphur content to an acceptable level and removal of the mineral matter (for example, by filtration). The product is a low-sulphur, low-ash solid (typically less than 1 per cent sulphur, 0.1 per cent ash) and can be used to substitute directly for heavy fuel oil without further refining.

SRC is in competition both with coal burned directly (for example, in a fluidised bed) and low calorific value gas. Although SRC is considerably more expensive than coal, the additional fuel costs are offset by a number of advantages (also shared by low calorific value gas).

(1) SRC can be used in compact, high output boilers, similar to those in use for fuel oil.
(2) No ash removal system is required.
(3) The low sulphur content may avoid the need for sulphur retention during or after combustion.

Compared with low calorific value gas, SRC has the advantage of low transport costs and would therefore be manufactured in large central plants rather than near the point of consumption.

5.11 Comparison of direct and indirect liquefaction

As noted earlier, there exist two separate technologies for converting coal into liquid fuels: the direct route described in this chapter and the indirect route via synthesis gas described in chapter 4. These technologies are generally regarded as being in competition and the purpose of the present section is to assess relative merits and future prospects.

The initial patents on both routes date from the second decade of the twentieth century and both were used on a commercial scale during the second world war. At the present time, however, the only large-scale coal liquefaction plants are those based on the Fischer–Tropsch process in South Africa. Only this type of indirect liquefaction can therefore be regarded as commercially available.

Development programmes for both routes are in progress although, until recently, the emphasis of research and development outside South Africa was overwhelmingly towards the direct route. Direct liquefaction still accounts for the majority of liquefaction research and development expenditure but renewed interest, stimulated by developments in catalysis such as the Mobil process, is now being shown in the indirect route.

In spite of the comparatively modest research and development expenditure, it is widely thought that the 'second generation' of indirect processes (such as Mobil) present fewer development problems and have a greater potential for near-term commercialisation than the direct processes. It remains to be seen whether this is true and, if so, whether the direct route offers sufficient advantages to replace the indirect route in the longer term.

The research and development programmes on direct liquefaction arose because of the general view that this approach offers substantially higher efficiencies and lower capital costs than existing, commercial indirect liquefaction technology (that is, Fischer-Tropsch). However, the development of second-generation indirect processes incorporating new catalysts and gasifiers has improved markedly the economics of this route. The economic prospects of the two approaches are discussed further in chapter 8.

Although the new indirect processes are still less efficient than the direct processes (typically by 5 to 10 percentage points), the product spectra are not directly comparable. Direct processes always produce a wide range of products and, as discussed earlier, these require refining to meet the market specifications (although this provides some flexibility in the product slate produced). By contrast, indirect processes can be highly specific with almost all of the product being obtained as a single marketable fuel that does not require further refining. In the case of the Mobil process, for example, the product is a high-octane gasoline. The lower efficiency of the indirect processes may therefore be offset by the higher value of the products obtained.

A number of other factors affect the comparison between direct and indirect processes. They include the following.

Coal type

The range of coals that can be used in direct liquefaction is not yet fully established but may be subject to some limitations on reactivity, mineral matter content and mineral type. With appropriate choice of gasifier, however, indirect liquefaction processes can accept virtually any coal.

Range of products

At present, specific catalysts for indirect processes exist only for methanol and gasoline. Improved catalysts for manufacturing jet fuel, diesel and chemical feedstocks are therefore required before indirect processes are able to compete with direct processes over the full range of products. Recent developments in catalysis have been rapid and present research programmes are aimed at extending the range of high-performance catalysts available.

Integration

Since the crude liquids produced by direct liquefaction processes require refining using present oil-refinery techniques, it has been suggested that coal liquefaction plants, at least initially, may be confined to manufacturing a synthetic crude oil ('syncrude') which would then be refined, perhaps together with crude oil, in existing oil-refineries. This possibility may give the direct route some advantage in terms of ease of integration with present liquid fuel processing and marketing operations.

Environment

Concern has been expressed about the carcinogenic properties of the polycyclic aromatic compounds formed as intermediates in the direct liquefaction of coal. This aspect is discussed further in chapter 7 but it may be noted here that such considerations need not apply to the indirect route.

Although both direct and indirect processes are being developed rapidly, it will probably not be until the 1990s that sufficient is known about their performance and economics on a large scale for commercial decisions to be made with confidence. It may be that the different approaches fulfil complementary roles and that both are required to meet the range of economic and market conditions that will exist in the future.

CHAPTER 6

OTHER TECHNOLOGIES

Although the major part of research and development effort on coal utilisation is directed to the technologies of fluidised bed combustion, gasification and liquefaction, a number of other developments that could have an impact on the future pattern of coal use are also taking place. Such developments are concerned not only with the coal utilisation technologies themselves but also their interactions with the processes of coal extraction, transport and handling, and other major energy sources such as crude oil and nuclear power. The purpose of the present chapter is to review briefly the most important of the developments in order to provide a context for the major technologies considered in the preceding three chapters.

6.1 Remote extraction

Underground gasification

The possibility of converting coal reserves *in situ* so that their energy can be obtained as a clean gas without men having to go underground has attracted widespread interest.

The concept was first suggested in 1868 by Sir William Siemens and later by Mendeleev in Russia. The first practical experiments were carried out in 1912 by Sir William Ramsey in County Durham but the work was stopped by the first World War. In 1913, Lenin reviewed the possibilities for underground gasification in an article in *Pravda* foreseeing the prospect of releasing the labour of millions of miners. In the 1930s, a large development programme was initiated by the USSR and considerable experience was obtained.

After the second World War, a number of other countries, including the United States, the United Kingdom and Belgium, also began experimental work. By the early 1960s, however, all these programmes had been abandoned (or, in the case of the USSR, reduced to a low level of effort) in the face of cheap oil and natural gas.

The different energy situation of the 1970s has led to a re-awakening of interest in underground gasification, particularly in the United States where a number of large-scale trials have been undertaken. Other countries that have begun experimental programmes include Belgium, West Germany, France and Canada, and work is continuing in the USSR.

An underground gasification scheme is usually started by drilling a series of vertical holes from the surface down into the coal seam. These boreholes then have to be linked within the seam to provide passages along which the gas can flow. Three main methods of linkage are available.

Pneumatic

The generally preferred method of linkage is to apply compressed air to one borehole and to allow the pressure difference to create a path for the air to the other borehole. When the air is flowing, a fire can be started at the exit borehole and the passageway enlarged by reverse combustion. This method may be unacceptably slow in coals that have a low permeability, or where linkage over a long distance is required, and other techniques may be preferred.

Hydraulic

An alternative to air is to use water under high pressure to create a linkage by fracturing the seam along planes of weakness (referred to as 'hydrofraccing').

Drilling

Linkage between vertical holes can be achieved directly by preliminary mining to construct roadways and galleries. This approach has the disadvantage that men and machinery still have to work underground. However, this may be avoided by employing the advanced directional drilling techniques developed by the oil industry. Starting vertically at the surface, the direction of the borehole is made to change as drilling progresses until the borehole is able to follow the coal seam at the appropriate depth.

The importance of drilling and linkage costs is such that a number of other techniques such as electrical heating and lasers have also been considered and these approaches may provide the basis for future improvements.

When linkage has been established, the reactant gases are pumped down one borehole, the coal is ignited and a combustible product gas obtained from another borehole. Two main types of gasification have been investigated.

Air gasification

If air is used as the gasifying agent, a low calorific value gas of 3 to 5 MJ/m^3 is obtained because the combustible components are diluted by the nitrogen in the air.

Oxygen–steam gasification

A higher calorific value gas of generally 7 to 12 MJ/m^3 can be obtained by using oxygen and steam as the gasifying agents since this avoids dilution with nitrogen.

In order to obtain a stable, controllable system producing a gas of adequate calorific value, the geometry of the gasifier must be designed with care. One of the simplest arrangements is shown in figure 6.1. The coal near the inlet boreholes is burned to carbon dioxide which is then reduced to carbon monoxide

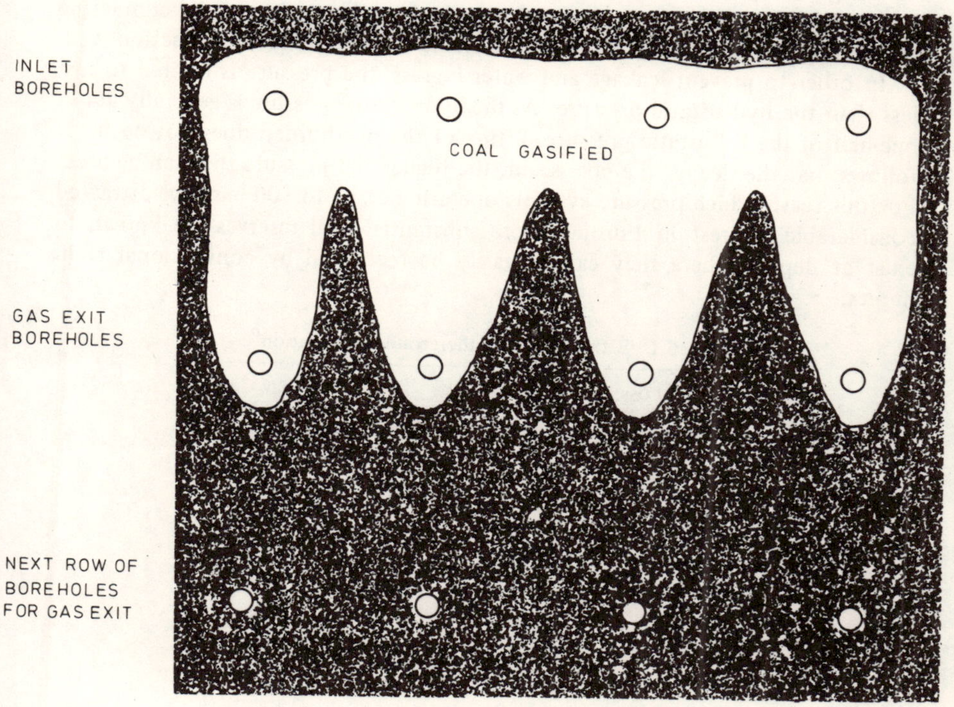

Figure 6.1 Layout of an underground gasifier

before leaving the seam via the exit holes. Usually, several inlet holes would be linked together as shown in the figure. The gasifier is made to progress through the coal field, converting the coal panel by panel. In practice, a number of such gasifiers operating in parallel would be employed.

Typically, the overall efficiency of underground gasification using air is 45 to 57 per cent (see table 6.1). Somewhat higher overall efficiencies appear to be possible using oxygen–steam gasification.

Underground gasification can take place using elevated pressures; experimental tests at up to 20 bar have been carried out. Operation at elevated pressure has some important advantages over low-pressure systems

(1) Equilibrium considerations favour methane formation (see chapter 4) which improves the calorific value of the gas.

(2) Gas-solid contacting (and therefore percentage conversion) are improved by the increased turbulence and longer gas residence times obtained at elevated pressures.

(3) The size of well required for a given flow rate of gas is smaller.

The gasifier geometry required for high-pressure operation differs significantly from that for low-pressure systems. In particular, the good gas–solid contacting obtained leads to radial growth of the flame front around the injection well.

In order to prevent leakage and water ingress, the pressure is limited to just less than the hydrostatic pressure. As the hydrostatic pressure is generally about one-half of the lithostatic pressure, lifting of the overburden does not occur. It follows that the deeper the coal seam, the higher the pressure that can be used. For this reason, high-pressure systems operating at up to 100 bar have attracted considerable interest in Europe where substantial coal reserves are known to exist at depths where they cannot easily be recovered by conventional techniques.

Table 6.1 Efficiency[a] of underground gasification[b]

Loss	Energy Content (% coal in place)
Coal unburned	20
Heating of rocks	10
Water evaporation	0–12
Unrecovered sensible heat in the gas	3
Gas leakages	5
Thermal energy for compression (including linkage)	5
Total	43–55
Efficiency[a]	45–57

[a] Efficiency is defined as the chemical energy (gross calorific value) of the gas expressed as a fraction of that of the coal in place.
[b] Source: J. Patigny and V. Chandelle, *Economic Prospects of Underground Gasification of Coal at Great Depth and High Pressure*, Paper to IEA Coal Working Party Meeting on Underground Gasification, Paris, 1977.

The possibility of high-pressure, deep-seam gasification has led to the suggestion that hydrogen (perhaps with steam) could be used as the gasifying agent. The hydrogenation reactions are exothermic (see chapter 4) and could therefore be self-sustaining. The main product would be SNG (methane) which could be separated from the unconverted hydrogen cryogenically. Some of the product would be reformed to provide the hydrogen consumed in the gasification process. No experimental work on this concept appears to have been carried out although the overall process efficiency may be low because of poor conversion of the hydrogen underground. For these reasons, the remainder of this section will concentrate on air and oxygen–steam systems.

Since the gases produced by underground gasification have a low or medium calorific value, their transport by pipeline is expensive compared with natural gas or SNG. The most economic use of the gases is therefore bulk markets near the point of production. The principal options are as follows.

Power generation

Probably the only economic market for air-based underground gasification systems is power generation. Trials have confirmed that the product gas can be used in boilers for conventional steam cycles with combined cycle operation being a possible future option. Oxygen–steam-based underground gasification systems can also be used for power generation.

Synthesis

If oxygen and steam are used as the gasifying agents, the product gas is suitable for upgrading by synthesis to SNG or liquid fuels (for example, methanol or hydrocarbon liquids by the Fischer–Tropsch process). The conversion technology is essentially the same as that required for surface gasification systems (see chapter 4).

The main cost in underground gasification is that of drilling the boreholes. The economics of gas production therefore depend critically on the following factors

seam thickness
seam depth
the spacing between boreholes

Underground gasification is generally cheapest in those locations where thick seams lie near to the surface. As these conditions also favour conventional mining, the comparative economics of this and underground gasification depend largely on the spacing between the boreholes. At present, the spacing between boreholes is limited by the distance over which linkage can be achieved. So far, experimental work has been confined to borehole spacings of less than 25 m. However, current programmes are directed towards extending the distance over which linkage can be achieved, possibly up to 100 m. It is not known yet whether efficient gasification would be obtained at this borehole spacing.

One of the attractions of underground gasification is that it may be able to exploit reserves that are unworkable or uneconomic with conventional mining techniques. Such reserves may include seams that are deep, thin or remote (for example, undersea), reserves that are too small to be extracted economically by conventional mining or situations where the geological conditions (for example, faulting or water) make conventional extraction difficult or dangerous. However, reserves suitable for underground gasification have to be in areas of low population density because of the large number of boreholes required and because subsidence is not easily controlled.

At first sight, underground gasification appears to have a number of environmental advantages when compared with conventional mining. In particular, coal handling, preparation and transport are avoided as are ash and spoil disposal. These advantages may, however, be offset by the following factors:

(1) Land use is considerable (a 500 MW thermal site requires 5 to 10 hectares in use at any one time and advances at a rate of about 0.2 ha/d). The visual impact would include drilling rigs, borehole cappings, pipelines, compressors and utilisation plant (for example, a power station).
(2) Excessive gas leakage could cause environmental problems, particularly near residential areas, and careful control will be necessary to limit this impact.
(3) Similarly, water in contact with an underground gasifier will become polluted and careful planning and control is necessary to avoid such pollution spreading through underground aquifers.

The advantages of underground gasification may be summarised as follows:

(1) Men do not have to work underground and labour intensity is low.
(2) Reserves that are impractical or uneconomic to work by conventional mining may be exploited (for example, deep seams or relatively small reserves).
(3) Many of the environmental impacts of conventional mining are avoided (although other environmental disadvantages may exist).
(4) The energy in the coal is obtained directly as a combustible gas.

The main requirement for more widespread use of underground gasification is an improvement in the economics. Although the trials that have been carried out are encouraging, further development of the technology is necessary, particularly with regard to techniques for reliable in-seam linkage of wells at commercial spacings. At present, however, underground gasification seems unlikely to displace conventional mining although it may find application for accessing reserves that otherwise could not be exploited.

Other remote extraction technologies

A number of other systems of extracting coal remotely (that is, without men going underground) by chemical conversion have been proposed. These include

pyrolysis by electrical heating
complete combustion
liquid solvent extraction

hot aqueous alkali oxidation
chemical comminution
microbiological degradation

Little practical work has been carried out on these systems although theoretical studies have revealed a number of problem areas. In many cases, the economics would be dependent on the same factors affecting underground gasification, particularly drilling costs and the spacing between boreholes. It therefore seems likely that research and development priorities in this area will remain on underground gasification until the technology has become more firmly established.

6.2 Coal preparation

The traditional methods of coal preparation are described briefly in chapter 1. Although coal preparation is a mature technology, developments in coal utilisation are providing incentives for departures from normal washery practice, particularly for the purposes of increased sulphur removal and production of low-ash boiler fuels.

Sulphur removal

The ash content of the clean coal from a conventional washery is generally 5 to 10 per cent and the pyritic sulphur content is reduced by up to 40 per cent of that in the raw coal. The loss of combustible material is usually of the order of one or two per cent. Interest in increasing the degree of sulphur or ash removal centres on the application of existing washing methods in optimised configurations as discussed below. However, such schemes invariably lead to an increased loss of combustible material.

Part of the sulphur in coal is present as pyrites in the mineral matter and a significant reduction in sulphur content can therefore be obtained by the removal of pyrites during coal preparation. This is discussed in the context of other options for sulphur control in section 7.2.

The most suitable technique for reducing the pyritic sulphur content of coals appears to be dense medium cyclones (operating on coal sizes of 25 to 0.5 mm). These can reduce the pyritic sulphur content by up to 60 per cent but at the expense of an increase in the combustibles loss to 5 to 35 per cent.

Froth flotation (operating on coal smaller than 0.5 mm) is not usually considered effective in separating the pyrites from the coal because the surface properties (on which the separation depends) are similar in the two cases. The United States Bureau of Mines, however, is developing a two-stage froth flotation system for this purpose. In the first stage, coal is floated and mineral matter removed whereas in the second stage the mineral matter is floated and the coal depressed using a chemical additive. Approximately 80 per cent of the pyrites can be removed but the loss of combustible material is about 35 per cent.

Low-ash boiler fuels

Interest in the preparation of a low-ash boiler fuel stems from the disadvantages arising from the presence of ash when coal is used directly in industrial boilers. In particular, the conversion of existing oil or gas-fired boilers to coal firing is generally not possible without substantial derating (see chapter 3).

In order to increase the ash removal, and thereby to minimise the derating on conversion, 'deep washing' is necessary. This involves pulverising the coal to a finely divided state to maximise the 'release' of the mineral matter trapped in larger coal particles. The pulverised coal can then be cleaned using a conventional froth flotation system. However, only coals with a low inherent ash content are suitable for treatment by this method.

The deep washing approach described above has been employed in the preparation of coal-water mixtures (see section 6.6). It has proved possible to remove 70 to 90 per cent of the pyritic sulphur and to reduce the ash content to 1 to 4 per cent with a combustibles loss of 1 to 3 per cent.

An alternative and novel approach to deep washing is the 'spherical agglomeration' technique being developed by the National Research Council, Canada. In this case, coal is finely ground in a petroleum fraction; the mineral matter is removed as agglomerates that are formed when an aqueous bridging liquid is added, leaving the clean coal suspended in the liquid. This process appears to have particular relevance to the preparation of coal-oil mixtures (see section 6.6).

Micronised fuel

Micronised fuel is coal that has been reduced to a maximum particle size of about 5 microns.

With pulverised fuel, the maximum particle size is about 100 microns and some of the finely divided inherent mineral matter remains embedded within the coal particles and cannot be removed by washing. Micronising the fuel releases much of the inherent mineral matter and, used as the first stage of a 'deep washing' process, enables the ash content, at least in theory, to be reduced to less than one per cent.

The additional mechanical energy required to produce a micronised fuel rather than a pulverised fuel is equivalent to the consumption of less than 5 per cent of the coal for power generation. Existing washing techniques, such as froth flotation, can be used for cleaning the micronised fuel. Micronised fuel may be prepared dry, as a coal–oil mixture or as a coal–water mixture, depending on the method of use.

The main interest in micronised fuel appears to be as a possible means of fuelling a coal-fired diesel engine (see below). In addition to permitting a low-ash content to be achieved, micronised fuel has the advantage of a short combustion time for this application.

6.3 Transport

Conventional systems

The conventional systems of coal transport include the following.

Rail

This is the world's main method of coal transport. The railway networks of the industrialised countries expanded along with coal production during the industrial revolution and the existence of this infrastructure is likely to ensure that rail retains its dominant position. The cheapest form of rail transport is unit trains (or merry-go-round trains), that is, trains dedicated to moving coal between two locations. At present, unit trains account for less than half of the coal transported by rail although their use is likely to increase substantially in the future. The size of unit trains varies from country to country; in the United Kingdom, unit trains carry up to 1000 tonnes each whereas in the United States up to 10 000 tonnes may be moved by a single train.

Inland waterways

Coal transport by barge on inland waterways can be the cheapest method of moving coal inland but only two waterway systems at present handle significant quantities of coal (the Mississippi river basin in the United States and the mainland system of north-west Europe). Individual barges usually carry a maximum of 1350 tonnes although a number of barges (sometimes up to 30 in the United States but usually less than 4 in Europe) may be fastened together.

Road

The most flexible coal transport system is lorries and, for this reason, they are used extensively although mainly for local deliveries of coal to relatively small consumers. The capacity of a coal lorry is typically 20 tonnes. Where convenient, rail is usually preferred for the bulk movement of coal.

Conveyor belts

These are usually used only for transport over short distances (for example in a mine or plant). However, a conveyor 16 km long transporting 1500 t/h coal has been built at Camp Breckenridge, Kentucky, in the United States.

Ships

Per tonne-kilometre, the transport of coal by ship is substantially cheaper than any other method of coal transport and, where there is a choice, is the preferred option. At present, coal is transported mainly in vessels of up to 125 000 deadweight tonnes. Vessels of up to 40 000 deadweight tonnes (referred to as 'handy size' vessels) are used for short and coastal routes. These are mainly self-unloading ships such as those that dominate coal transport on the Great Lakes. Although more expensive to build and operate than larger vessels, they have more flexibility in not requiring conventional port unloading facilities. For long ocean routes, the most widely used vessels are those of the 'Panamex' class (60 000 to 80 000 deadweight tonnes). However, as port facilities develop, it is expected that larger dry bulk carriers will be used increasingly in the future.

Slurry pipelines

Although the first patent was issued in 1891, slurry pipelines are generally regarded as a new technology and indeed the only major development in the field of coal transport in recent times. The pioneer installation was the 175 km long, 250 mm diameter Consolidation Coal pipeline at Cadiz, Ohio, commissioned in 1957. This moved 1.25 Mt/a of coal until it was closed in 1963 for commercial reasons.

The only long-distance pipeline at present operating in the Western World is the 435 km long, 460 mm diameter Black Mesa pipeline in Nevada. This has a capacity of 4.5 Mt/a and was completed in 1971.

The main stages in coal transmission by slurry pipeline are as follows.

Slurry preparation

The coal is first crushed and ground, typically to a maximum particle size of 2.5 mm. The choice of particle size is a compromise between segregation (particles too coarse) and difficulties in dewatering (particles too fine). The particles are then mixed with water to form a slurry containing about 50 per cent by weight of coal.

Transmission

The slurry is pumped by reciprocating pumps into the pipeline with booster pumping stations at 100 to 200 km intervals. The pumping pressure is up to 100 bar and is sufficient to move the slurry at 1 to 2 m/s. Velocities in this range are necessary to ensure that segregation of the coarse particles is avoided as this may lead to blockages.

Coal-slurry pipelines are usually buried underground for protection against freezing and are subject to a maximum gradient of about 16 per cent to prevent coal settling. A further difference between coal-slurry pipelines and oil/gas pipelines is the abrasive and corrosive nature of coal–water slurries. Methods of protection against these effects include allowing additional pipe thickness and adding corrosion inhibitors.

Dewatering

On arrival at the destination, the coal slurry is fed to a dewatering plant which removes most of the water. Efficient dewatering is particularly important if the coal is to be used in a power station (the main application so far) or transported further because of the detrimental effect of moisture on the calorific value.

The slurry is fed first to continuous vacuum filters or centrifuges which produce a dewatered cake of coal containing about 20 per cent water and a waste water stream containing about 5 per cent coal. The dewatered cake is suitable for direct combustion or further transport and the waste-water stream is passed to clariflocculators for purification. These produce a clear overflow stream that can be used as cooling water or disposed of and an underflow stream ('ink') containing 10 to 25 per cent solids. The disposal of this 'ink' is the main environmental problem of coal-slurry pipelines. The maximum particle size of the solids is less than 40 microns, making further dewatering impractical with existing technology and combustion difficult (although not impossible). The most convenient solution is to dump the 'ink' in large and unsightly ponds.

Slurry pipelines are best suited to the transport of large quantities of coal over long distances and, as such, compete mainly with rail transport. The relative economics of slurry pipelines and existing rail systems are unclear although slurry pipelines appear attractive when compared with new railway lines.

A possible constraint on the growth of slurry-pipeline transport in the United States is the reluctance of rail companies to grant a right of way to slurry-pipeline companies with whom they may be in direct competition. This problem may not apply elsewhere and is not expected to be a long-term constraint in the United States.

A further problem is water supplies. A slurry pipeline requires large quantities of water and this may restrict its application in areas that do not have plentiful water supplies.

Provided that these difficulties can be overcome, and an environmentally acceptable method of disposal found for the 'ink' residue, then slurry-pipeline transport may be expected to become more widespread in the future. A number of new pipelines of up to 1 m diameter (capacity 25 Mt/a) have been proposed, particularly in the United States.

Future developments

A number of coal-pipeline transport systems other than the coal–water slurry system have been suggested although no demonstration-scale units have yet been built.

For example, if oil is produced near to the location of the coal-mine, then coal could be transported by pipeline as a coal–oil slurry. The coal–oil slurry could be burned directly in a power station, thereby avoiding the costs and problems of the dewatering stage necessary for coal–water slurries.

Alternatively, coal–methanol slurries have been suggested as a medium for the pipeline transport of coal (for example, the proprietary 'Methacoal' system). A proportion of the coal is gasified and converted into methanol at the mine and the methanol is then mixed with the remainder of the coal to form a slurry. The mixture contains 50 to 75 per cent by weight of coal particles with the maximum particle size being about 150 microns. Under these conditions, the slurry is a stable pseudo-thixotropic suspension that requires only minimal agitation to prevent separation. At the point of consumption, the coal–methanol mixture can either be burned directly or can be separated to a high efficiency by distillation. In the latter case, the methanol can either be marketed as a fuel or chemical feedstock or returned to the mine for re-use in the slurry. The overall water requirements of the system are low compared with those of coal–water slurries and, as for coal–oil slurries, the costs and problems of the dewatering stage necessary with coal–water slurries are avoided. The main disadvantage of the concept, however, is the cost and efficiency loss for the conversion of coal into methanol, particularly if the slurry is used directly as a power station fuel.

A further slurry-pipeline concept under investigation uses liquid carbon dioxide as the transport medium. The operating pressure is 50 to 150 bar and therefore lies within the range of existing gas-pipeline technology. Studies by Arthur D. Little and W. R. Grace indicate that coal concentrations of 70 to 80 per cent (by mass) can be achieved with sufficiently good flow characteristics for throughputs to be up to three times those for coal–water slurries at comparable pipeline diameters. Separation at the point of consumption is achieved by pressure let-down with fabric filters being favoured for the final stage of recovery of the coal. The separation efficiency is high and the efficiency loss arising from the high moisture content of the fuel in coal–water slurry systems is avoided. Both open loop and carbon dioxide recycle configurations have been considered with the latter being favoured because of the difficulty in finding markets for large quantities of carbon dioxide. It is envisaged that the carbon dioxide would be obtained from coal combustion; in the open loop configuration, it would be necessary to burn about 9 per cent of the coal at the point of production. However, only make-up quantities are required in the recycle configuration and these could be supplied from the power station combustion gases at the point of consumption.

6.4 Handling

For large-scale applications such as power stations and future gasification and liquefaction plant, extremely efficient handling systems (for example, rail delivery and conveyor belt transport) have been developed. It therefore seems unlikely that there is much scope for further improvement in this area.

In the industrial sector, however, coal suffers in comparison with oil and gas because traditional handling methods are less convenient, dirtier and more labour intensive. There is, therefore, a considerable incentive to develop improved handling techniques for industrial applications.

The traditional method of industrial coal handling is based entirely on mechanical systems. These are briefly described below, together with recent developments in the conventional technology and other approaches, such as pneumatic conveying, that are now becoming widespread. In all these cases, it is assumed that the coal is fired as delivered and that no further coal preparation is required.

Traditional systems

The most common method of coal delivery to industrial consumers is by road although, for large installations, rail, ship or barge may be used instead.

Road deliveries are usually made by conventional tipper lorries that tip the coal into reception bunkers at (or below) ground level. If the combustion system is also sited at ground level, the coal has to be lifted by a conveying system before it can be burned. Usually, the conveying system delivers coal into a service hopper from which it flows by gravity into the combustor feed system. The main stages in a traditional industrial coal handling system are illustrated schematically in figure 6.2.

It can be seen that gravity plays an important part in determining the requirements for coal handling facilities. Indeed, the conveying system for transferring the coal from the reception bunker to the service hopper is one of the major items of cost in the configuration described above.

For this reason, if the reception bunkers can be sited at a higher level than the combustors (for example, on a sloping site), then the conveying system and service hoppers can be dispensed with and the combustors fed directly by gravity from the reception bunkers. For small installations, it is often economic to site the boilerhouse underground so that this arrangement is possible.

In most cases, however, a conveying system is required to transfer the coal to the service hopper and a number of mechanical devices are available. These include belt elevators, bucket elevators, drag-link conveyors, vibratory conveyors, screw conveyors and grab-skip-hoists.

From the service hopper, the coal flows under gravity into the feeding system

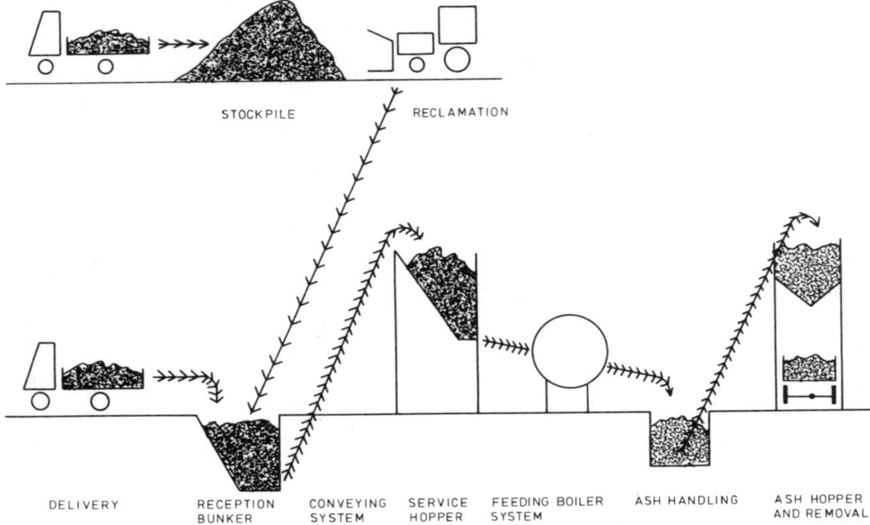

Figure 6.2 Traditional coal handling system

which delivers it at the required rate to the combustion bed. The feeding system is an integral part of the combustor and the type of feeding system used therefore depends on the combustor design. The main options are summarised in table 6.2.

Reception bunkers usually have a capacity of up to 30 tonnes. Where more storage than can conveniently be provided by reception bunkers is required, open stockpiles may be used. Reclamation from open stockpiles may be achieved most simply by a tractor and bucket although this method is not amenable to automation. The only automatic method of stockpile reclamation is the grab-skip-hoist which is operated from an overhead monorail. Coal reclaimed from stockpiles by a tractor and bucket would be delivered to the reception bunkers

Table 6.2 Industrial combustor coal feeding systems

Combustor type	Feeding system
Sprinkler/spreader stoker	Gravity via mechanical sprinkler or pneumatic conveyor (from reception hopper)
Chain-grate stoker	Self-feeding travelling grate
Coking stoker	Ram
Underfeed stoker	Screw conveyor or ram
Fluidised bed combustor (lump coal)	Rotary valve, screw conveyor or pneumatic conveyor
Fluidised bed combustor (crushed coal)	Pneumatic conveyor

as for new deliveries. If, however, a grab-skip-hoist has been installed, this would normally deliver coal from stock directly into the service hopper and would also be used to transfer coal from the reception bunker to the service hopper.

Since the quantity of ash to be removed from an installation is typically only one-tenth of the quantity of coal feed, ash handling is of less importance in the overall economics. On small systems, manual removal and storage in bins may be justified whereas, on larger systems, the traditional approach is to use mechanical devices (for example, belts, vibrators or skip hoists) similar to those used for coal conveying.

Recent developments

Although mechanical handling systems have a long history, improvements in their performance and reliability continue to be made. These arise partly from the use of modern materials and control technology and partly from evolutionary developments in design and system concepts.

One such development is the provision of a belt conveyor on the delivery lorry. This permits delivery directly into an above-ground bunker and may remove the need for a separate service hopper and on-site conveying system. Although a conveyor lorry is more expensive and less versatile than a conventional tipper lorry, this approach can be economic in appropriate circumstances.

Another recent approach by the NCB and manufacturers is the tipping bunker. Coal is delivered from a tipping lorry into a horizontal ground-level bunker which is then progressively tipped by hydraulic jacks to provide a controlled feed to a conveying system.

Also under consideration is the possibility of adapting the container-lorry system to coal delivery. Environmentally, this approach is particularly attractive when compared with open tipping.

When long-term storage is required, but insufficient space exists for open stockpiles, storage silos may be used. Individual silos of up to a capacity of 1000 tonnes are now becoming available. Delivery to silos is usually by a pneumatic conveying system and reclamation from silos is usually by a screw conveyor.

Pneumatic conveying systems

Despite the continued improvements in mechanical handling systems, the requirements for a higher level of amenity and lower costs in the industrial sector have led to a steady growth in the popularity of pneumatic conveying systems since their introduction during the 1960s.

There are two main types of pneumatic conveying systems for coal.

Dilute phase (or lean phase)

In a dilute phase system, the coal is metered into an air stream, usually by means of a rotary valve or screw feeder. The coal moves along the pipeline at relatively high velocities, typically 25 m/s. However, the pressure difference required is less than 1 bar and the system can therefore operate with a fan under either positive pressure (forced draught) or negative pressure (induced draught). The particles of coal are suspended in the air flow and have to be disengaged at their destination by means of cyclones or a vented hopper. The transport of coal as a suspension requires a comparatively large volume of air per unit mass of coal as implied by the term 'dilute phase'.

Dense phase

A dense phase pneumatic conveying system is essentially a batch-operated process with the cycle of operations being as follows. A pressure vessel under the bunker is filled by gravity and then sealed. Compressed air, typically at 3 to 10 bar, is introduced at the top of the pressure vessel and this forces the coal into the pipeline in the form of a slug. Under the action of the compressed air, the slug of coal travels along the pipeline at a relatively low velocity (typically 2.5 m/s). Owing to the low velocity, degradation of the coal and abrasion of the pipework are less than with a dilute phase system. When the slug reaches the service hopper, the compressed air is turned off and the cycle repeated. The volume of air per mass of coal transferred is low compared with that required by a dilute phase system and the problem of separating the air from the coal is virtually eliminated.

Dilute phase systems have been used for many years to transport pulverised coal in pulverised fuel power stations. For the industrial market, however, their application is limited at present to singles because the pipelines are prone to blockages when using smalls. Dense phase systems can handle both singles and smalls but are more expensive than dilute phase systems. In general, therefore, dilute phase systems are preferred for singles and dense phase systems for smalls.

Pneumatic conveying systems can now be used at every stage of coal handling.

Delivery

Pneumatic delivery lorries employ positive pressure dilute phase conveyors. If required, these can deliver the coal directly into a service hopper thereby eliminating the need for a reception bunker.

Conveying

The main application of both dilute and dense phase systems is for conveying coal from reception bunkers to service hoppers.

Feeding

Positive pressure dilute phase conveyors can be used to feed coal directly from the reception bunker to the combustion chamber as, for example, in the Vekos sprinkler stoker.

Ash handling

Negative pressure dilute phase conveyors are sometimes used for ash handling. However, the abrasion rates on pipework and other components are high. Dense phase systems can also be used, but other methods (for example based on hydraulic systems) are often preferred.

Pneumatic conveying systems are of interest because they offer the industrial consumer a high level of amenity in coal handling. In particular, the coal is conveyed in a closed pipeline so that dust is suppressed and the pipeline can be routed to reduce its prominence. A disadvantage compared with mechanical systems is that power consumption is higher (although still a small fraction of the total boilerhouse consumption).

Further improvements of pneumatic conveying systems are however, possible. For example, a recent development of the negative pressure dilute phase conveying system is the NCB suction nozzle. This eliminates the need for the conventional mechanical feeding systems at the pick-up point of a dilute phase system so that retrieval of coal from a reception bunker requires only that the nozzle is immersed into the coal.

Hydraulic systems

Although a logical alternative to pneumatic conveying is hydraulic conveying, the introduction of water as a transport medium for coal has important implications for the method of use and this aspect is therefore discussed separately (see slurry pipelines, section 6.3, and coal–water mixtures, section 6.6).

Hydraulic systems are commonly employed for ash handling as they serve both to quench the hot ash and to reduce dust. In the simplest system, the ash falls into a sluice along which water is pumped. The ash is carried to a storage pit where it settles and awaits periodic removal. The water is decanted, filtered and recirculated. Other systems employ a conventional mechanical conveyor (for example, a drag-link conveyor, belt conveyor or vibratory conveyor) submerged in the water. This approach has the advantage of enabling the ash to be delivered to an above-ground ash bunker.

6.5 Domestic market

Smokeless fuels

When a bituminous coal is heated above 400 °C it releases volatile matter (see section 5.3), typically equivalent to one-third of its initial dry, ash-free weight. If such a coal is used on a conventional domestic appliance, the volatile matter does not burn completely. Not only is this a source of inefficiency but, more seriously, it results in the emission of smoke in the flue gases. For environmental reasons, smoke emissions from domestic appliances are controlled in many countries and a number of smokeless fuels are marketed.

For present purposes, smoke is defined as unburned products entrained in the flue gases, and consists of two components: soot (particles of carbon) and tar fog.

Both of these arise primarily from the tar component of the volatile matter. Chemically, tar is mainly aromatic and, characteristically, tends to burn with a smoky flame.

To qualify as a domestic smokeless fuel, a solid fuel must have the following attributes:

(1) It must burn 'smokelessly' (that is, smoke emissions must be below a specified standard) and therefore produce little or no tar when heated.
(2) It must be in 'lump' form because a solid fuel containing a significant proportion of fines is not suitable for burning on domestic appliances.

The properties required for metallurgical coke are also those of a lump fuel having a low volatile matter content. For this reason, many of the processes for domestic smokeless fuel manufacture may be applicable to the production of formed coke, as discussed in section 6.8. However, the manufacture of formed coke is less well-established than that of smokeless fuels and further development work is necessary before the full possibilities for this application can be assessed.

The manufacture of domestic smokeless fuel involves the two main technologies of carbonisation and briquetting described briefly below.

Carbonisation

Coals containing up to about 15 per cent volatile matter produce little or no tar when heated and may be regarded as smokeless fuels. These coals include anthracites (0 to 10 per cent volatile matter).

Smokeless fuels may be manufactured from coals containing more than 15 per cent volatile matter by heating in the absence of air to remove the tar component of the volatile matter. The process is essentially one of pyrolysis (see section 5.3) although, in the context of smokeless fuel manufacture, is usually referred to as carbonisation.

In the case of coking coals, the coal particles soften and coalesce to form a coherent semi-coke when heated to 400 to 500 °C. A strong lump coke suitable for use as a smokeless domestic fuel can therefore be manufactured from a crushed coking by carbonisation alone, for example, in a slot-type oven.

For coal fines other than for coking coals, carbonisation can reduce the volatile content to produce a smokeless coke or char having approximately the same size distribution as that of the original coal.

Tar is one of the first components to be released. Its evolution commences at about 400°C and is substantially complete by 600 °C. For these reasons, carbonisation to produce smokeless fuels is usually carried out in this temperature range. Temperatures above 800 °C are preferably avoided otherwise the resulting product will not be sufficiently reactive (that is, will not burn easily) on some domestic appliances.

Carbonisation processes may be divided into two main categories: indirectly heated processes and directly heated processes.

Indirectly heated processes are generally batch-operated with the most common type being the slot ovens used for metallurgical coke manufacture (see section 6.8) although the design of this equipment is far from ideal for operating within the preferred temperature range discussed above.

The designs of directly heated processes favoured for lump coal or briquettes are fixed bed kilns or vertical retorts and, for crushed coal, are fluidised bed systems. The processes are usually continuously operated. The heat is generally supplied by passing hot combustion gases through the bed although steam may also be used. Alternatively, air can be passed through the bed to permit some combustion.

Briquetting

As noted above, carbonisation of a coking coal, whether in lump form or crushed, produces a lump smokeless fuel directly. In order to produce a domestic smokeless fuel from fines using a non-coking coal, mechanical compaction, usually to form briquettes, is necessary. This may occur in two ways: with the coal particles or with char particles obtained by carbonising the coal.

The possible approaches to briquetting are further multiplied by the existence of two different techniques: briquetting with a binder and briquetting without a binder.

The most flexible method of producing briquettes is to employ a binder. This acts as an adhesive and holds the particles together. The pitch fraction of coal tar is commonly used as a binder, although coal tar itself, petroleum-derived bitumen or other materials such as sulphite liquor are also used.

Under high pressures, briquetting can sometimes be carried out without using a binder. The feed material must be capable of plastic deformation; the ease with which deformation occurs affects both the pressure required and the strength of the resulting briquette. Suitable materials include hot char produced by the rapid carbonisation of some coals under precisely controlled conditions, lignites, after crushing and drying to an optimum moisture content, and bituminous coal pulverised to less than 0.1 mm.

Various types of briquetting equipment are available; the most common are listed in table 6.3. In general, the higher the maximum operating pressure, the more costly the equipment. It is therefore usual to employ the lowest pressure type of equipment suitable for the feedstock. For completeness, pelletising processes are included in the table. These rely on agglomeration of a 'soft' feedstock without the imposition of direct mechanical pressure but produce a product that is comparatively weak. Although a number of such processes have been employed in the past, there appears to be little current interest in this approach for the production of smokeless fuel.

Table 6.3 Types of briquetting equipment

Description	Pressure (MN/m^2)	Typical applications
Ring roll press	150–200	Bituminous coals and lignites without a binder
Reciprocating extrusion press	100–150	Lignites and hot char without a binder
Double roll press	About 30	Hot char without a binder and any feed with a binder
Pelletising	Not applicable	Hot char

Following their manufacture, the 'green' briquettes may require further heat treatment to improve their strength or to make them smokeless. In the present context, the term 'heat treatment' is taken to exclude a simple heat soak. If the briquetting of a smokeless coal or char has been carried out with a binder, heat treatment to about 300 °C is always necessary to obtain optimum strength and to render the binder smokeless. The briquettes may be carbonised to higher temperatures instead but this reduces the binder strength and so some further cohesion of the particles in the briquette is generally required in compensation. If briquetted without a binder, the briquettes may be suitable for direct use if the coal or char is smokeless; otherwise, carbonisation of the briquettes is usually employed.

The various options for obtaining smokeless domestic solid fuels are summarised in figure 6.3. As shown in the figure, these options differ mainly in the order of the briquetting and carbonisation stages. For simplicity, processes involving the crushing of lump coal, or coke manufactured from lump coal, are considered with the equivalent process based on coal fines.

The choice of process is, however, related closely to the properties of the coal feed, as illustrated in figure 6.4. The figure indicates (as shaded areas) process/coal combinations that are either unnecessary or not technically possible at present. The main classes are as follows.

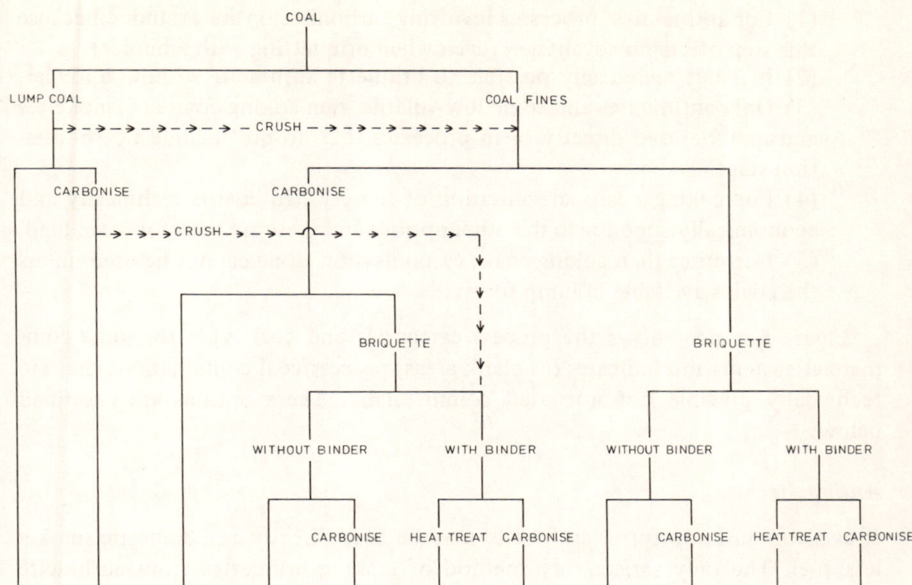

Figure 6.3 Classification of processes for smokeless fuel and formed coke manufacture. Shading indicates those process/coal combinations that are impractical or uneconomic

PROCESS OR MATERIAL	ANTHRACITE	LOW VOLATILE NON-COKING	COKING	HIGH VOLATILE NON-COKING	LIGNITE	TWO COMPONENTS
LUMP COAL	DIRECT USE	DIRECT USE (SOME COALS)				
LUMP COAL CARBONISED				REXCO		
CARBONISE			COALITE SUNBRITE			
CARBONISE BRIQUETTE WITHOUT BINDER				HOMEFIRE		ANCIT
CARBONISE BRIQUETTE WITHOUT BINDER CARBONISE						
CARBONISE BRIQUETTE WITH BINDER HEAT TREAT						
CARBONISE BRIQUETTE WITH BINDER CARBONISE						
BRIQUETTE WITHOUT BINDER						
BRIQUETTE WITHOUT BINDER CARBONISE					LURGI SPULGAS	
BRIQUETTE WITH BINDER HEAT TREAT	MULTIHEAT EXTRAZIT					
BRIQUETTE WITH BINDER CARBONISE		PHURNACITE HBNPC				

Figure 6.4 Process/coal combinations used for the manufacture of domestic smokeless fuel

(1) For anthracites, processes involving carbonisation are excluded because this step offers no advantages (even when briquetting with a binder).
(2) It is not technically possible to briquette anthracite without a binder.
(3) Only anthracites and some low-volatile, non-coking coals are smokeless and may be used directly or in processes that do not include a carbonisation stage.
(4) For coking coal, carbonisation of the crushed coal is technically and economically superior to the other options and these are therefore excluded.
(5) For other than coking coals, carbonisation alone cannot be used unless the coal is available in lump form.

Figure 6.4 also shows the process categories and coal types for some commercial systems and indicates (as blank areas) process/coal combinations that are technically possible but not used commercially. These options are described below.

Anthracite

If available in lump form, anthracite may be used directly as a domestic smokeless fuel. The only satisfactory method of making briquettes from anthracite fines (known as anthracite duff) is to use a binder. In general, the briquettes have to undergo heat treatment to achieve maximum strength and, in some cases (for example with a pitch binder), to render the binder smokeless.

Low-volatile, non-coking coal

Some of the lower-volatile matter content coals in this category (for example, dry steam coals) may be used directly in lump form as domestic smokeless fuel. For coal fines, the main approach employed is to briquette with a binder and then to carbonise the briquettes. A number of other 'technically possible' options arise but have not been implemented as commercial processes. In principle, many of the processes used for high-volatile, non-coking coals and lignites may be applicable but there is little interest in such developments because of the comparatively low production and reserves of low-volatile non-coking coals.

Coking coal

The only method used for manufacturing smokeless fuel from coking coal is carbonisation of the crushed coal, the process usually being carried out in indirectly heated slot ovens or retorts.

High-volatile, non-coking coal

Smokeless fuel can be manufactured by carbonising lump high-volatile, non-coking coal in retorts. This approach has been employed in processes (Rexco, for example) designed for the production of smokeless fuel.

For manufacturing domestic smokeless fuel from fines of a high-volatile, non-coking coal, interest has been directed mainly towards processes such as Homefire in which the coal is carbonised and the char is briquetted hot without a binder. Other approaches are also possible, such as the application of the HBNPC process to high-volatile, non-coking coals and the application of the FMC formed coke process to domestic smokeless fuel manufacture. In the latter case, however, processing of the briquettes would probably be confined to heat treatment rather than to carbonisation.

Lignites

Owing to their high-volatile matter, lignites produce a relatively weak coke if carbonised in lump form although the product may be acceptable in some cases. The main method of producing a smokeless fuel from lignites is, however, to crush and briquette. Since lignites are comparatively soft, briquetting is carried out without a binder, after the moisture content has first been reduced to 15 to 20 per cent. The green briquettes require carbonising to produce a smokeless fuel.

Two components

In the options discussed above, it is implicitly assumed that the feed material is either a single coal type or a homogeneous blend of coals of different types that behaves in the processes like a single coal type. However, processes have been developed which may, for simplicity, be regarded as having two separate feed coals. These are subjected to different treatments in the early stages and the streams are combined later in the system. Such processes are not conveniently included in the same categories as those for single-component feeds.

Two-component processes generally employ a coking coal and a non-coking component. The non-coking component may be either anthracite (for example, the Ancit process) or bituminous coal (for example, the BFL process for formed coke manufacture).

The non-coking component is heated or carbonised to about 700 °C and then mixed with the coking component. The temperature of the mix is about 500 °C and is sufficient for the coking component to become plastic and to act as an adhesive for the non-coking component. The mixture is briquetted hot, usually in a roll-press. The green briquettes may require heat treatment or carbonisation to increase their strength.

Special appliances

The manufacture of smokeless fuels represents one approach to meeting the need for environmentally acceptable solid fuel combustion in the domestic sector. An alternative adopted by the NCB is the development of a new range of domestic combustion appliances that are able to burn bituminous coal with sufficient smoke reduction to satisfy emission regulations (in the United Kingdom).

In conventional domestic appliances, the air enters at the bottom of the bed and passes upwards in counter-current flow to the coal which is fed to the top. Such a design is called an 'updraught' appliance. As the coal descends through the bed, it is heated and, above about 400 °C, begins to release volatile matter. The volatile matter is swept upwards by the air into the cooler parts of the bed and is chilled to below its ignition temperature. Combustion of the volatiles is therefore incomplete, resulting in the emission of smoke if a bituminous coal is being burned.

In domestic appliances developed by the NCB, the coal is burned by 'downdraught' combustion, as illustrated in figure 6.5. Air is fed to the top of the firebed and passes downwards through the bed in co-current flow with the coal. The volatile matter released as the coal is heated up is carried by the air into the region at the bottom of the bed where the temperature is highest. Here, additional air is injected to ensure that combustion of the volatile matter is completed and the resulting flue gases are free from smoke.

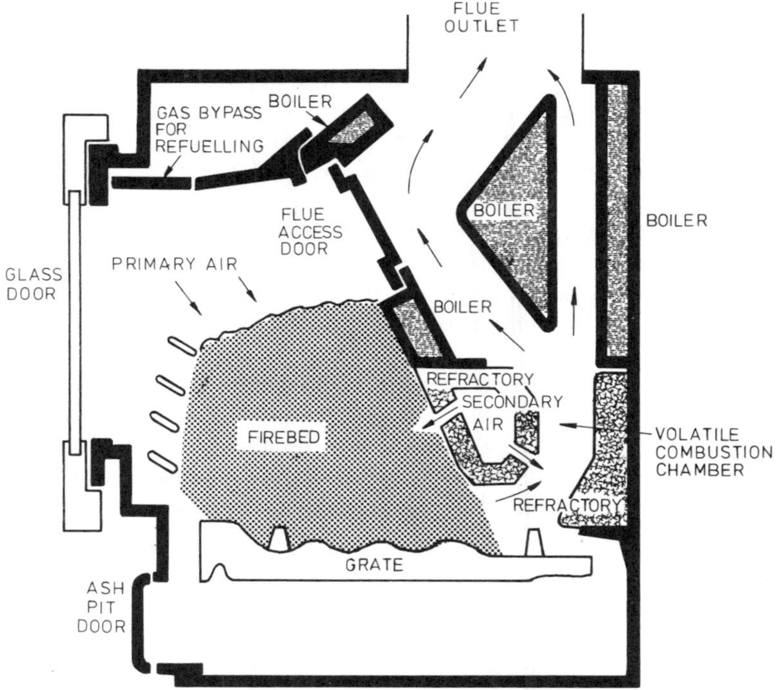

Figure 6.5 A bituminous coal-burning domestic smokeless appliance

The design of the combustion chamber and the choice of air flows are both important to the successful operation of the appliance; the requirements for efficient volatiles combustion are time, temperature and turbulence. In practice, it is found that a mixture of volatile matter and air has to be maintained above 600 °C for at least 0.5 s to achieve complete combustion.

The concept of downdraught combustion is not new and dates from Dalesme's heating machine of 1680 (see figure 6.6). The principle was later rediscovered and patented by James Watt in 1785.

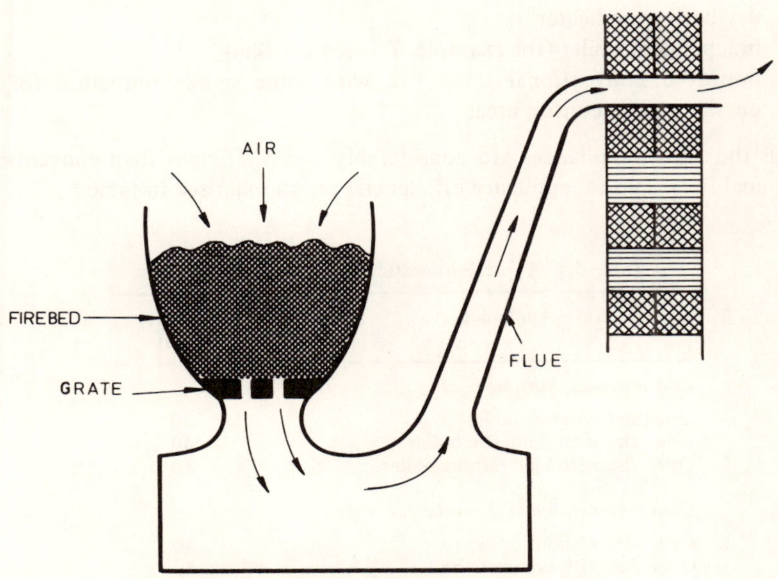

Figure 6.6 Dalesme's heating machine (1680)

The first generation of smoke-reducing appliances became available in the early 1970s. These were continuous-burning closed roomheaters with high-output back boilers and were capable of providing full central heating for a small two or three bedroom house. The design was, however, complicated and expensive and, in particular, incorporated a fan to provide the combustion air. The grade of coal was restricted to singles (12.5 to 25 mm) preferably of weakly caking or non-caking quality.

A range of second-generation appliances was subsequently developed to overcome some of the disadvantages of the early models. These later appliances are of a simpler and cheaper design and operate solely on natural chimney draught thereby avoiding the need for a fan. All domestic grades of coal (singles, doubles and trebles) can be burned and the appliances are not as sensitive to the caking properties of the coal as the earlier designs.

The first model of the second generation of smoke-reducing appliances, the Rayburn Prince 76, became available in 1975. There are now appliances of this type available, or in an advanced stage of development, for all of the main categories of domestic solid fuel systems, including

open roomheater with high-output back boiler (for example, Rayburn Prince 76)
closed roomheater with high-output back boiler (for example, Rayburn Coalglo C30)
dry back roomheater
independent boiler (for example, Trianco Coalking)
improved conventional open fire with some smoke reduction for use outside smoke control areas

All the above appliances are considerably more efficient than conventional open coal fires. Typical appliance efficiencies are summarised in table 6.4.

Table 6.4 Typical domestic appliance efficiencies[a]

Appliance	Efficiency (%)
Conventional, burning coal	
Open fire without boiler	30
Open fire with domestic boiler	40
Open fire with high output boiler	50
Conventional, burning smokeless fuel	
Open fire without boiler	40
Open fire with domestic boiler	50
Open fire with high output boiler	60
Closed roomheater without boiler	65
Closed roomheater with boiler	75
Independent boiler	65–75
Smoke-reducing appliances burning coal	
Openable roomheater with boiler	55–60
Closed roomheater with boiler	70–75

[a] Sources: *Centrepiece No. 13*, published by the Solid Fuel Advisory Service, Hobart House, Grosvenor Place, London, 1975; R. C. Payne, Developing Coal Burning Appliances for Clean Air, *Build. News*, 1979.

District heating

In district heating schemes, a large number of individual dwellings, often many thousands, are heated by hot water (or, in a few cases, steam) piped from a central heat source. The heat may be obtained from a boiler (a 'heat only' system) or may be heat rejected from a power generation system. The latter case is known as 'combined heat and power' and is discussed further in section 6.7.

Commercial district heating schemes have existed on a small scale since the end of last century; the New York steam supply system was started in 1882, followed by Hamburg in 1894. Some major district heating schemes were initiated between the two World Wars, including the systems in Moscow, Leningrad and West Berlin. However, most of the present capacity was built after the second World War, including all the systems in Sweden, Finland and Poland. The popularity of district heating continues to increase, particularly in continental Europe.

Despite the improvements in domestic fuels and appliances described in the preceding sections, coal (used directly) remains at a disadvantage when compared with other methods of domestic heating because of its comparative inconvenience and low level of amenity. For this reason, the use of coal as the fuel for district heating schemes has the attraction that it permits coal to supply the domestic heating market with levels of amenity and convenience similar to those of the competing forms of energy. Although district heating schemes may also be fuelled by oil, natural gas or even nuclear power, coal's increasing competitiveness in industry may be expected to lead to its acquiring a substantial share of this potentially important part of the domestic heating market.

6.6 Industrial market

Improvements to conventional stokers

Over recent years, the development of new combustion systems for the industrial market has been dominated by fluidised bed combustion (see chapter 3). However, conventional stoker systems possess some inherent advantages over fluidised bed combustion. These are summarised below.

Start-up

A stoker can be started using a simple kindling system (or electrical ignitor, see later) whereas a fluidised bed combustor requires an oil or gas burner, generally rated at more than 20 per cent of the total boiler output. This adds significantly to the cost and complexity of the fluidised bed combustor and its associated control system.

Fan power

Since the pressure drop across a stoker grate and bed is small compared with that across a fluidised bed, distributor plate and cyclones, fan power requirements are low.

Gas clean-up

The high temperature in a stoker bed causes sintering of the ash with the result that entrainment in the gas stream is low. Whereas a cyclone system is always required for gas cleaning with a fluidised bed combustor (and other devices such as bag filters or electrostatic precipitators may also be necessary), stokers can be operated with simple grit arrestors or, in some cases, no gas clean-up.

These advantages are of particular value in small installations where the additional complexity of a fluidised bed system represents a proportionately greater capital cost penalty. For this reason, there is an incentive to develop improved conventional stokers in parallel with the work on industrial fluidised bed combustors. The main interest appears to be in the United Kingdom where the NCB have initiated a programme concentrating on systems of up to 2 MW and covering the following aspects.

Improved underfeed stokers

Traditionally, underfeed stokers are employed for systems of up to 2 MW but do not offer automatic de-ashing and are comparatively inefficient. The development of an automatic de-ashing system has both improved the level of amenity and resulted in an increase in efficiency by permitting operation at a lower excess air level. Automatic start-up using electrically heated air is also now available.

New reciprocating grate stokers

Based on existing combustion techniques for larger applications, two new reciprocating grate stokers (the ESCOM stoker and the Proctor Mini-Coking stoker) have been designed for boilers up to 2 MW. In both cases, the aim is to produce a compact, high combustion intensity system in order to reduce the capital cost. This is achieved by

- optimisation of the pattern of air distribution
- restriction of the coal feed to weakly caking and non-caking washed singles
- the use of high-chrome stainless steel grate bars for good high-temperature properties

With these features, operation at burning rates of 3 MW/m^2 is possible. This corresponds to an increase of 50 per cent compared with the burning rate commonly achieved using the traditionally conservative design of chain grate stoker and is similar to the burning rate for industrial fluidised bed combustors.

Automatic ignition

Automatic ignition systems using small electrical air pre-heaters have been developed for application to a wide range of conventional stokers. These, together with improved control systems based on microprocessors, are capable of contributing significantly to a higher degree of operational amenity for stoker systems.

Coal-liquid mixtures

Coal-liquid mixtures (often abbreviated to CLMs) are of two main types: coal-water mixtures (CWMs) and coal-oil mixtures (COMs).

Coal-liquid mixtures were first proposed as long ago as 1879 when a wide-ranging United States patent (covering both coal-water and coal-oil mixtures) was awarded to Munsell and Smith. Since then, coal-liquid mixtures have enjoyed periods of interest, notably during the two World Wars, when security of oil supplies was in doubt.

Coal-liquid mixtures now have two main attractions: they are cheaper than fuel oil on an energy content basis (because coal is cheaper than fuel oil) and they are more convenient to use than coal because some of the advantages of liquid fuels (for example, ease of handling and combustion) are retained. For the immediate future, the main incentive for developing coal-liquid mixture technology is to provide a means of converting existing oil-fired boilers to use a cheaper fuel.

Interest in coal-liquid mixtures revived during the 1970s following the oil crisis, and a number of research and development programmes were started.

The initial emphasis was on coal-oil mixtures as these were regarded as presenting fewer technical problems, both in preparation and combustion. Several production plants of more than 100 t/d have been built and numerous demonstration trials carried out. One of the largest programmes was that of Florida Power and Light; coal-oil mixtures were produced in a 1700 t/d plant and burned in a 400 MW(e) boiler designed for oil firing. Coal-oil mixture technology may therefore be regarded as an established and proven technology.

Development work on coal-water mixtures began more recently. The main incentive is that coal-water mixtures are widely expected to be substantially cheaper than coal-oil mixtures. This is simply because they avoid using the relatively expensive fuel oil component.

The potentially attractive economics of coal-water mixtures have encouraged the establishment of development programmes in several countries, notably Sweden, the United States and Canada. So far, production plants exist only up to 300 t/d and boiler trials up to 30 MW(th), with larger facilities being planned in the future.

At present, therefore, coal-water mixture technology is at an earlier stage of development than that of coal-oil mixtures. With limited large-scale operating experience, coal-water mixtures cannot yet be regarded as an established commercial option, although demonstration programmes are progressing rapidly. If these are successful, it may be that coal-water mixtures will largely supersede coal-oil mixtures on economic grounds in the market for coal-liquid mixtures.

The technologies of producing and using coal-water and coal-oil mixtures are outlined in the following sections.

Coal-water mixtures

The main stages in the preparation of coal-water mixtures are as follows.

Pulverisation

The coal is pulverised to standard PF sizes, often by a wet-milling process. In order to obtain a high solids content in the final coal-water mixture, the size distribution can be carefully controlled to provide a high particulate packing density. This usually involves selective grinding to increase the proportion of fine material, giving what is referred to as a 'bimodal' size distribution.

Screening

The pulverised coal is screened, typically at 200 microns, to remove oversized particles which might otherwise block the burners.

Washing

The coal is washed in multi-stage froth flotation cells to reduce the ash and sulphur contents of the fuel. Typical product specifications are an ash content of 1 to 4 per cent (dry basis) and the removal of 70 to 90 per cent of the pyritic sulphur, with a combustibles loss of 1 to 3 per cent.

Dewatering

Dewatering is carried out to reduce the water content to that required in the final fuel. Typically, coal-water mixtures contain about 70 per cent coal and 30 per cent water. With the appropriate choice of size distribution (see above), the solids content can be increased to about 80 per cent.

Fuel preparation

Surfactant additives are mixed with the dewatered slurry in a homogenising vessel. The main purpose of the additives is to stabilise the slurry, that is to inhibit sedimentation during storage. The additives also improve the viscosity of the mixture (generally less than 1000 centipoise in coal-water mixture specifications) and assist atomisation in burners.

Although coal-water mixtures have a comparatively low sulphur content owing to the use of 'deep washing' in their preparation, it may be possible to reduce sulphur emissions further by incorporating limestone into the fuel. This possibility is still at the research stage, but could be attractive because the flame temperature when firing coal-water mixtures is lower than for conventional oil or PF flames.

It may be noted that coal-water mixtures for industrial use, as described above, differ from those used for slurry pipeline transport (see section 6.3) in employing pulverised coal rather than crushed coal. Also, the coal/water ratio is higher in a coal-water mixture and the fuel is fired directly, without dewatering.

The main market for coal-water mixtures in the near-term is the conversion of existing oil-fired boilers to coal-firing, and most of the current development effort is being directed towards this objective. The most important sectors of the market are power generation and industrial water-tube boilers, where the costs for replacement boilers are high, the lifetime of the existing boiler stock is generally long and the load factor (and therefore fuel costs) is high.

Coal-water mixtures may also have a long-term market as a fuel for new boiler installations. One perspective is to regard coal-water mixtures as a convenient method of storing, transporting and handling pulverised fuel. They may therefore be attractive in situations where on-site preparation of pulverised fuel has to be avoided for reasons of amenity or the economies of scale, and where other considerations (safety, for example) preclude the central preparation and distribution of pulverised fuel.

In particular, centralised preparation and reduced handling costs may make coal-water mixtures attractive for use at smaller boiler sizes than is the case for pulverised fuel. For large fire-tube boilers, there may be an additional advantage. Conventional methods of coal firing (both stokers and fluidised bed combustion) suffer from a reduction in heat transfer to the fire-tube because of shielding by the grate or distributor plate. In common with oil or gas flames, heat transfer from a coal-water mixture flame would take place over all the available fire-tube area.

Coal-water mixtures are also a convenient method of feeding coal continuously into pressurised reactors, and are employed for this purpose in the Texaco gasifier. Other applications of this type, such as pressurised fluidised bed combustion, may also prove to be attractive.

With suitable preparation, the storage, handling and combustion characteristics of coal-water mixtures are, in many ways, similar to those of fuel oil. However, the conversion of existing oil-fired boilers to coal-water mixtures involves some changes, particularly in the following respects.

Fuel storage and feeding

In most cases, existing storage equipment can be used, although it may be necessary to introduce modifications to protect against corrosion. New pumping equipment is generally required, and precautions against freezing in winter may be needed.

Burners

It is necessary to install new burners when changing from oil-firing to firing with coal–water mixtures. Several development programmes are in progress and the combustion of coal–water mixtures has been demonstrated with both air-atomised and steam-atomised designs.

Some decrease in efficiency when burning coal–water mixtures is inevitable because of the additional latent heat loss associated with the water in the mixture. This loss is, however, small, being less than 5 per cent of the gross calorific value for a 70 per cent coal, 30 per cent water mixture.

Boiler derating

The presence of coal ash in a coal–water mixture fuel means that some derating may be necessary for boiler applications. There are two main reasons for this (see also chapter 3).

(1) The gas exit temperature from the combination chamber has to be reduced to minimise fouling and slagging on the convective section of the boiler.

(2) Gas velocities through the convective section of the boiler have to be reduced in order to avoid erosion of the heat transfer surfaces by the ash.

The amount of derating depends on a number of factors, the most important being the ash content, the ash abrasiveness and the boiler design. For a fuel having a low content of non-abrasive ash, it may be possible to avoid derating with an oil-fired boiler of conservative design. Under adverse conditions, derating by 30 to 40 per cent may be required.

The effect of the ash content on derating emphasises the importance of the 'deep washing' stage of preparation. For the same reason, only coals that have a low inherent ash content may be regarded as suitable for use in preparing coal-water mixtures.

Even under conditions where a significant derating is required, this may not represent an obstacle to the introduction of coal–water mixtures. Many industrial sites have excess boiler capacity, sometimes because of conservation measures. In other cases, dual firing may be possible, so that oil can be used to meet peak demands.

Ash collection

In order to satisfy atmospheric particulate emission requirements, coal–water systems require the installation of equipment such as electrostatic precipitators or bag filters to remove the ash from the combustion gases. This can add significantly to the capital cost of conversion. Soot blowers are also required to permit the periodic removal of ash deposits from heat transfer surfaces and an ash offtake must be provided from the bottom of the combustion chamber.

Coal–oil mixtures

The preparation of coal–oil mixtures is usually a simpler process than that for coal–water mixtures described in the previous section. The 'deep washing' stage is normally omitted and either conventional dry PF is prepared and then mixed with the oil, or the feed coal is mixed with the oil and milled in suspension.

Typically, a coal–oil mixture contains 30 to 50 per cent coal, compared with 70 to 80 per cent coal for a coal–water mixture. As with coal–water mixtures, stability is an important consideration. Water appears to be a good stabilising agent at concentrations up to about 5 per cent (mass basis). Dispersing agents, emulsifiers and thickeners are sometimes also used. A stable fuel can be produced more easily with a fine particle size distribution.

The presence of oil in the fuel makes the combustion properties of coal–oil mixtures nearer to those of fuel oil than is the case for coal–water mixtures. For this reason, combustion technology for coal–oil mixtures is usually regarded as easier. Also, the relatively high calorific value per unit volume of coal–oil mixtures means that storage requirements are less than for coal–water mixtures.

For similar reasons, the ash content of a coal–oil mixture may be lower than that of the corresponding coal–water mixture, when expressed per unit of energy content. Derating is still generally necessary, although the amount of derating may be somewhat reduced. However, the installation of ash collection and handling equipment is usually required when converting from oil to coal–oil mixtures.

If a low sulphur coal is available, a coal–oil mixture may have the advantage of a lower sulphur content than the fuel oil. Conversely, the use of a high-sulphur coal could result in an increase in the sulphur content of the fuel.

Coal-fired diesel engines

More than 50 years ago, Pavolikowski, an associate of Diesel, experimented with powdered coal as an engine fuel. However, the greater convenience and lower costs of liquid fuels led to development work on diesel engines being concentrated in this direction. Although interest has remained, the problem of erosion and corrosion of mechanical components arising from the presence of ash in the fuel has so far prevented commercial exploitation using coal.

With present fuel prices, however, there is now an economic incentive to change from oil to coal-firing of diesel engines and development efforts have been renewed over recent years.

Technically, the best prospect appears to be the conversion of slow-speed, two-stroke diesels. This is partly because the long residence times permit the efficient combustion of a slow-burning fuel such as coal and also because the use of exhaust ports, rather than valves, reduces the effects of corrosion and erosion.

Slow-speed, two-stroke diesels usually burn heavy residual fuel oils and are available in the larger size ranges (from about 6 to 40 MW) for marine and stationary land-based power sources.

Development work on coal-fired diesels was carried out in Germany in the 1930s. It was found that changes in the piston ring/cylinder liner metallurgy were necessary in order to reduce the wear rates on mechanical components. Much of the work concentrated on designs in which pulverised coal was injected into a pre-chamber early in the compression stroke by using a low-pressure injector.

More recently, the advantages of using micronised fuel (that is, coal that has been reduced to a maximum particle size of about 5 microns, see section 6.2) have gained recognition for the following reasons:

Using micronised particles makes it easier to obtain complete combustion within the short residence time of about 0.1 s available (in slow speed diesels).

Micronising the coal releases much of the inherent mineral matter as discrete particles (see above). This enables washing to reduce the ash content to a low level, less than the maximum value of 1.5 to 3 per cent thought to be acceptable for diesel engines. Furthermore, the fine particles of ash remaining in a micronised fuel may be expected to be less erosive than coarser particles.

The micronised fuel may either be used dry, or as a coal–oil or coal–water mixture. The use of coal–oil or coal–water mixtures appears to be favoured because of their advantages in handling and injection into the engine.

Engine modifications thought to be necessary include

improved design of fuel injectors, particularly to operate at the high pressure part of the stroke

changes in piston ring lubrication and exhaust port (or valve) design to limit corrosion and erosion

the introduction of particulate collection equipment to limit emissions to the atmosphere

Although interest in coal-fired diesel engines has revived comparatively recently, there is already some optimism that technical solutions can be found. The use of micronised fuel for medium-speed diesels (and small gas turbines) is less attractive than for slow-speed diesels because of the greater technical problems although these applications remain longer-term possibilities. Pressurised fluidised bed combustion is the preferred approach for direct coal firing of large gas turbines (see chapter 3).

If the current development programmes for coal-fired diesels are successful, the comparatively low fuel costs of these engines will make them more competitive with gas turbines, particularly for industrial applications (including combined heat and power). Coal-fired diesels for marine applications are a further possibility although these would be in competition with conventional boilers and steam turbines for coal-fired ships (see next section).

Coal-fired ships and locomotives

The general interest in the replacement of oil by coal for bulk combustion has resulted, naturally, in attention being directed towards the traditional markets for coal, two of the most important of which were ships and locomotives. Clearly, the traditional technologies could be re-introduced if the economic conditions were appropriate and environmental constraints could be met. However, new technology may be more attractive.

For ships, the principal method of propulsion at present is diesel engines which are used in about 90 per cent of the world's shipping fleet. The remainder are powered mainly by oil-fired steam cycles or gas turbines. For coal-fired ships, steam cycles are the only option available in the short term. These typically have an efficiency of 25 per cent for marine applications as compared with over 40 per cent for modern marine diesels.

The first of the new generation of coal-fired ships is expected to employ conventional stokers burning smalls. In particular, spreader stokers are widely favoured because of their high combustion intensity, good load-following capability, flexibility to coal type and the considerable past experience of their use in marine service.

Other coal-based combustion technologies, including the following, may subsequently be employed.

Pulverised fuel

Although this is a well-developed technology for land-based applications, there are disadvantages for marine applications in storing and handling the fuel if prepared on shore (particularly in minimising the risks of fire or explosion). Also, the costs of accommodating pulverising equipment on board may be prohibitive.

Coal–liquid mixtures

These offer a convenient and safe method of storing, handling and firing coal for marine boilers. In particular, reconversion to fuel oil may be possible at ports where coal–liquid mixtures are not available.

Fluidised bed combustion

The main advantage of fluidised bed combustion for marine applications is flexibility to coal type. This is particularly important because of the variations in the quality of the coal available at ports around the world. Special design features would be required, for example, the installation of vertical baffles in the bed to reduce the effect of ship motion on the fluidised bed.

In the future, synthetic liquid fuels manufactured from coal may be used in marine diesels. Conventional diesel fuels could be manufactured for high-speed and medium-speed diesel engines but, for the slow-speed engines used in the larger ships, it may be more economic to produce a low-ash solvent-refined coal. In either case, the high efficiency that can be obtained from a diesel engine as compared with a steam cycle is offset by the efficiency losses in the manufacture of synthetic fuels. A further option, particularly for slow-speed two-stroke engines, may be direct coal firing (see above). This may involve the preparation of a micronised fuel, possibly as a coal–water mixture, but would avoid the losses inherent in coal conversion processes.

The economics of coal-fired ships depend on a trade-off between running costs and capital costs (see section 8.5). At present, it appears that coal may have some advantage for the larger vessel sizes. The main applications are expected to be bulk carriers, particularly of coal and grain.

Direct coal firing appears less favourable for locomotives than for ships principally because of the need for an extremely compact unit. The traditional design of steam locomotive had an efficiency of only 5 to 7 per cent compared with about 30 per cent for a diesel locomotive. The economics of reverting to the traditional steam locomotives are, therefore, unlikely to be favourable even with a substantial diesel/coal price differential. It may be possible to increase the efficiency of steam locomotives using modern technology; efficiencies of 10 to 14 per cent have been claimed for new systems. Proposals include the use of cyclone combustors, fluidised bed combustors and two-stage gasification/combustion systems. However, even if such improvements can be achieved, the following other ways of employing coal to power the railway system exist and may be more attractive.

Electrification

The overall efficiency of an electric locomotive based on coal-fired power generation would be about 30 per cent.

Synthetic fuels

Diesel fuel can be manufactured from coal (see chapter 5) and could be used in conventional diesel locomotives. An overall efficiency of about 20 per cent should be possible.

6.7 Power generation market

Flue gas desulphurisation

Interest in advanced power generation systems such as fluidised bed combustion and gasification derives partly from the prospect of an increase in efficiency and partly from their ability to reduce emissions of sulphur dioxide to the atmosphere. As noted in chapter 3, the pulverised fuel systems used for almost all coal-fired power stations are a highly developed technology but the only commercially available method of reducing sulphur dioxide emissions is flue gas desulphurisation (FGD). Pulverised fuel combustion with FGD therefore represents an alternative to the advanced power generation systems discussed above for sulphur retention in new stations, and retrofitting FGD plant is one of the main options for reducing sulphur emissions from existing stations.

The first major FGD plant was built at Battersea power station, London, in 1932 and employed a simple aqueous scrubbing system using water from the River Thames. The effluent was returned to the river after processing. This plant, together with the subsequent plants built at Fulham, Bankside and North Wilford, has not been operated for a number of years.

Recent interest in FGD can be traced to the late 1960s when Detroit Edison demonstrated the technology on two full-scale power station boilers. There are now several hundred FGD plants in operation, although these are confined largely to Japan and the United States.

The process that has dominated recent commercial experience is lime/limestone scrubbing. This is sometimes referred to as the 'first generation' FGD technology (the installations in the 1930s being excluded). A typical design is illustrated in figure 6.7.

After dust removal, the combustion gases are passed upwards through a scrubbing tower in counter-current flow to a spray of limestone/water slurry or lime/water slurry. The choice between limestone and lime depends on the location and the degree of sulphur removal required. Lime is cheaper to transport (less mass per unit of calcium) and more reactive but the costs of calcination and hydration are significant. Recent designs generally favour the use of limestone.

The slurry is removed from the bottom of the scrubbing tower to a hold tank to permit completion of the reaction between sulphurous acid and the sorbent. Some of the slurry is recycled to the scrubbing tower and a purge stream is taken to a dewatering plant and then to a settling pond for disposal. Excess water from the dewatering plant and the settling pond is recycled to the scrubbing tower. As discussed in section 7.2, there is concern about the environmental problems associated with the disposal of the thixotropic calcium sulphite sludge which results from lime/limestone scrubbing processes.

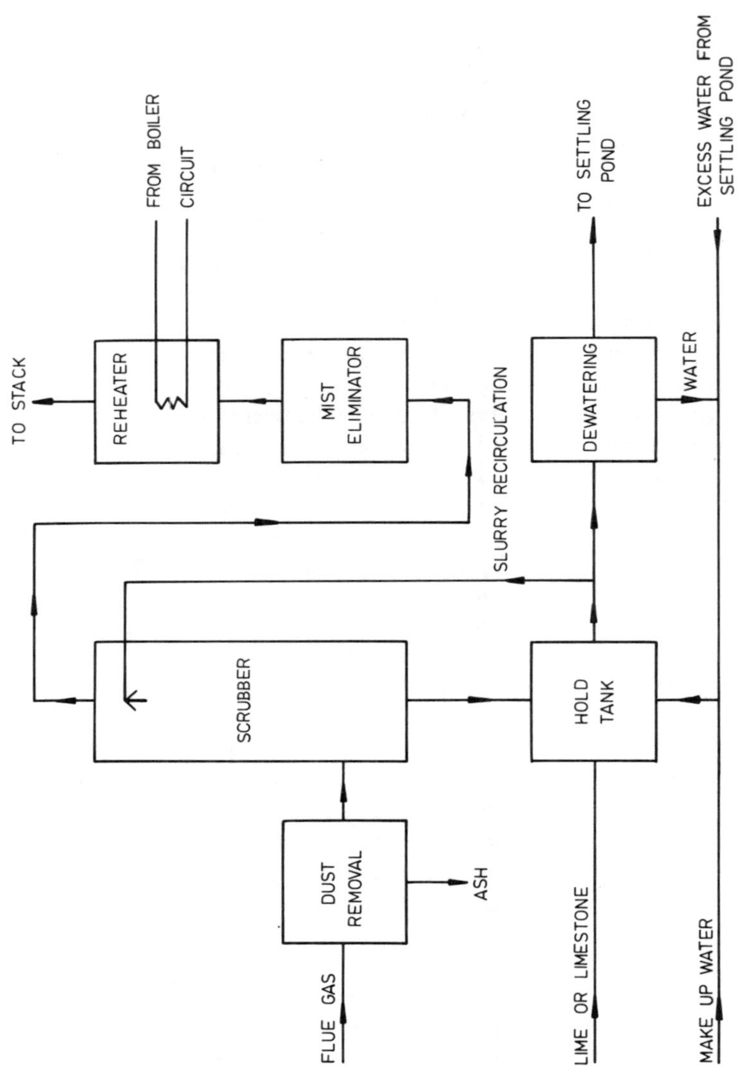

Figure 6.7 Lime/limestone scrubbing plant

The desulphurised gases leave the scrubber at about 50°C and, after passing through a mist eliminator, are reheated to provide the plume with sufficient buoyancy before emission to the atmosphere. Usually, reheating is accomplished by placing tube heaters in the gas stream and using water or steam from the boiler circuit.

In early designs of lime/limestone scrubbing FGD plant, plugging, scaling and corrosion in the slurry recirculation circuit were important problems and led to poor availability of the plant. However, evolutionary developments have much improved the performance for recent systems.

A large number of variations on the above design are possible and include no pre-removal of dust, omission of the hold tank, no recycle of water from the settling pond (this is referred to as the 'open-loop' system) and no reheating. Reheating is particularly important in the economics of FGD because of its high energy requirements. Opinions differ on the amount of reheating required to avoid the environmentally undesirable phenomenon of the plume returning to ground level too quickly.

The disadvantages of the lime/limestone scrubbing concept have prompted work on numerous other designs of FGD plant. Over fifty processes are in various stages of development or commercial application. Although most of these are at an early stage, several are in use commercially and may therefore be regarded as 'second generation' alternatives to the first generation of lime/limestone scrubbing processes.

The developing FGD technologies can be classified in several ways, the most common being whether the process is non-regenerable or regenerable, and whether the sulphur removal stage results in a wet or dry product. The main approaches are discussed below.

Non-regenerable processes

Non-regenerable processes operate by consuming a sorbent which is then disposed of with the sulphur. This approach is therefore economic only if the sorbent is cheap. For this reason, attention has concentrated on the use of limestone (or lime). Processes have been developed to use as sorbent sodium carbonate or bicarbonate where naturally occurring deposits exist, or fly ash where this has good sulphur retention properties. However, such processes are of limited applicability.

In addition to lime/limestone scrubbing, non-regenerable processes involving wet sulphur removal include gypsum processes, double alkali processes and sea water scrubbing.

For gypsum processes, the first stage is similar to lime/limestone scrubbing processes in that the sulphur dioxide is removed by scrubbing with a lime or limestone slurry. The calcium sulphite is then oxidised with air to form calcium sulphate (gypsum) which is removed as a slurry and dewatered. The main advantage of gypsum processes is that they avoid the production of the unstable sulphite sludge, giving instead a product that can be sold or disposed of as landfill. This approach is of particular interest in Japan which has no indigenous supplies of gypsum. Oxidation units can be added to conventional lime/limestone scrubbing systems, and many such units are in operation in Japan. Integrated processes designed for gypsum production include the Chiyoda Thoroughbred 121 process (using a limestone slurry as sorbent and operational on a 500 MW scale) and the Saarberg–Hoelter process (using a lime slurry as sorbent and operational on a 175 MW scale).

Double alkali processes use sodium salts (mainly sodium sulphite) as alkalis for scrubbing the combustion gases. The sodium salts are then regenerated by reaction with lime in a second stage, the sulphur being removed as a calcium sulphite/calcium sulphate sludge. The main advantages claimed for this two-stage approach are avoidance of plugging and scaling problems, reduced energy consumption and a better waste product. These features are offset by the cost penalties associated with the complexity of the process and the loss of small quantities of the relatively expensive sodium salts.

For coastal power station sites, the use of sea water for aqueous scrubbing with discharge back into the sea after neutralisation with limestone would be possible, following the approach of the first FGD plants built in the 1930s. Although the costs would be low, the environmental effects of the effluent on marine life are questionable.

Non-regenerable dry removal processes employ the technique of spray drying. This generally involves scrubbing the combustion gases with a concentrated lime slurry. The water content is sufficiently low for the sensible heat in the combustion gases to dry the sorbent completely before reaching the exit of the spray tower. The sorbent is collected at the bottom of the tower and in subsequent bag filters. In some designs, the combustion gases are cleaned using electrostatic precipitators prior to the FGD plant in order to recover fly ash separately as a saleable product. The costs of spray drying systems appear attractive compared with those of conventional lime/limestone scrubbing, and the product is more easily disposable. However, a disadvantage of spray drying is that the sorbent utilisation is lower than for lime/limestone scrubbing, and this may mean that it is uneconomic for high sulphur coals.

Regenerable processes

In regenerable processes, the sulphur extracted from the combustion gases is converted into a saleable product, generally elemental sulphur, sulphuric acid or liquid sulphur dioxide. The main advantages compared with non-regenerable processes are that, except for make-up quantities, the costs of sorbent supply are avoided and the disposal problems of the spent sorbent are eliminated.

Regenerable wet removal processes employ an aqueous solution of an effective sorbent to scrub the combustion gases. The sorbent is then regenerated in a separate reactor to produce a concentrated stream of sulphur dioxide for purification or conversion into other chemicals.

The most widely used system of this type is the Wellman–Lord process, which has commercial operating experience in over 30 installations. The process uses sodium sulphite as the sorbent, and regeneration is accomplished by steam stripping in an evaporator–crystalliser.

An alternative approach is the magnesium oxide process in which an aqueous slurry of slaked magnesium oxide is used as sorbent. This is regenerated by calcination in a reducing atmosphere at about 800°C. Other regenerable wet removal processes are being developed but are currently at a comparatively small scale. These include the citrate process (using an aqueous solution of sodium citrate as sorbent) and the ammonia scrubbing process (using aqueous ammonia as sorbent).

Non-regenerable dry removal processes are at an early stage of development. They involve contacting the combustion gases with a fixed or moving bed of solid sorbent which can be regenerated to produce a concentrated sulphur dioxide stream. Sorbents include supported metal oxides (copper oxide on alumina in the Shell–UOP process) and active carbon (the Bergbau-Forschung process).

The various types of FGD plant are summarised in table 6.5.

Table 6.5 Summary of flue gas desulphurisation processes[a]

Non-regenerable	Wet	Lime/limestone scrubbing Gypsum Double Alkali Sea water scrubbing	 Chiyoda Thoroughbred 121 Saarberg–Hoelter
	Dry	Spray drying	
Regenerable	Wet	Wellman–Lord Magnesium oxide	
	Dry	Supported metal oxides Activated carbon	Shell–UOP Bergbau-Forschung

[a]Based on *Control of Sulphur Oxides from Coal*, Technical Information Service Report ICTIS/TR21, IEA Coal Research, London.

Magnetohydrodynamics

If a partially ionised (and therefore electrically conducting) gas is passed through a magnetic field, an electric current is generated and may be extracted by placing electrodes in the gas stream at right angles to the magnetic field. This simple and elegant concept for power generation is known as magnetohydrodynamics (MHD) and was suggested by Faraday in 1832. However, the process remained of academic interest until the 1950s when materials technology had progressed sufficiently for potentially practical generation schemes to be considered.

Although only 0.1 per cent of the gas need be ionised to obtain an adequate electrical conductivity, achieving this just by heating the gas would require a temperature of several thousand degrees. It is therefore necessary to 'seed' the gas with an element having a low ionisation potential. Potassium is generally preferred because of its good availability (caesium has a lower ionisation potential but is prohibitively expensive for commercial plant). The seed is usually introduced as the carbonate since this permits retention of the sulphur in the fuel as potassium sulphate. Typically, the amount of seed required is about 35 per cent of the mass of coal burned and gives a potassium concentration of 1 per cent (mass basis) in the combustion gases.

Even with the addition of the potassium carbonate seed, it is necessary to operate at a temperature of about 2500 °C for sufficient ionisation to occur. Such high temperatures cannot be achieved by coal combustion under conventional conditions. The preferred approach appears to be to pre-heat the combustion air to about 1400 °C although oxygen enrichment is also a possibility.

In addition to electrical conductivity, two other factors have important effects on the output from an MHD generator.

Gas velocity

A high gas velocity through the MHD generator is required; a typical value is 1000 m/s in present design concepts.

Magnetic field

The magnetic field strength also has to be high and, for this reason, superconducting magnets are generally specified.

Although a number of MHD configurations have been considered for coal-based applications (including closed-cycle systems and open-cycle systems employing coal gasification), the direct combustion, open-cycle concept appears to be the preferred option. A typical such system is illustrated in simplified form in figure 6.8.

Coal is fed to a combustion chamber and burned in air compressed to about 10 bar and pre-heated to about 1400 °C. Most of the ash in the coal is removed from the combustion chamber as slag. The combustion gases are then mixed

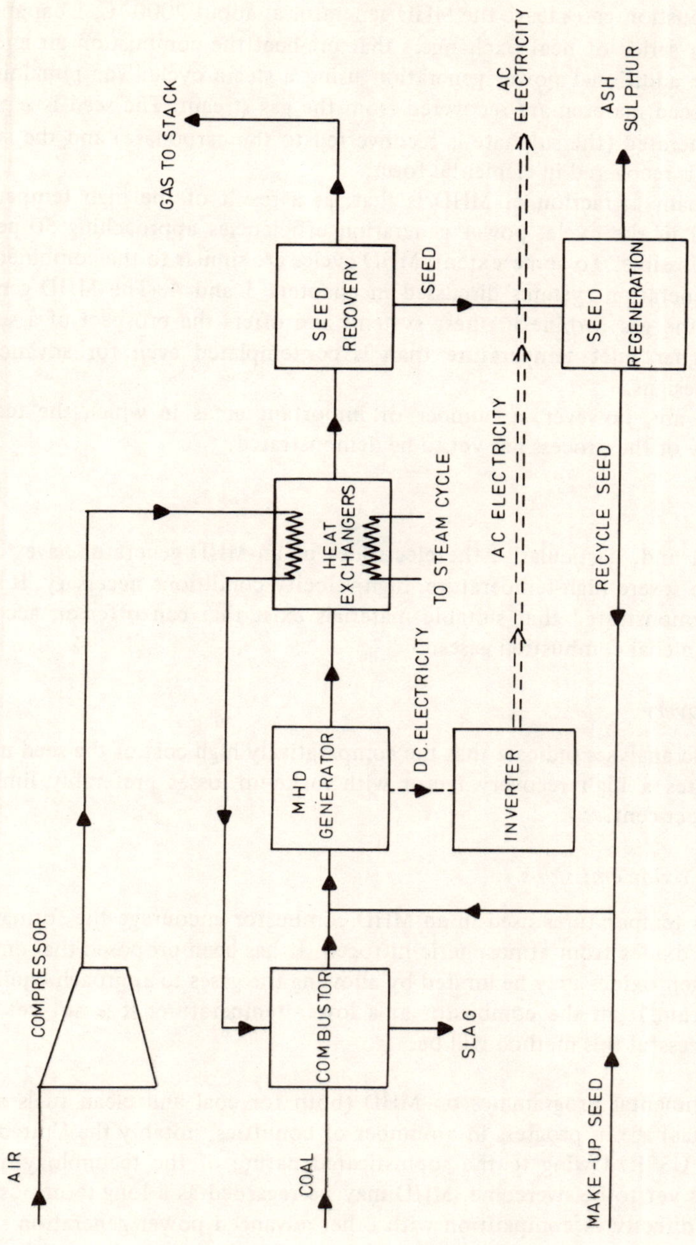

Figure 6.8 MHD power generation system

with the seed material and passed through the MHD generator. Direct-current electricity is generated and is converted to alternating current by an inverter. The combustion gases leave the MHD generator at about 2000 °C, 1 bar and pass through a series of heat exchangers that pre-heat the combustion air and raise steam for additional power generation using a steam cycle. The remaining ash and the seed material are recovered from the gas stream. The seed is separated and regenerated (the sulphate is reconverted to the carbonate) and the sulphur captured is recovered in elemental form.

The main attraction of MHD is that, as a result of the high temperatures employed in the cycle, power generation efficiencies approaching 50 per cent may be possible. To some extent, MHD cycles are similar to the combined cycle power generation systems discussed in chapters 3 and 4. The MHD generator replaces the gas turbine in these systems and offers the prospect of a substantially higher inlet temperature than is contemplated even for advanced gas turbine designs.

There are, however, a number of important areas in which the technical feasibility of the process has yet to be demonstrated.

Materials

The duct and, particularly, the electrodes of an MHD generator have to withstand the severe high-temperature, high-velocity conditions necessary. It has yet to be demonstrated that suitable materials exist that can offer an acceptable lifetime in coal combustion gases.

Seed recovery

Economic analyses indicate that the comparatively high cost of the seed material necessitates a high recovery factor with make-up losses preferably limited to about 1 per cent.

Nitrogen oxide emissions

The high temperatures used in an MHD combustor encourage the formation of nitrogen oxides from atmospheric nitrogen. It has been proposed that emissions of nitrogen oxides may be limited by allowing the gases to approach equilibrium downstream from the combustor at a lower temperature. It is not yet known how successful this method will be.

Experimental programmes on MHD (both for coal and clean fuels such as natural gas) are in progress in a number of countries, notably the United States and the USSR. Owing to the sophisticated nature of the technology and the problems yet to be overcome, MHD may be regarded as a long-term possibility and not directly in competition with other advanced power generation systems such as combined cycles.

Fuel cells

All processes that involve the conversion of heat energy into mechanical power are limited in their efficiency by the theoretical maximum of the Carnot cycle. In particular, the Carnot cycle efficiency represents an upper bound for each of the methods of power generation discussed so far (including MHD).

In a fuel cell, as in a conventional electrochemical cell, the chemical energy of the fuel is converted directly into electricity (that is, without using a working fluid and a heat engine) thereby avoiding the Carnot cycle efficiency limitation. Unlike a conventional cell, however, a fuel cell operates on a continuous supply of fuel.

The principle may be illustrated by reference to the hydrogen–oxygen fuel cell (the 'hydrox' fuel cell) operating with an acidic electrolyte. At the fuel electrode (anode), the hydrogen loses electrons to produce hydrogen ions (in chemical terms, an oxidation reaction). The electrons flow through the external circuit (doing work) and re-appear at the oxygen electrode (cathode). The hydrogen ions migrate through the electrolyte to the cathode where they combine with oxygen and the electrons to produce water. The process for an alkaline fuel cell is similar except that the internal transfer of charge is achieved by the migration of hydroxyl ions from the oxygen electrode to the fuel electrode. In both cases, the overall reaction may be written as

$$H_2 + \tfrac{1}{2}O_2 = H_2O + \text{electricity} + \text{heat}$$

The fuel cell was first demonstrated in 1839 by Sir William Grove and an improved design was built by Mond and Langer in 1889. However, it was not until 1959 that a system suitable for practical application was demonstrated by Bacon and his colleagues in Cambridge following three decades of dedicated effort. The commercial development of fuel cells was taken up in the United States as part of the space programme and, sponsored by NASA, hydrox cells were used on the Gemini and Apollo spacecraft.

Effort in the United States is now being directed towards the use of this technology for commercial power generation. Although, in principle, a range of feedstocks (including methanol, hydrazine, ammonia and methane) can be used in fuel cells, the two major developments that appear likely to achieve early commercialisation are both of the hydrox type.

Phosphoric acid cells

These typically have carbon-based electrodes impregnated with a platinum catalyst separated by a silicon carbide matrix impregnated with the phosphoric acid electrolyte. The operating temperature and pressure are about 200 °C and 3 to 10 bar respectively.

The phosphoric acid fuel cell is likely to be the first type to become available commercially. A 4.8 MW (electrical) demonstration unit constructed by United

Technologies for Commonwealth Edison in New York is nearing completion and the manufacture of packaged 11 MW (electrical) units is expected to begin by the mid-1980s.

Molten carbonate cells

Molten carbonate fuel cells typically have nickel-based electrodes separated by an alumina matrix impregnated with a mixture of alkaline carbonates. The operating temperature and pressure are 600 to 700 °C and 3 to 10 bar respectively.

The molten carbonate cell is at an earlier stage of development than the phosphoric acid cell and commercial availability is forecast for the 1990s. The advantages claimed for the molten carbonate cell are principally

the high operating temperature permits the waste heat to be recovered as high-grade steam, possibly for use in a steam cycle to improve the overall efficiency

adequate reaction rates can be achieved at the high temperatures used without expensive electro-catalysts

the cell will consume both carbon monoxide and hydrogen thereby avoiding the need to carry out a shift reaction to provide a pure hydrogen feed

Development problems still to be overcome include electrolyte loss and sensitivity to poisoning by sulphur compounds.

Commercial power generation concepts employing the above fuel cells are based on a three-stage process outlined in figure 6.9. The hydrocarbon fuel is first converted into hydrogen and purified. The hydrogen is then reacted with oxygen (as air) in the fuel cell and, finally, the resulting direct current electricity is converted into alternating current.

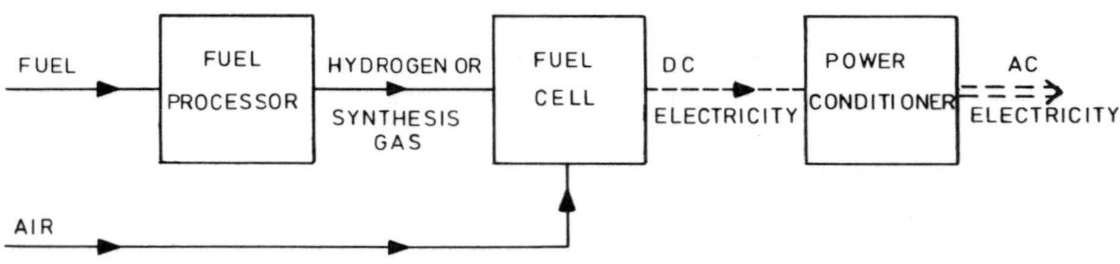

Figure 6.9 Fuel cell power generation system

Initial applications of fuel cells have concentrated on clean fuels (natural gas and naphtha) with hydrogen being produced by steam reforming. However, hydrogen can also be produced by coal gasification (using oxygen and steam as gasifying reactants) and fuel cells therefore represent an alternative to advanced combined cycle technology for high-efficiency power generation systems based on coal gasification.

If successful, molten carbonate fuel cells may be best suited to coal gasification applications because of their ability to utilise carbon monoxide directly and because high efficiencies are possible in large central plants by recovering the waste heat using a steam cycle.

The advantages claimed for fuel cells of all types include

environmental impact is low
manufacture in small packaged units suitable for local power generation is possible
load following and part-load characteristics are excellent

The overall efficiency of the present generation of phosphoric acid systems is about 40 per cent with efficiencies of about 50 per cent being expected with molten carbonate technology. However, the capital costs are relatively high and the extent to which fuel cells displace more conventional approaches to power generation, particularly coal-based systems, depends on the reduction in capital cost that can be achieved in commercial production.

Combined heat and power

As existing power generation efficiencies are only about 35 per cent and, even with advanced concepts such as MHD or fuel cells, efficiencies in the future are unlikely to exceed 50 per cent, there is at first sight a considerable incentive to utilise for heating purposes the 50 to 65 per cent of the input energy normally wasted. This is the objective of combined heat and power (CHP) systems.

For convenience, two main cases can be distinguished, depending on whether the operator is an electric utility or an industrial concern.

Electric utility

Owing to the economies of scale, electric utilities generally employ large central power stations based on steam cycles. The heat rejected by these stations, although large in quantity, is at a fairly low temperature (typically 30 °C) and, as such, is directly suitable only for agriculture or aquaculture applications.

Higher-temperature heat can be obtained by extracting all of the steam ('back pressure' turbines) or part of the steam ('pass out' turbines) at an intermediate temperature and pressure. This reduces the electricity generation capacity of the station but increases the 'overall efficiency' (that is, electricity plus useful heat) substantially, 75 per cent being typical. A further factor affecting the economics of CHP schemes is that the need to maintain heat supplies may necessitate a CHP station operating out of the power station 'merit-order' (see chapter 8), thereby involving the costs of operating more expensive power plant than would otherwise be the case.

The use of large coal-fired power stations for combined heat and power schemes is particularly attractive. Coal-fired stations can be sited near centres of population and therefore nearer than nuclear stations to potential heat demands, and can offer lower operating costs than oil-fired stations at present fuel oil prices.

The output of a large central station is such that the corresponding heat demand is usually distributed over a considerable area. As a result of the distances involved, it is usually considered economic to distribute only low-temperature hot water in order to minimise losses. For this reason, interest in utility CHP schemes has in the past been directed mainly towards applications to residential heating. However, the possibilities for smaller, local CHP power stations, connected to a utility grid but supplying mainly higher-temperature heat (hot water or low-pressure steam) to industrial consumers are now increasingly being recognised.

Industrial

Many industrial processes require both electricity and heat (for example, as process steam) and combined heat and power systems offer one method of meeting these demands.

For large-scale applications, coal-fired steam cycle systems, as described above for utility applications, may be employed. Below about 20 MW (electrical), however, the economies of scale make such systems comparatively expensive and other solutions are generally sought.

The main choices for smaller-scale industrial power generation applications are gas turbines and diesel engines, both of which produce hot exhaust gases from which heat can be recovered and so are suitable for CHP operation. These systems burn fuels (for example, light distillate and heavy fuel oil, respectively) that are considerably more expensive than coal. There is, therefore, an incentive to develop technologies that will permit coal to be used, directly or indirectly, for industrial power generation (and CHP) by these methods. The principal options are:

fast fluidised bed air heaters (for gas turbines)
pressurised fluidised bed combustion (for gas turbines)
low calorific value gas manufacture (for gas turbines)
coal–oil and coal–water mixtures (for diesels)

Finally, it may be noted that the general prospects for combined heat and power depend, to some extent, on institutional factors. For example, the willingness of a utility to buy surplus power from an industrial concern at a satisfactory price has an important influence on the economics of industrial CHP.

Compressed air energy storage

An energy storage system that could assist utilities to meet peak electricity demands is compressed air energy storage (CAES). This involves using off-peak electricity to compress air for storage underground. At times of peak electricity demand, the compressed air is withdrawn from storage and heated by combustion for expansion through a gas turbine.

The compressed air energy storage concept was patented during the 1940s but it was not until 1978 that the world's first (and only) commercial unit began operation. The plant, which is sited at Huntorf in West Germany, is oil-fired and has a capacity of 290 MW.

A number of design studies have been carried out, concentrating mainly on oil-fired systems. The storage pressure is typically 40 to 80 bar and storage facilities include solution-mined salt caverns (as at Huntorf), hard rock caverns and aquifer formations.

The feasibility of coal-fired compressed air energy storage systems based on pressurised fluidised bed combustion has been investigated by the United Technologies Research Center. The results indicate that such systems could be attractive for intermediate load factor duties. However, oil-fired gas turbines and oil-fired compressed air energy storage systems would be preferred for the lowest load factor duties where their comparatively low capital cost is the decisive factor.

Refurbishing

The useful lifetime of a fossil-fuel power station is usually regarded as being 30 to 40 years. After this period of time, stations have generally been retired; the economics of continued operation are unattractive because of the low design efficiency as compared with modern stations. However, stations built from the 1950s onwards have efficiencies similar to those of the most recently installed units. For such stations, 'refurbishing' to provide an additional period of operational life may be considered.

Refurbishing is not a well-defined term but can apply to a range of actions from 'heavy' maintenance to the renewal of major equipment items such as boilers and turbines. Refurbishing has attracted interest over recent years because, under suitable conditions, it can be more economic than the construction of the equivalent new generation capacity.

Retrofitting

The term 'retrofitting' is usually applied to a change in the method of fuelling a boiler before the end of its useful life and often arises in the context of changing from oil-firing to coal-firing. Apart from conventional combustion systems, methods of retrofitting oil-fired boilers for coal-firing include

fluidised bed combustion
gasification
coal–oil mixtures
coal–water mixtures

It is also possible to retrofit an existing boiler (burning either coal or oil) by using the exhaust gases from a gas turbine (fired, for example, by pressurised fluidised bed combustion or a coal-derived gas) as the source of heat instead of the combustion chamber. This option is sometimes referred to as 'repowering'.

6.8 Coke manufacture

Developments of conventional practice

The manufacture of coke for metallurgical processes was introduced by Abraham Derby in Shropshire during the early 1700s and has expanded with the coal industry through the industrial revolution and up to recent times. In spite of this, the coking process has remained basically unchanged for a century. Coal, crushed so that about 80 per cent of the particles are less than 3 mm in diameter, is charged from the top into slot-type ovens. The largest of these are 14 m long, 6.5 m deep and 0.45 m wide. A battery can contain up to 100 ovens. The ovens are heated indirectly through the side walls which are usually made of silica refractory brick. Between the walls of adjacent ovens are flues through which the combustion products of the fuel gas pass maintaining the oven at a temperature of 1100 to 1300 °C. A typical oven is illustrated in figure 6.10.

The coal near to the oven walls is heated rapidly and heat is then gradually transmitted through the remainder of the charge. When the temperature reaches 350 to 400 °C, a coking coal will soften and then begin to decompose. As heating continues, the particles coalesce to produce a coherent porous structure. This plastic stage generally lasts for little more than 100 °C after which resolidification takes place. The rigid semi-coke formed then contracts, producing fissures which affect the size and strength of the final coke. When the centre of the charge reaches 900 °C, a further period of up to 3 hours is allowed as a heat soak. The total carbonising time ranges from 12 to 30 hours.

There are two main incentives to depart from traditional coking practice.

(1) Only certain coals, known as prime coking coals, possess the special combination of properties necessary to produce good coke. In countries with a long history of coke manufacture, for example the United Kingdom,

the most easily won reserves of prime coking coal have already been exhausted. In addition, progress in blast furnace technology has led to increasingly stringent quality specifications for coke.

(2) Coke ovens are expensive and consume substantial quantities of energy. There are, therefore, the usual economic incentives to improve productivity and to reduce energy consumption.

In response to the above factors, a number of developments in the conventional methods of oven operation are occurring and include the following.

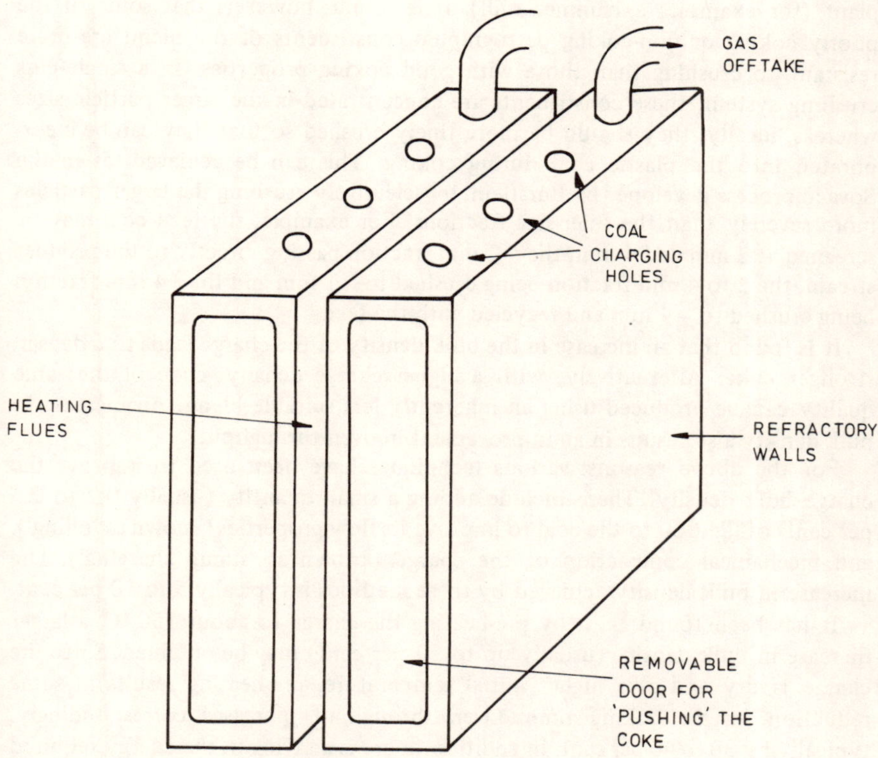

Figure 6.10 Metallurgical coke oven design

Blend formulation

The art of blending to produce satisfactory coke from component coals that are otherwise unsuitable for coke manufacture has advanced considerably over recent years. Indeed, cokes have been produced commercially at the NCB's Lambton works and at the BSC's Scunthorpe works from blends that contain no prime coking coal.

Good coking coals possess caking properties and volatile matter contents lying within a narrow range and an effective blend must therefore contain components that are complementary in these respects.

Charge preparation

The main areas of development interest in charge preparation are selective crushing, pre-heating and partial briquetting.

Traditionally, a blend of coking coals would be crushed together to the required size distribution of 0 to 3 mm in a single pass through the crushing plant (for example, a hammer mill). It is found, however, that some of the poorly coking or non-coking petrographic constituents of the blend are more resistant to crushing than those with good coking properties. In a single-pass crushing system, these constituents are concentrated in the larger particle sizes whereas, ideally, they should be more finely crushed so that they can be incorporated into the plastic coal during coking. This can be achieved, as in the Sovaco process developed by Burstlein, by selectively crushing the larger particles more severely than the finer-size fractions. For example, the feed coal may be screened at 2 mm and 4 mm, the −2 mm fraction passing directly to the product stream, the 2 to 4 mm fraction being crushed to −1 mm and the +4 mm fraction being crushed to −4 mm and recycled with the feed.

It is found that an increase in the bulk density of the charge leads to a denser, stronger coke. Alternatively, with a higher charge density, coke of the same quality can be produced using an inherently less suitable blend. An increase in bulk density also results in an improvement in oven throughput.

For the above reasons, various techniques have been used to improve the charge bulk density. These include adding a small quantity (usually 0.2 to 0.5 per cent) of light oil to the coal to improve its flow properties (known as 'oiling'), and mechanical compaction of the charge (known as 'stamp charging'). The increase in bulk density achieved by these methods is typically 5 to 10 per cent.

It has been found that, by pre-heating the charge to about 150 °C, a larger increase in bulk density (usually up to 20 per cent) may be obtained. Since the charge is dry and at a higher initial temperature, pre-heating results in some reduction in the coking time. Oven throughput increases correspondingly, typically by 30 to 40 per cent. In addition, there is a saving in the energy required by the coke oven of about 25 per cent. However, this energy has to be supplied to the pre-heater although to do so may be more convenient or cheaper.

The possibility of combining pre-heating and stamp charging is also under consideration; initial results indicate that further increases in bulk density can be achieved.

An alternative method of increasing the bulk density of the charge is partial briquetting. This involves briquetting up to one-third of the feed coal with a binder usually derived from the by-product tar. The briquettes are then mixed with the remainder of the coal and charged to the oven in the usual way. An increase in the bulk density of up to 20 per cent may be obtained with associated benefits for coke quality and oven throughput. The resulting coke is homogeneous, there being no indication of the original presence of briquettes in the charge.

Carbonising conditions

The main departures from conventional carbonising conditions of current interest are fast carbonising rates, high carbonising temperatures and programmed heating.

One of the important factors affecting the carbonising rate is the rate of heat transfer across the oven wall. This may be increased by reducing the wall thickness and by employing refractory materials (such as magnesite and silicon carbide) having a higher thermal conductivity than the conventional silica refractory. The main incentive for such changes is to obtain higher oven throughputs with increases of about 20 per cent being possible (both for thin walls and improved refractories).

An alternative method of increasing the throughput is operation at higher carbonising temperatures. Typically, increasing the flue temperature from 1100 to 1350 °C increases the throughput by 50 per cent and at 1600 °C the throughput doubles. For this reason, the use of higher carbonising temperatures of up to about 1350 °C has gradually become more widespread. Although higher carbonising temperatures bring additional gains in throughput, these are offset by significant disadvantages including the cost of the special refractory wall materials (fused mullite, for example), a deterioration in coke quality and a substantial increase in energy requirements. It therefore seems unlikely that there will be a movement towards further increases in carbonising temperature in the immediate future.

The increased importance of energy consumption has led to the investigation of programmed heating systems in which the heat input is varied according to the requirements of the charge. It appears that there is no advantage for coke quality in heating the charge to above about 1000 °C. However, when high carbonising temperatures are used in order to increase throughput, it is found that, towards the end of carbonisation, much of the coke may reach temperatures that are unnecessarily high and wasteful of energy. In a programmed heating system, the energy input is reduced towards the end of carbonisation so that the final coke temperature is lower than would otherwise be the case and energy is saved. Depending on the carbonising conditions, a reduction of about 10 to 15 per cent in energy consumption can be obtained.

Post-carbonisation treatment

Conventionally, when carbonisation is completed, the coke is pushed from the oven into a 'coke car' in which it is rapidly quenched by a water spray to prevent combustion. The sensible heat in the coke is wasted.

Recovery of the sensible heat may, however, be accomplished by dry quenching. This involves delivering the hot coke from the 'coke car' into a vessel that is sealed so that the sensible heat of the coke can be transferred to a recycled inert gas (for example, combustion gases and steam). In present plants, the heat recovered, which typically corresponds to half of the heat supplied to the oven, is passed to boilers for steam raising. Other uses, such as pre-heating the coal, may be possible.

Although dry quenching was first introduced in the 1920s, there were few installations until comparatively recently. Dry quenching is, however, now used extensively in the USSR and Japan. Current interest derives only partly from energy savings; there are substantial environmental advantages in avoiding the emission of the quench gases to the atmosphere and some improvement in coke quality is also claimed.

Formed coke

The term 'formed coke' refers to the production of briquettes of suitable quality to be used in a blast furnace (or foundry) as a replacement for, or in conjunction with, conventional oven coke. As discussed earlier, the coal types that may be used in conventional coking are restricted to a relatively narrow range by the special properties required, even with the recent advances in blend formulation. The main advantage of formed coke therefore is that, in principle, almost any type of coal can be used to produce metallurgical coke by the choice of a suitable briquetting process.

Other advantages that have been suggested are as follows.

(1) The possibility of improving the performance of the blast furnace and reducing the amount of coke consumed by using a fuel of uniform shape and size, optimised for the conditions in the furnace.
(2) Control of pollutant emissions is simpler and cheaper than for conventional coking plants.
(3) Formed coke processes lend themselves more easily than coke ovens to automation and flexible operation.

The technology of formed coke manufacture is related closely to that of domestic smokeless fuel and, for this reason, only the major differences will be outlined in the present section. In comparing the two applications, it should however be noted that formed coke manufacture is at a much earlier stage of development than smokeless fuel manufacture and therefore the full potential of smokeless fuel processes in the metallurgical field has yet to be realised.

The specification for formed coke is generally more stringent than for domestic smokeless fuel, particularly in the following respects.

Lower reactivity

In order to obtain a sufficiently low reactivity for metallurgical use, it is generally necessary to carbonise to a high temperature at some stage of the process (the direct use of lump anthracite is an exception). Even so, the reactivity of formed coke is generally greater than that of conventional coke.

Stronger briquettes

Abrasion and degradation in metallurgical processes are severe (particularly in a blast furnace) and the briquettes therefore have to be stronger than the minimum acceptable for domestic smokeless fuel.

For comparison with figure 6.4, the main process routes for manufacturing formed metallurgical coke are summarised in figure 6.11. A number of options available for domestic smokeless fuel manufacture are excluded as possible methods of formed coke manufacture because of the more demanding specifications described above.

(1) Low-volatile non-coking coal is generally considered too reactive for direct metallurgical use (without prior carbonisation).

PROCESS OR MATERIAL	ANTHRACITE	LOW VOLATILE NON-COKING	COKING	HIGH VOLATILE NON-COKING	LIGNITE	TWO COMPONENTS
LUMP COAL	DIRECT USE					
LUMP COAL CARBONISED						
CARBONISE			CONVENTIONAL COKE OVENS			
CARBONISE BRIQUETTE WITHOUT BINDER						ANCIT
CARBONISE BRIQUETTE WITHOUT BINDER CARBONISE				SAPOZHNIKOV		B F L
CARBONISE BRIQUETTE WITH BINDER HEAT TREAT				I ChPW *		
CARBONISE BRIQUETTE WITH BINDER CARBONISE				F M C		
BRIQUETTE WITHOUT BINDER						
BRIQUETTE WITHOUT BINDER CARBONISE						
BRIQUETTE WITH BINDER HEAT TREAT						
BRIQUETTE WITH BINDER CARBONISE		HBNPC INIEX DKS				

* LUMP COAL IS CARBONISED, THEN CRUSHED

Figure 6.11 Process/coal combinations for formed metallurgical coke trials. Shading indicates those process/coal combinations that are impractical or uneconomic

(2) The coke product obtained by carbonising lump high-volatile non-coking coals or lignite is, in general, too weak for use as a metallurgical coke.

(3) Briquetting anthracite with a binder and heat treating gives a product that is too weak for use as a metallurgical coke.

As for domestic smokeless fuel, there remain a number of coal/process options that appear not to have been investigated either because of limited availability of the coal or because the technology is thought to be difficult.

In some of the processes, notably Ancit and BFL, a proportion of coking coal is required in the feed. This may be regarded as a disadvantage since the main incentive for formed coke manufacture is to reduce the dependence of the iron and steel industry on coking coal supplies. The coking coal required in these processes is, however, a comparatively small percentage (less than 30 per cent) of the total coal processed.

A number of formed coke plants of 100 to 300 t/d capacity have been built and have provided formed coke in sufficient quantities for blast furnace trials. In general, the results have been satisfactory but there is little indication of any significant advantages over conventional coke. Optimisation of the properties of formed coke for the blast furnace is likely to be hindered by the absence of a detailed understanding of the processes occurring within the furnace and the large quantities (more than 5000 t per trial) of the fuel required for blast furnace tests.

Formed coke technology is still at an early stage of development. Where coking coal and coke ovens are available, the conventional technology seems likely to continue to provide the most economic solution. The main opportunity for formed coke occurs where new plant is required and large reserves of a cheap, non-coking coal are available.

Special carbons

Although the dominant markets for the solid products of carbonisation are metallurgical coke and domestic smokeless fuel, a number of small-scale applications exist for special, high-quality carbons. The main prospects for the future expansion of coal use in this market appear to be as follows.

Adsorbents

Adsorbents (or 'active carbons') can be made from coal by a wide range of processes all of which rely on enhancing its internal porosity. This is accomplished by a combination of carbonisation and limited gasification (termed 'activation') using steam or carbon dioxide. Non-caking coals with a low ash content are preferred. Bituminous coals and lignites are generally used in briquetted form whereas anthracites may be used directly.

World production of active carbons is 0.57 Mt/a (1980), most of which is manufactured from coal.

Electrode carbons

At present, electrode carbons are manufactured principally by coking petroleum feedstocks in a delayed coker to produce a 'green' (or 'needle') coke. This is then extruded with coal tar pitch as a binder and subjected to a range of heat treatments (including, in some cases, graphitisation), depending on the application.

World electrode carbon production amounts to about 6 Mt/a (1976). Of this, about 70 per cent is provided by petroleum coke and 20 per cent by pitch binder with the remainder being attributable to coal-based materials (principally pitch coke, bituminous coal and anthracite).

It has been shown that high-quality electrode cokes can be manufactured by coking a coal solution. This may be prepared (see chapter 5) by dissolving coal in a liquid solvent and filtering the digest to remove mineral matter and insoluble carbon. The solvent is recovered when the solution is coked. Although the market size for electrode coke is relatively small, the product values are high and this process may therefore provide an attractive application of direct coal liquefaction technology.

Carbon fibres

At present, carbon fibres are manufactured mainly by high-temperature (approximately 2700 °C) treatment of artificial fibres such as polyacrylonitrile (PAN). However, coal tar pitch can also be used as a starting material and commercial production using this approach is now beginning (for example, Mitsui and Mitsubishi).

It has also been shown that suitable starting fibres may be obtained by spinning a coal solution; the final carbon fibre product has properties similar to the best commercially available PAN-derived fibres. However, the present size of the carbon fibre market is small and the introduction of a process based on a coal solution is unlikely to be worthwhile unless there is a substantial increase in world carbon fibre demand.

6.9 Combined processes

In addition to processes based on coal alone, hybrid processes using both coal and another energy source are being investigated. Three examples of such processes are discussed below.

Coal–oil co-processing

A natural extension of coal's displacement of fuel oil for power generation and industrial bulk steam raising is its introduction into oil-refineries to meet internal process demands. In particular, coal could be used to produce process steam, electrical power and hydrogen.

At present, these are provided by heavy fuel oil or fuel gas. As oil prices rise, however, it may become preferable to upgrade such materials to produce high value products and to substitute using coal.

In addition to this indirect use of coal in oil refining, various methods of introducing coal as a feedstock into refinery processes have been proposed. These generally involve processing a mixture of heavy oil and coal and fall into two categories.

Thermal cracking of coal–oil mixtures

Coal may be mixed with the heavy oil feed to thermal cracking processes. The resulting product spectrum compares favourably with that obtained from heavy oil alone. Indeed, it has been claimed that some interaction between the coal and oil occurs and that the presence of the coal tends to reduce coke formation. The Cherry P process is of this type.

Hydrogenation of coal–oil mixtures

Alternatively, coal can be mixed with the feed to processes for upgrading heavy fractions by hydrogenation. In order to obtain a satisfactory yield, it is necessary that the heavy oil feedstock has a high content of aromatics. Where this is the case, co-processing offers the possibility of introducing a limited amount of coal hydrogenation within an existing oil-refinery.

Current interest in the hydrogenation of coal–oil mixtures appears to be concentrated in the USSR.

Finally, it has been suggested that the installation of coal liquefaction plants at, or adjacent to, refinery sites may be advantageous so that sharing of facilities and joint processing of products may take place. A number of arrangements are possible, and can include direct, indirect or pyrolysis processes. A particular case is the limited introduction of pyrolysis to provide sufficient char for process steam, power generation and hydrogen production; the pyrolysis liquids are refined together with the appropriate petroleum fractions.

Nuclear heat gasification

On the assumption that nuclear energy will become significantly cheaper than fossil fuel energy, attention has been directed towards opportunities for nuclear heat to replace fossil fuels in fields other than power generation. One such possible application is coal gasification.

For many coal gasification processes, the carbon conversion (that is, the proportion of the input carbon reporting in the product) is low. This arises partly because carbon is burned to carbon dioxide within the process in order to support endothermic reactions taking place. For example, simplified energy and carbon balances for a typical SNG process are shown in table 6.6. Although the overall thermal efficiency is approximately 70 per cent, SNG contains nearly twice the energy per mass of carbon as coal and the carbon conversion is, therefore, only about 40 per cent.

Table 6.6 Energy and carbon balances for SNG production

	Energy (taking coal ≡ 1.0)	Carbon balance (taking coal ≡ 1.0)
SNG	0.7	0.4
Carbon dioxide	–	0.6
Losses	0.3	–
Total	1.0	1.0

If, therefore, nuclear heat were substituted for the combustion of fossil carbon as a means of supporting the endothermic reactions within the gasification process, a substantial increase in carbon utilisation would be possible.

The effects on the energy supply system of providing nuclear heat for coal gasification include:

(1) Nuclear energy is able to contribute to meeting those energy demands for which the properties of hydrocarbon fuels are particularly advantageous.

(2) Fossil fuel carbon is conserved by reducing the coal required to meet a given demand for SNG. There is also a proportionate reduction in the generation of fossil fuel pollutants.

The concept of using nuclear heat to support endothermic gasification reactions has been the subject of a substantial research and development programme in West Germany. Two configurations, differing in the type of endothermic reaction employed, have been proposed.

Coal/steam

Nuclear heat is transferred indirectly to a fluidised bed coal gasifier operating with steam as the reactant gas.

Steam reforming of methane

Hydrogen for hydrogasification of coal is produced by steam reforming part of the methane product using nuclear heat.

Both of these processes require heat at a higher temperature than can be obtained from a conventional nuclear reactor and the development of a new type of reactor, the high temperature reactor (HTR), is therefore necessary. Experimental reactors using helium as the coolant and refractory nodules of pyrocarbons or silicon carbide to contain the fissile material have been operated. At present, the maximum core exit temperature achieved is 950 °C which, after heat exchange differentials, provides a maximum delivered temperature to the reactor of 900 °C.

There exist a number of problems that have to be solved before nuclear heat gasification can be regarded as a commercial option.

Temperature

As steam reforming takes place at 600 to 825 °C, the temperature at which the nuclear heat is available should not be a limiting factor in this case. Coal gasification usually takes place at over 1000 °C, however, in order to obtain sufficiently high reaction rates. Even if adequate reaction rates can be achieved at the lower temperatures available from an HTR, the choice of coals is likely to be restricted to the more reactive types (such as lignites) and residence times will be long, requiring large, expensive gasification reactors.

Materials

The choice of heat exchanger materials for heat transfer to a fluidised bed coal gasifier at temperatures of 900 °C or above is a critical problem area. The reducing conditions and the presence of sulphur compounds both accelerate corrosive attack. The availability of materials suitable for commercial applications has yet to be demonstrated.

Integration of plants

Problems of safety and security, while not insuperable, combine to militate against the operation of nuclear and coal systems in close proximity.

Economics

HTR technology is still at an early stage of development. There is, therefore, a considerable margin of uncertainty associated with the costs of high-temperature nuclear heat supply.

Nuclear heat gasification may be regarded as a long-term option for using nuclear energy to meet premium heating demands in the domestic and commercial sectors. However, there exist a number of other, nearer-term technologies that are capable of satisfying the same objective.

Electrical heating

Nuclear electricity can be used instead of SNG for space heating, either by resistance heating or by heat pumps.

Nuclear district heating

Low-temperature nuclear heat may be used directly to meet space heating demands via a district heating network.

Hydrogen for distribution

Nuclear electricity may be used to manufacture hydrogen (by electrolysis) for distribution as a fuel.

Hydrogen for coal conversion

Hydrogen manufactured by nuclear electrolysis may be used in coal conversion processes (including gasification) to increase the carbon conversion. This possibility is discussed further in the following section.

Non-fossil derived hydrogen

Since coal is deficient in hydrogen, all coal conversion processes for which liquid or gaseous hydrocarbon fuels are the main product involve the manufacture of hydrogen.

In conventional coal conversion processes, the hydrogen is manufactured internally by decomposing steam using a carbonaceous feedstock, the oxygen in the steam ultimately being rejected as carbon dioxide.

The overall carbon conversion of the process is, therefore, low but can be increased by providing hydrogen from an external source, such as electrolysis using nuclear electricity.

The use of non-fossil derived hydrogen has the advantage of simplifying conversion processes significantly and therefore of reducing their capital cost. For example, most coal conversion processes have a requirement for oyxgen. If non-fossil derived hydrogen is produced, co-product oxygen is also available to meet plant demands without recourse to air separation equipment.Also, the shift reactor may be eliminated from gasification/synthesis processes as the carbon monoxide to hydrogen ratio can be adjusted simply by hydrogen addition.

In some coal conversion processes, however, gasification for hydrogen production is a convenient method of utilising low value by-products. If external hydrogen is introduced, alternative uses for these by-products have to be found, for example, combustion to provide process steam.

The supply of hydrogen produced from nuclear electricity to coal conversion processes shares the following advantages with nuclear heat gasification (see previous section).

(1) Nuclear energy can contribute to meeting the demand for hydrocarbon fuels.
(2) Fossil carbon is conserved by increasing the carbon utilisation of coal conversion processes.

There are, however, a number of differences between the two approaches:

(1) Non-fossil derived hydrogen can be produced from any source of electricity including hydro and the renewables. In particular, existing nuclear reactor technology (the PWR, for example) may be employed.
(2) The nuclear and coal plants may be separated geographically.
(3) Storage of the hydrogen to smooth out supply/demand variations is possible.
(4) Both liquefaction and gasification processes can be provided with external hydrogen.

The economics of using non-fossil derived hydrogen in coal conversion processes clearly depend on the price at which the hydrogen can be supplied. In the future, off-peak electricity from nuclear stations may become available at marginal production cost. This advantage is, however, offset by the low utilisation factor for the electrolyser when operating with off-peak power.

For hydrogen production either from a dedicated nuclear station or from off-peak nuclear electricity, a real increase in European coal price levels by a factor of about three (more in low-cost coal regions) would be required before the use of nuclear electrolytic hydrogen would become economic. In the long term, the economics may be assisted by the development of advanced electrolysers.

CHAPTER 7

ENVIRONMENT

During the 1960s and 1970s, a growth in the public awareness of, and interest in, the environmental effects of industrial activities was evident. As a result, coal utilisation practices have come under closer scrutiny and there has been a trend towards increasingly severe controls almost everywhere.

Meeting more stringent environmental controls generally requires a greater plant investment. Also an energy input is usually needed to operate the pollution control equipment, thus resulting in the less efficient overall use of energy. Although the cost implications of controls can, therefore, be estimated, the benefits are less readily quantified. This is because considerable uncertainty and controversy surrounds the effects of coal-derived pollutants, particularly in the following respects:

(1) Comparatively little is known about the long-term effects of exposure to coal-derived pollutants at low concentration levels. In particular, threshold concentrations (that is, pollutant concentration levels below which there is no health hazard) have not, in general, been determined for coal-derived pollutants.
(2) Coal processes usually result in the release of a number of different pollutants simultaneously. Little is known about the impact of multiple coal-derived pollutants and their possible synergistic effects.

The appropriate balance between the costs of pollution-abatement controls and the resulting benefits to the 'quality of life' is, therefore, a matter of judgement. The controls that are in force at present vary widely from country to country and it is particularly difficult to forecast whether the general trend towards stricter environmental controls will continue in the future or whether there will be some relaxation as energy prices increase.

It is already clear, however, that environmental considerations are having an important influence on the development of new coal utilisation technologies; design concepts for new processes usually assume that conformity to strict environmental standards is required.

There are a number of ways of classifying the environmental effects of coal utilisation, including

(1) by type of process (for example, combustion, carbonisation, gasification, liquefaction)
(2) by the form of the pollution (for example, atmospheric emissions, liquid effluents, solid residues)

(3) by the scale on which the pollution may have an effect (that is, local, regional or global)
(4) by the nature of the environmental impact or health hazard

Since present markets for coal are dominated by combustion processes, attention has naturally been concentrated on the environmental impacts of these technologies. Of particular concern has been the effects of emissions to the atmosphere of combustion products, and various aspects of this field (particulate matter, sulphur dioxide, nitrogen oxides, carbon dioxide and trace species) are discussed in the first five sections of this chapter. The environmental effects associated with the solid residues of combustion are considered in section 7.7.

The other major market for coal at present, carbonisation, produces trace organic species (section 7.5) and aqueous effluents from gas cleaning (section 7.6).

The environmental effects of future gasification and liquefaction plants appear likely to be similar to those of existing combustion and carbonisation processes. Indeed, many of the emissions from these conversion plants are expected to arise from combustion to supply internal power and process steam requirements. Liquid effluents (similar to those from carbonisation) and solid residues (similar to combustion residues) will also be produced.

The main environmental concerns are summarised in simplified form in table 7.1, and the similarities and relationships between the impacts associated with the main classes of coal utilisation technology are illustrated in figure 7.1.

Although the present chapter is concerned specifically with the effects of coal utilisation, many of the same pollutants also arise from the use of petroleum-

Table 7.1 Principal environmental concerns arising from coal utilisation

Pollutant	Human health	Material damage	Ecology	Haze and smog	Global climate
Particulate matter	✓			✓	
Sulphur	✓	✓	✓	✓	
Nitrogen oxides	✓		✓	✓	
Carbon dioxide					✓
Trace species	✓			✓	
Liquid effluents			✓		
Solid residues	✓				

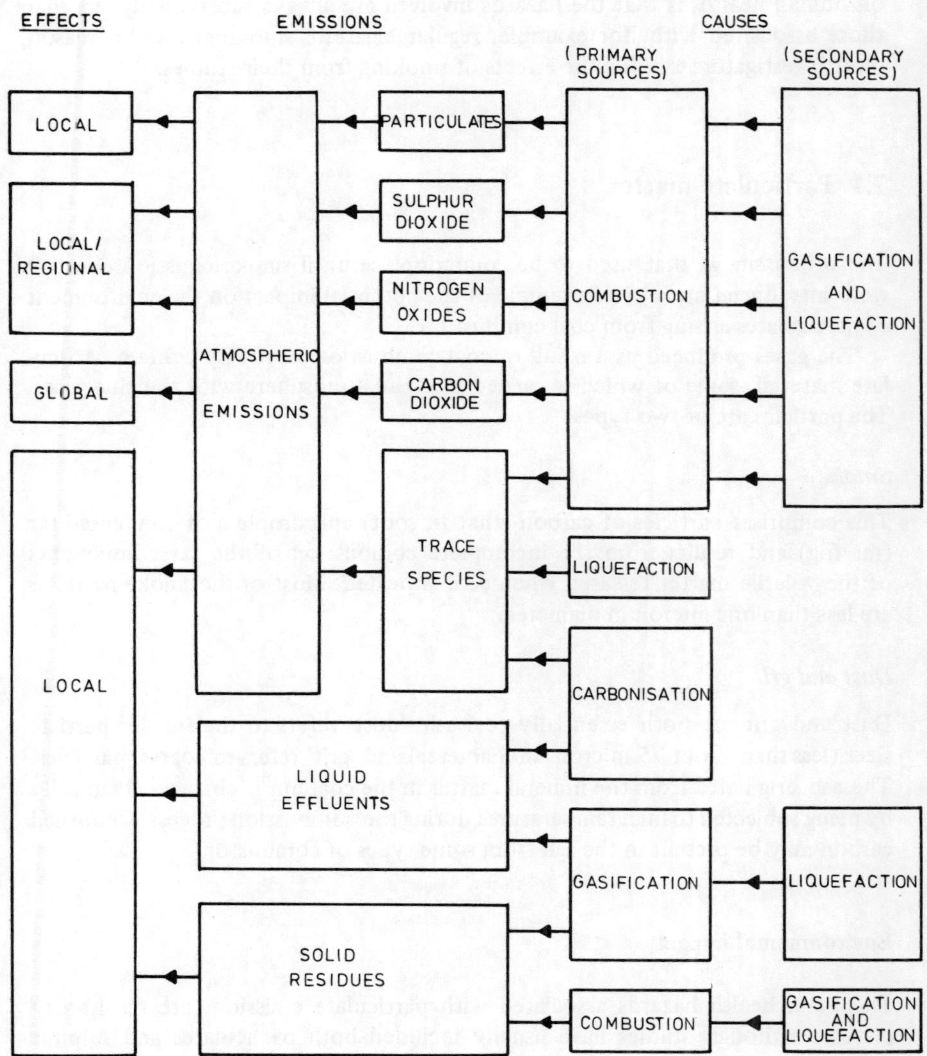

Figure 7.1 Main environmental concerns for coal utilisation. 'Secondary' sources are defined as contributing to environmental emissions indirectly, by virtue of incorporating a primary source. For example, a liquefaction plant includes a gasification system and therefore produces gasification solid residues

derived fuels. The overall environmental impact of coal combustion and conversion must, therefore, be considered in the general context of energy supply and demand.

A further consideration, particularly when assessing the impact of air pollution on human health, is that the hazards involved are always substantially less than those associated with, for example, regular cigarette smoking; for this reason, most investigators exclude the effects of smoking from their studies.

7.1 Particulate matter

The urban smogs that used to be commonplace until smoke emission controls were introduced are a vivid example of the potential impact on the environment of particulates arising from coal combustion.

The gases produced as a result of coal combustion inevitably contain particulate material, some of which is carried into the atmosphere with the flue gases. The particles are of two types.

Smoke

This comprises particles of carbon (that is, soot) and droplets of condensed tar (tar fog) and results from the incomplete combustion of the tar components of the volatile matter released when coal is heated. Most of the smoke particles are less than one micron in diameter.

Dust and grit

Dust and grit are both essentially coal ash; 'dust' refers to the smaller particle sizes (less than about 75 microns in diameter) and 'grit' refers to coarser particles. The ash originates from the mineral matter in the coal but is changed chemically by being subjected to high temperatures during the combustion process. Unburned carbon may be present in the ash from some types of combustor.

Environmental impact

The main health hazards associated with particulate emissions are respiratory diseases although studies have usually included both particulates and sulphur dioxide.

The worst effects of particulates on health have been the result of smoke emissions. Under certain weather conditions, extensive smoke emissions lead to the formation of urban smogs that can increase mortality, especially among the old and chronically sick. The London smog of 1952, for example, has been estimated to have caused 4000 additional deaths.

The extent of the hazard associated with dust and grit depends on the particle size.

(1) Dust particles larger than about 15 microns, and grit particles, tend to be deposited from the atmosphere under their own weight in due course. These particles are too large to be respirable.

(2) Dust particles smaller than 15 microns remain suspended in the atmosphere for long periods and can be inhaled.

In general, dust and grit particles are chemically inert and mainly consist of stable oxides. Trace quantities of a large number of elements can, however, be found and their environmental implications are discussed further in section 7.5. Apart from this aspect, the larger sizes of dust and grit that settle out are a nuisance rather than a direct health hazard. The smaller, respirable dust particles may, in principle, lead to increased respiratory disease, although the concentrations resulting from coal combustion are extremely low and no evidence of adverse health effects has been established.

Sources

Particulate emissions arise principally from two sources.

Domestic combustion

Particulate emissions from domestic appliances are mainly as smoke while dust and grit emissions are comparatively low.

Industrial installations and power stations

A properly designed, maintained and operated industrial or utility furnace or boiler does not normally emit significant amounts of smoke and, when smoke emission does occur, it can usually be remedied by attention to these aspects. Some emissions of dust and grit are inevitable.

Control

Control techniques for particulate emissions from solid fuel combustion are well-known.

In order to reduce the smoke emissions from domestic heating appliances, two main options are available:

(1) The combustion of a smokeless fuel (that is, anthracite, coke, or a smokeless briquette, see section 6.5).

(2) The use of smoke-reducing combustion appliances (also see section 6.5).

The dust and grit emissions from industrial installations and power stations can be controlled by cleaning the combustion gases. A number of devices are available, some of which have been described in previous sections. For completeness, the main types are summarised below.

Cyclones

The particles are separated from the gas stream by centrifugal forces which are induced by creating a vortex type of motion.

Aqueous scrubbers

These use a water spray to wet the particles and thereby assist their separation from the gas stream by inertial or gravitational means.

Electrostatic precipitators

A high voltage is applied to electrodes between which the combustion gases pass. Particles in the gas become electrically charged and attracted to the electrodes from which they are removed periodically by mechanical methods.

Bag filters

The combustion gases are passed through fabric bag filters in which the particles collect. The particles are removed by reversing the air flow and flexing the fabric with a 'pulse jet' or by shaking.

The main method of dust collection in industrial installations is cyclones. These generally remove about 80 per cent of the particles and the most efficient units are effective down to about 10 microns.

For power stations, electrostatic precipitators are usually employed. Collection efficiencies of over 99.5 per cent can be achieved with substantial removal of particles in the size range 0.1 to 10 microns which mostly escape collection in cyclones. The overall performance of bag filters is similar to that of electrostatic precipitators although removal can be better for the smaller particle sizes. Bag filters have been installed at some power stations. Both precipitators and bag filters are normally used after the air heater, at a temperature of about 150 °C, although 'hot' precipitators operating upstream at temperatures of up to 375 °C are sometimes used for low sulphur coals because of the poorer electrical conductivity of the gas in this case.

High-efficiency aqueous scrubbers can remove up to 98 per cent of the particles in a gas stream and are effective for particles down to about one micron. Aqueous scrubbers are not generally favoured as a means of particulate removal in combustion applications, however, mainly because of scaling and corrosion problems although they are the preferred choice for carbonisation plant and gasification systems (see chapters 4 and 6).

The performance and applications of the main technologies for dust and grit removal are summarised in table 7.2.

Table 7.2 Dust and grit removal methods[a]

Method	Typical removal efficiency (%)	Minimum effective particle size (microns)	Main applications
Cyclones	80	10	Industrial boilers/furnaces (main method used)
Aqueous scrubbers (high efficiency)	98	1	Carbonisation and gasification
Electrostatic precipitators	99.5	0.1–0.3	Power stations (main method used)
Bag filters	99.5	0.1–0.3	Power stations and industrial boilers/furnaces (some)

[a]Sources: Selection of Gas Cleaning Equipment: a Study of Basic Concepts, *Conference of the Filtration Society*, London, 1969; *Coal and the Environment*, Report by the Commission on Energy and the Environment, H.M.S.O., London, 1981.

7.2 Sulphur

The emission of sulphur compounds to the atmosphere has attracted more attention and controversy over recent years than almost any other form of pollution.

Sulphur is present in coal in two forms.

Organic

Some of the sulphur is part of the organic coal substance.

Inorganic

Sulphur also exists as inorganic compounds in the mineral matter, particularly as sulphides (for example, pyrites).

When coal is burned, both the organic and inorganic sulphur are released. A proportion of the sulphur (usually about 15 per cent) is retained in the coal ash; almost all of the remainder is emitted as sulphur dioxide in the combustion gases.

Sulphur dioxide may be removed from the atmosphere by two mechanisms.

Dry deposition

This occurs by the chemical reaction or absorption of sulphur dioxide on surface features such as buildings and vegetation.

Washout

Sulphur dioxide is soluble in water, forming sulphurous acid, and is therefore readily removed by rainfall (and other forms of precipitation).

In parallel with the above removal mechanisms, sulphur dioxide can also be oxidised in the atmosphere. The principal product is ammonium sulphate, which usually accounts for more than 60 per cent of the sulphur, with the remainder comprising mainly calcium and sodium sulphates (arising from the particulate matter present) and free sulphuric acid. The reactions that occur are extremely complex and the detailed chemistry is not yet fully understood.

The mechanism of dry deposition is far less effective for sulphates than for sulphur dioxide (typical deposition velocities are 1 mm/s and 8 mm/s respectively). For this reason, the removal of sulphates depends mainly on 'washout' by precipitation. In the absence of precipitation, sulphates can travel long distances, resulting in what is known as the 'long range transport' (LRT) problem. Since the sulphates are slightly acidic, their presence in the atmosphere contributes to acidity in the rainfall. The environmental implications of 'acid rain' are discussed later.

The overall sulphur system for coal combustion is illustrated, in a simplified form, in figure 7.2. Also shown in the figure is a typical percentage breakdown of sulphur deposition by the various mechanisms.

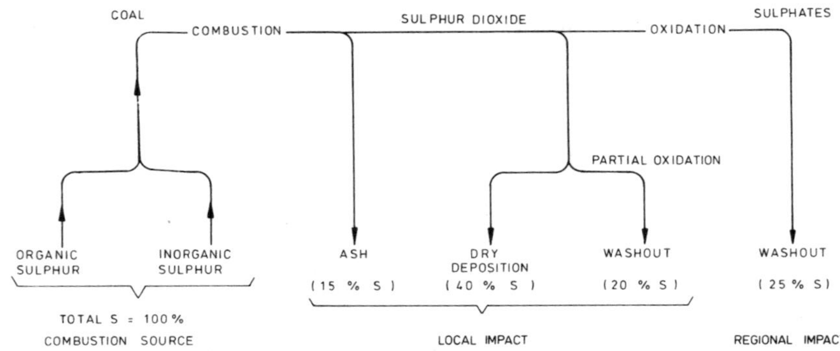

Figure 7.2 Interactions of sulphur with the environment (sources: R. A. Barnes, *The Long Range (Transboundary) Transport of Air Pollution*, Report for the ECE senior advisors on environmental problems, 1978; *Coal and the Environment*, Report by the Commission on Energy and the Environment, H.M.S.O., London, 1981)

Environmental impact

The environmental impact of sulphur emissions can be divided into two categories.

Local

Since sulphur dioxide is oxidised slowly in the atmosphere to sulphates, the local impacts of sulphur emissions are confined principally to atmospheric sulphur dioxide concentrations.

Of primary concern is the effect of sulphur dioxide on respiratory disease. Although this field has been the subject of intensive study, dose–response relationships are still uncertain.

High sulphur dioxide concentrations usually occur together with other pollutants (for example, particulates) which may also cause respiratory disease. Synergism of the effects of these pollutants is thought to be significant and complicates an overall assessment of the role of sulphur dioxide.

In addition to concern about the health hazards, unnaturally high concentrations of sulphur dioxide can also be responsible for two further adverse environmental effects: accelerated corrosion of materials and buildings and reduced growth rates of agricultural crops.

Regional

Because of the time taken to achieve dispersal on a regional basis, the main environmental effects in this case derive from sulphates and, in particular, the acidity associated with them. The main areas of concern are respiratory disease, effects on ecosystems and atmospheric haze.

It has been suggested that acid sulphates, even if present at low concentrations in the atmosphere, may contribute to respiratory disease. At present, the evidence is uncertain and conflicting and more work needs to be carried out on this aspect.

The effects of sulphates on ecosystems occur as a result of long range transport and 'acid rain' (that is the deposition of acid sulphates in rainfall). The main concerns are for the impact of this additional acidity in remote rural areas with non-calcareous rocks and soils and are directed particularly towards reduced tree growth in forests and the decline in fish populations in freshwater systems.

However, the extent to which 'acid rain' contributes to the above effects is the subject of considerable controversy, as are the possible mechanisms involved.

Of less prominence in the current debate on the environment, but less disputed, is the contribution of particulate sulphates in the atmosphere to a reduction in visibility. The effect referred to is not the severe reduction in visibility in urban areas on moist winter days (that is, smog) but the occurrence of a general haze that reduces visibility at times when meteorological conditions would otherwise have been clear. Such sulphate hazes are probably most noticeable during summer in remote rural areas where previous generations would probably have enjoyed the 'amenity' of clear visibility. In the United Kingdom, for example, sulphate aerosols probably account for about half of the light extinction in dry conditions.

Sources

It is convenient to distinguish three types of emission source for sulphur dioxide: low level (domestic combustion), medium level (industrial and commercial combustion) and high level (power stations).

When combustion gases are emitted from a chimney, they are carried downwind and are diluted by turbulent diffusion with the atmosphere. For a domestic chimney, the maximum ground-level concentration occurs within a few metres whereas for power stations, where the stack may be 200 m high, the maximum ground-level concentration occurs typically 5 to 10 km from the source.

It follows that high-level emitters contribute relatively little to ambient sulphur dioxide concentrations; the main contributors to localised sulphur dioxide pollution are low and medium-level emitters. Indeed, reductions in urban sulphur dioxide concentrations have been largely achieved as a result of the general decline in the domestic market for coal.

Since high-level emissions remain airborne before contact with the ground for longer than those from low and medium-level sources, there is less opportunity for removal by dry deposition. Power stations may, therefore, be expected to contribute rather more to sulphate formation and long-range transport than domestic and industrial/commercial combustion.

Meteorological conditions play an important role in determining the ambient sulphur dioxide concentrations and the deposition pattern resulting from sulphur emissions. For example, in stagnant air systems such as anticyclones, high concentrations of sulphur dioxide, and hence sulphates, may build up. Inversion layers may also result in an accumulation of pollutants, in this case by inhibiting the dispersal of low and medium-level emissions. Other weather conditions, particularly steady moist airflows, promote the long-range transport of sulphates.

Control

The main options available for controlling the effects of sulphur emissions from coal combustion are as follows.

Dispersal

It is possible to limit ground-level concentrations of sulphur dioxide by employing tall stacks thereby ensuring efficient dispersal of the combustion gases. Although most effective for power stations and large industrial installations, a 'tall stack policy' often also applies to smaller industrial and commercial plant and is effective in reducing local sulphur dioxide concentrations arising from these sources and the consequent possible hazard to human health (respiratory disease).

Dispersal does not reduce sulphur emissions and consequently does not alleviate the concerns (such as acid rain) associated with the long-range transport of air pollutants (indeed, by minimising deposition near the source, some additional sulphate formation may occur).

Retention during or after combustion

This option is practical only for industrial and power generation systems. Flue gas desulphurisation is described in section 6.7. By using this technology, up to 90 per cent of the sulphur can be removed from the stack gases before emission to the atmosphere.

Alternatively, fluidised bed combustion offers the opportunity of removing up to 90 per cent of the sulphur during combustion by adding limestone or dolomite to the bed (see chapter 3).

Normally, in the above cases, the sulphur is retained as calcium sulphite or sulphate (see the descriptions of the technologies given earlier). Unless regeneration is employed, the residue has to be disposed of, and this introduces other environmental disadvantages to be discussed later.

Coal cleaning

Since a significant proportion of the sulphur in coal is present as inorganic compounds in the mineral matter (for example, pyritic sulphur), washing the coal to reduce the mineral matter content also reduces the sulphur content. Conventional coal preparation systems (see section 1.1) can rarely remove more than about 30 per cent of the pyritic sulphur although advanced systems (see section 6.2) designed for optimal sulphur reduction are now available. These can remove up to 60 per cent of the pyritic sulphur but at the expense of a significant increase in the loss of combustible material in the discard. None of the organic sulphur is removed by coal preparation processes.

As a result of the high loss of combustible material, it is rarely worthwhile to introduce additional washing alone as a means of sulphur control. It may, however, be useful to prepare a low-sulphur (clean coal) stream for direct use, together with a high-sulphur (middlings) stream for use with sulphur control technology (for example, flue gas desulphurisation).

The limited amount of sulphur that can be removed by the physical separation techniques based on density differences used in coal preparation has prompted interest in the development of chemical cleaning systems. These generally involve the following steps

coal is reacted with a liquid leachant
the 'clean' (sulphur-free) coal and spent leachant are separated
the leachant is regenerated and recycled

Leachants include ferric sulphate (TRW Meyers process), sodium and calcium hydroxides (Battelle Hydrothermal process) and water with gaseous air or oxygen (Ledgemont and USBM/ERDA processes). In all cases, costs are dominated by the solids–liquid separation step.

These processes are all at an early stage and considerable development work is required before commercialisation is possible. Sulphur removal efficiencies of 90 to 95 per cent are claimed for pyritic sulphur with a thermal loss of about 5 per cent. The effect on the organic sulphur of chemical cleaning is not clear; some processes appear to give a reduction whereas other similar systems give no reduction.

Conversion

In order to remove both organic and inorganic sulphur with a high efficiency, the chemical conversion of coal into a gaseous or liquid fuel (see chapters 4 and 5) is necessary. If the conversion is carried out only as a means of producing a low-sulphur fuel for bulk combustion applications (rather than to produce a premium fuel), then the preferred products are clean fuel gas (low or medium calorific value gas) and clean fuel oil or solvent refined coal.

A clean fuel gas may be produced by air/steam or oxygen/steam gasification. After an initial particulate clean-up stage, the gas can be processed to remove over 99 per cent of the sulphur by using existing desulphurisation technologies (see chapter 4).

The manufacture of a clean fuel oil or solvent refined coal employs the same processes of coal solution, mineral matter separation and hydrogenation (see chapter 5) as liquefaction for premium products except that the degree of hydrogenation is kept to the minimum necessary to achieve the desired degree of desulphurisation. Sulphur removal efficiencies of 90 per cent are usually possible.

Alternatively, premium fuels such as SNG and light liquid fractions may be manufactured from coal for supply to markets in which a high value is placed on amenity. The efficiency of sulphur removal in the manufacture of these fuels is generally over 99 per cent.

The various sulphur control options discussed above are summarised in table 7.3, together with the corresponding removal efficiencies. With sulphur emissions, as with many forms of pollution, attitudes differ on the extent of control necessary.

For power stations and large industrial installations, the basic decision is whether dispersal to limit ground level concentrations is sufficient or whether emission controls are required to reduce both sulphur dioxide and sulphate levels on a local and/or regional scale. The divergence of views is exemplified by the difference in approach between the United Kingdom and the United States; the former adopts a tall stack dispersal policy and the latter enforces emission controls.

Retention of sulphur during or after combustion (by flue gas desulphurisation

Table 7.3 Sulphur control options[a]

Method	Typical sulphur removal efficiency (%)
Dispersal	0
Flue gas desulphurisation	Approx. 90
Fluidised bed combustion with acceptor addition	Up to 90
Conventional coal preparation	Up to 40, of pyritic S only
Advanced coal preparation	Up to 60, of pyritic S only
Chemical cleaning	90 to 95, of pyritic S and up to 70, of organic S
Gasification	At least 99
Liquefaction	At least 99

[a]Sources: *Coal and the Environment*, Report by the Commission on Energy and the Environment, H.M.S.O., London, 1981; *The Control of Sulphur Oxides emitted in Coal Combustion*, Economic Assessment Service report B1/77, IEA Coal Research, London.

or acceptor addition to fluidised bed combustors) is likely to be economic only for power stations and large industrial combustors. As indicated in table 7.3, however, each of these methods is capable of removing at most 90 per cent of the sulphur. If the use of 'the best available technology' is accepted as the guiding concept for sulphur control, and is interpreted literally, these options would be rejected in favour of conversion, for example, to a clean fuel gas, which permits 99 per cent of the sulphur to be retained.

For domestic and small industrial/commercial consumers, the only practical and economic methods of providing a coal-based energy supply with low sulphur emissions are likely to be as follows.

Conversion

The type of conversion will depend on the application. Where a high level of amenity is important (for example, the domestic sector), SNG or light liquid fuels will probably be preferable to clean fuel gas or the heavier liquid fractions.

Centralisation

The problem of controlling sulphur emissions from small-scale combustion equipment can be overcome by using a central installation. For example, the increased use of district heating or electricity would enable coal to supply the premium sectors (domestic and small industrial/commercial) from large central plants incorporating appropriate sulphur control technology.

7.3 Nitrogen oxides

Nitrogen oxides (commonly referred to as NO_x) are produced when coal, and all other fossil fuels, are burned. The principal components of concern are nitric oxide (NO), which typically accounts for about 95 per cent (by volume) of the nitrogen oxide emissions, and nitrogen dioxide (NO_2). Trace amounts of nitrous oxide (N_2O) are also produced. Nitrogen oxides arise from two sources: organic nitrogen (that is, nitrogen present in the coal) and atmospheric nitrogen.

The total NO_x content of combustion gases is usually in the range 400 to 1000 ppm (by volume) for coal combustion. In conventional combustion systems most of the NO_x (typically 80 per cent) is derived from the organic nitrogen.

In the atmosphere, nitric oxide undergoes a complex series of oxidation reactions. The main reaction pathway is to nitrogen dioxide which, in turn, is converted into nitric acid and nitrates. Although the reaction mechansims are not fully understood, it is known that sunlight, ozone and hydrocarbons can play important roles. Ozone may act directly as the oxidant for the conversion of nitric oxide to nitrogen dioxide and may also be produced by the photolytic decomposition of nitrogen dioxide to nitric oxide. Hydrocarbons (and derived oxygenates, for example, aldehydes) result in the oxidation of nitric oxide to organic nitrates such as peroxyacetylnitrate (PAN) and participate in a chain of reactions that result in abnormally high atmospheric ozone concentrations.

It is important to note that fossil fuel combustion accounts for only about 10 per cent of the total NO_x emissions to the atmosphere, with about half of this value being attributable to coal. A further 20 per cent arises from the use of fertilisers.

However, approximately 70 per cent of NO_x emissions are of biological origin. These occur mainly as nitrous oxide (N_2O) which is relatively inert in the lower atmosphere and not considered to be directly harmful. In the upper atmosphere, most of the nitrous oxide decomposes into nitrogen and oxygen but a proportion (about 5 per cent) is converted into nitric oxide which, in turn, is oxidised by the ozone present to form nitrogen dioxide.

The main reactions and products associated with atmospheric NO_x are illustrated in figure 7.3.

Environmental impact

The environmental impact of nitrogen oxides is, to some extent, similar to that of sulphur dioxide and may be considered in the same categories.

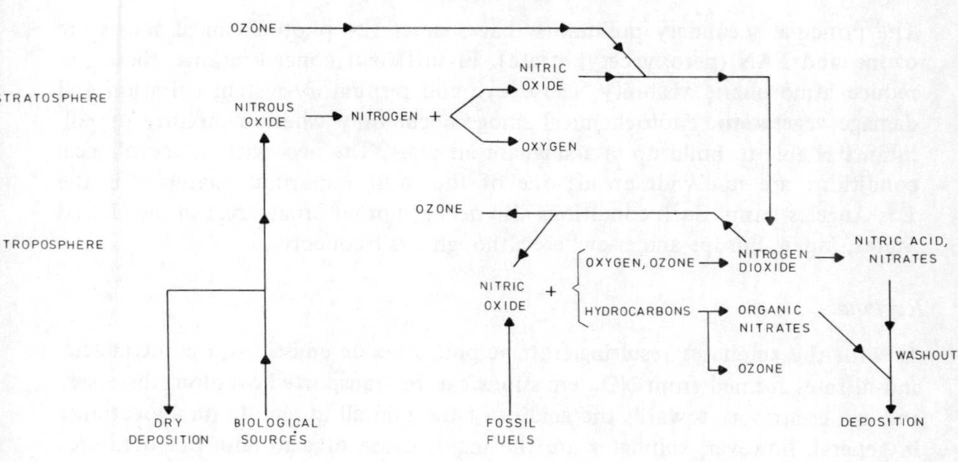

Figure 7.3 Interactions of nitrogen with the environment

Local

Ground level concentrations of NO and NO_2 in urban areas arise mainly from road-vehicle exhausts. Average concentrations generally lie in the range 0.02 to 0.06 ppm although peak concentrations can be several times higher. Rural concentrations are lower, usually in the range 0.005 to 0.02 ppm. These values compare with background concentrations of N_2O from biological sources of 0.25 to 0.4 ppm.

The main concerns associated with high local concentrations of NO and NO_2 are respiratory disease, damage to vegetation and photochemical smog (with associated health and vegetation effects).

At high concentrations, NO_2 is known to cause respiratory disease and NO impairs the oxygen transport function of haemoglobin. There is also some evidence that both NO and NO_2 adversely affect the immune system. Little is known about human dose-response relationships at low dosage levels or possible synergisms with other urban atmospheric pollutants such as sulphur dioxide, sulphates and particulates. It appears generally to be accepted, however, that any risks associated with atmospheric NO_x concentrations are much less serious than those of sulphur dioxide and sulphates.

Experiments have also demonstrated that relatively high concentrations of NO and NO_2 (that is, more that 0.5 ppm) can damage vegetation. As with human health risks, there is little evidence to imply any direct effects at typical current urban concentrations. However, recent work indicates that NO_2 and SO_2 acting in synergism at low concentrations may reduce the yield of agricultural grasses.

Photochemical smogs are formed when nitrogen oxides react with oxygen and hydrocarbons in the presence of sunlight to form secondary pollutants.

The principal secondary pollutants that characterise photochemical smogs are ozone and PAN (peroxyacetylnitrate). In sufficient concentrations, these can reduce atmospheric visibility, cause eye and respiratory-system irritation and damage vegetation. Photochemical smogs occur only where a mixture of pollutants is able to build up in a stagnant air mass. The necessary meteorological conditions are not widespread; one of the most important examples is the Los Angeles basin. Such conditions also occur in other urban areas in the United States, Japan, Europe and elsewhere although less frequently.

Regional

As with the sulphates resulting from sulphur dioxide emissions, the nitric acid and nitrates formed from NO_x emissions can be transported over long distances and can contribute towards the acidity of the rainfall in remote rural locations. In general, however, sulphates are the major cause of acid rain. Furthermore, only a minor proportion of the total nitrogen-based acidity is attributable to coal (other sources, such as fertilisers and transport vehicles, are usually more important).

Global

Concern about the effects on the global climate of the emission of nitrous oxide has been expressed in two areas: the 'greenhouse effect' and ozone depletion.

It has been suggested that an increase in the nitrous oxide concentration in the upper atmosphere may absorb more infrared radiation, leading to a 'greenhouse effect' on the earth's climate (see section 7.4). Also, the conversion of nitrous oxide to nitrogen dioxide in the upper atmosphere may deplete the ozone layer and increase the amount of ultraviolet radiation reaching the earth's surface.

As almost all of the nitrous oxide emissions arise from biological processes (only trace quantities being produced by fossil fuel combustion), these areas of concern do not relate directly to coal utilisation.

Control

Most of the work on the control of nitrogen oxide emissions from coal combustion has been directed towards pulverised fuel systems. The two main approaches are as follows.

Burner design

The two principal consequences of burner design that affect NO_x production are temperature/residence time relationship and oxygen concentration.

At the short gas residence times (a few seconds) in most combustors, the formation of NO_x from atmospheric nitrogen can occur only at high temperatures in excess of about 1500 °C. If the product gases are cooled slowly from the peak combustion temperature, the NO_x can decompose at lower temperatures. In high intensity combustion systems, however, a high flame temperature is often followed by rapid cooling of the gases. This effectively 'freezes' the NO_x content at the high temperature level. Nitrogen oxide production from atmospheric nitrogen can, therefore, be controlled by limiting the combustion temperature and avoiding rapid cooling of the gases.

Temperature has comparatively little effect on the formation of NO_x from the organic nitrogen in coal. It appears that most of this NO_x is derived from the nitrogen released in the volatile matter. Although the mechanisms are not fully understood, NO_x formation from the organic nitrogen can be limited by restricting the oxygen supply in the region where the volatile matter is first released.

The techniques that have been used to modify the combustion conditions to reduce NO_x emissions often involve both reductions in combustion temperature and staged oxygen supply. The options include

burner redesign to delay mixing of the secondary air supply
two-stage combustion in which a fuel-rich flame is supplemented by a separate air supply or air-rich flames
low overall excess-air level operation
flue gas recirculation (this reduces both the combustion temperature and the oxygen content)
fluidised bed combustion (low combustion temperature operation)

Removal after combustion

A number of processes for removing NO_x from combustion gases are being investigated but none are yet commercially available for use on coal-fired combustors. Development interest is greatest in Japan, where stringent NO_x emission standards have been introduced. Processes can be subdivided into wet removal and dry removal systems according to whether or not aqueous scrubbing is employed.

Conventional flue gas desulphurisation systems do not reduce NO_x emissions because NO, which is the main form present in combustion gases, is relatively insoluble. One approach to wet removal, therefore, is to oxidise the NO to NO_2 before the FGD plant by adding ozone or chlorine dioxide to the flue gases. The NO_2 formed is sufficiently soluble to be removed with the SO_2 in the scrubber. The main disadvantages of this system are the costs of the oxidant supply and the waste water treatment required to remove nitrates.

The most advanced example of the dry removal approach is the reduction of NO to elemental nitrogen by reaction with ammonia and oxygen over a catalyst at 350 to 400 °C. Since the conversion reactions generally take place upstream of the particulate removal system in order to avoid the costs of hot gas cleanup or flue gas reheat, catalyst deactivation by fly ash is a potential problem area. Another possible disadvantage is the formation of ammonium sulphate and bisulphate by the reaction of ammonia with sulphur trioxide in the flue gas. These ammonium sulphates can cause fouling and corrosion of downstream equipment and may result in sulphate fume pollution in the stack emissions.

A number of other removal processes, both wet and dry, have been proposed but are at a less advanced stage of development than the examples described above. These include simultaneous absorption of SO_2 and NO in an aqueous solution of a chelating compound (for example, ferrous EDTA) and simultaneous adsorption of SO_2 and NO on active carbon.

Even if the difficult technical problems facing removal processes can be solved, they seem likely to be expensive. For this reason, combustion system design is the generally favoured approach to NO_x control, except where stringent emission standards are considered necessary.

7.4 Carbon dioxide

Carbon dioxide is the fifth most abundant gas in the atmosphere (after nitrogen, oxygen, water vapour and argon) and plays an important part in life processes. In particular, carbon dioxide is extracted from the atmosphere by photosynthesis and returned to the atmosphere by plant decomposition and animal respiration.

Although it cannot, therefore, be regarded as a pollutant in the conventional sense, carbon dioxide has attracted attention because of its possible effect on the global climate. Concern centres on the now well-established observation that the concentration of carbon dioxide in the atmosphere is increasing. When reliable measurements began in 1958, the carbon dioxide concentration was 315 ppm (by volume) whereas, by 1980, the concentration had increased to 338 ppm, an average increase of 0.3 per cent, or 1 ppm, per annum. For comparison, it has been estimated that the carbon dioxide concentration at the beginning of the industrial revolution was 290 ppm.

The reasons for concern for the global climate may be summarised as follows. Most of the solar radiation intercepted by the earth is in the visible and ultraviolet regions of the spectrum. Of the total 0.35 kW/m^2 incident radiation, about 35 per cent is reflected by the atmosphere or the earth's surface and 20 per cent (mainly ultraviolet) is absorbed by the atmosphere, leaving about 45 per cent (mainly visible light) to be absorbed by the earth's surface. Of this absorbed radiation, about 40 per cent is re-radiated from the surface. However, because the surface temperature is relatively low, this re-radiation occurs mainly in the infrared part of the spectrum. The main radiation fluxes are illustrated in figure 7.4.

The importance of carbon dioxide as a component of the atmosphere is that, in common with water vapour, it is transparent to visible radiation but absorbs

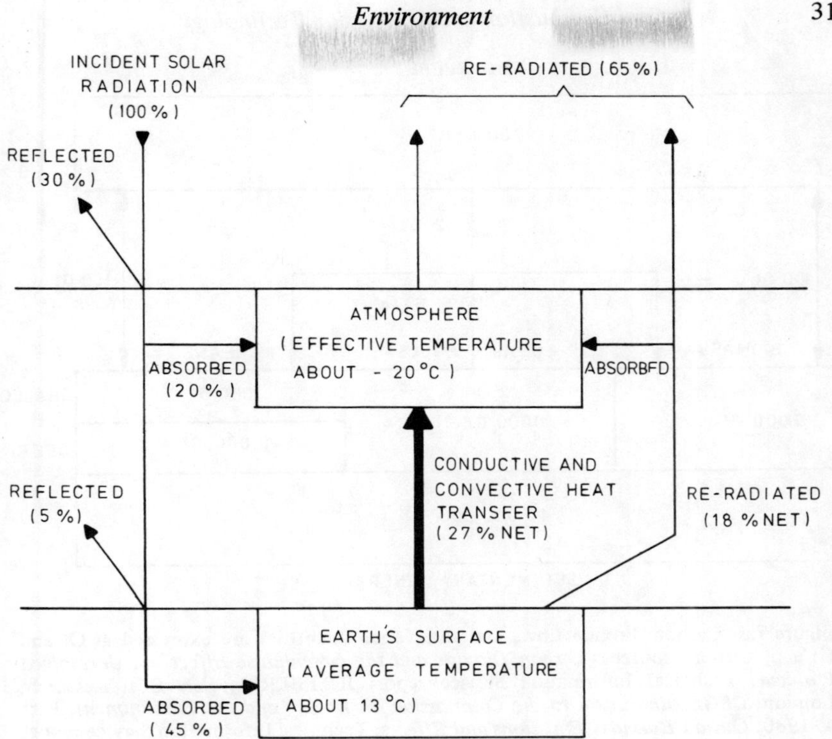

Figure 7.4 Main radiation fluxes in the earth–atmosphere system (source: J. T. McMullan, R. Morgan and R. B. Murray, *Energy Resources and Supply*, Wiley, New York, 1976)

infrared radiation. Not all of the re-radiated energy is, therefore, lost directly to space; some is trapped by absorption in the atmosphere. This phenomenon has come to be known as the 'greenhouse effect' (because the glass in a greenhouse performs a similar function) and makes the earth's surface significantly warmer than would otherwise be the case.

The possibility of changes in the carbon dioxide level affecting the climate was raised as long ago as 1863 by the British physicist Tyndall. However, the availability of more accurate data on carbon dioxide concentrations and computer-based simulation techniques have led to increased concern over recent years.

Reservoirs and flows

In order to place the above speculations in context, it is useful to consider the overall distribution of carbon dioxide. The main flows and reservoirs are shown in simplified form in figure 7.5.

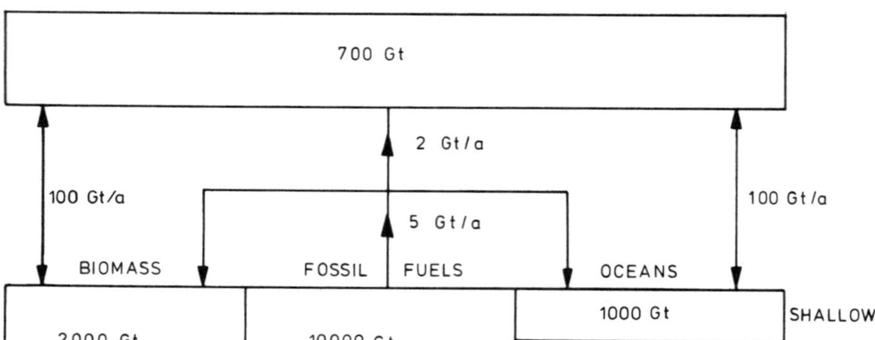

Figure 7.5 Carbon dioxide flows and reservoirs. Quantities are expressed as Gt and Gt/a of carbon (sources: *Carbon Dioxide and the Greenhouse Effect–an Unresolved Problem*, Technical Information Service report ICTIS/ER01, IEA Coal Research, London; *CEGB submission to the Commission on Energy and the Environment*, Part 4, 1980; *Carbon Dioxide–Emissions and Effects*, Technical Information Service report ICTIS/TR18, IEA Coal Research, London)

The amount of carbon stored as carbon dioxide in the atmosphere is about 700 Gt. This is small compared with the 40 000 Gt of carbon in the ocean depths (mainly as bicarbonates), and with the 10 000 000 Gt of carbon in sedimentary minerals (mainly as calcium carbonate). The potential of the ocean depths for absorbing the carbon dioxide released by fossil fuel consumption is, therefore, considerable; however, neither the rate of absorption nor the factors affecting it are yet known with any certainty.

It is thought that carbon dioxide in the ocean surface waters is approximately in equilibrium with that in the atmosphere, as is the exchange of carbon dioxide between biomass and the atmosphere. Expressed in terms of carbon, each of these fluxes is estimated to be approximately 100 Gt/a. The rate of carbon dioxide emission by fossil fuel combustion is comparatively small at 5 Gt/a (carbon), that is, about 2.5 per cent of the overall carbon flux. In addition, it has been suggested that deforestation may contribute an amount of carbon dioxide to the atmosphere similar to that of fossil fuel combustion.

The effect of absorption by biomass appears to be reflected in diurnal and seasonal variations in carbon dioxide concentrations at ground measuring stations of up to ± 20 ppm. However, the annual average carbon dioxide concentration varies by only ± 1 ppm at different locations on the earth's surface and by 2 ppm from the troposphere to the stratosphere (the northern hemisphere and the troposphere have the higher concentrations).

The observed increase in the atmospheric carbon dioxide concentration corresponds to a net carbon flux to the atmosphere of about 2 Gt/a. This is about 40 per cent of the rate of carbon dioxide production from fossil fuels and indicates that at least 60 per cent of the fossil fuel carbon dioxide emitted is absorbed by biomass or the oceans. As discussed above, however, not all of the increase in atmospheric carbon dioxide necessarily arises from fossil fuel combustion.

Since the ocean surface waters may be regarded as saturated with carbon dioxide and essentially at equilibrium with the atmosphere, an important factor affecting atmospheric carbon dioxide levels is the exchange between the ocean surface waters and the ocean depths. The rate of exchange is believed to be low, thereby reducing the effectiveness of the ocean depths as a sink for atmospheric carbon dioxide emissions. The mechanisms involved are not, however, well understood.

Climatic effects

Although the increase in the atmospheric carbon dioxide concentration is well established, no change in global temperatures has yet been measured. This is because any such change would be small and therefore masked by the 'noise' of normal climatic fluctuations. Indeed, over the last 40 years, the average global temperature has fallen by 0.4 °C.

All the predictions about the climatic effects of carbon dioxide are, therefore, based on mathematical models that attempt to simulate the main factors affecting the global climate. In spite of the uncertainties, most of the recent models lead to the conclusion that a doubling in the atmospheric carbon dioxide concentration would increase the mean global temperature by 1 to 4 °C. The increase would not, however, be uniform, a higher value of about 10 °C being expected for the polar regions.

A doubling of carbon dioxide concentrations would not occur for 300 years at the present rate of increase. However, with a continued growth in fossil fuel consumption, such levels could be reached by the second half of the next century.

Changes in temperature of the magnitude indicated by the models would, if correct, have a profound effect on the global climate and agriculture. In particular, the pattern of global rainfall is likely to change with some regions becoming wetter and others drier. Perhaps more serious, however, is the possibility of the polar ice caps melting with a consequent increase in sea level. This would take many centuries and, during this time, fossil fuel reserves would be exhausted at the rates of extraction contemplated.

Discussion of mechanisms

Most of the temperature increase predicted by the models occurs as a result of the following feedback mechanism. An increase in global temperature leads to more evaporation from the oceans and therefore higher water vapour concentrations in the atmosphere. Since water vapour, like carbon dioxide, is an efficient absorber of infrared, less energy is lost by re-radiation to space from the earth's surface. The balance of radiation fluxes summarised in figure 7.4 is therefore disturbed in the direction of enhancing the 'greenhouse effect' and making the climate still warmer than would be the case if the carbon dioxide concentration were the only factor.

The models on which the above estimates are made contain a number of approximations. In particular, little is known about the other feedback mechanisms that may be important. These may include the following:

(1) Increases in the atmospheric carbon dioxide level and the global temperature may lead to more plant growth which would tend to increase the rate of carbon dioxide removal.

(2) An increase in temperature would reduce the solubility of carbon dioxide in the ocean surface waters and may therefore impair the effectiveness of the oceans as a sink for carbon dioxide.

(3) An increase in temperature would increase evaporation and therefore cloud cover. This would reduce the amount of incident solar radiation.

Since some of the feedback mechanisms (including the main one studied so far, involving increased atmospheric water vapour concentrations) are positive, there is concern that the overall effect may be one of positive feedback. If this were the case, even relatively small increases in atmospheric carbon dioxide concentrations could destabilise the delicate natural balances of the global climate with possibly disastrous consequences.

The potential seriousness of the carbon dioxide/'greenhouse effect' problem and the extent of the uncertainties involved in the present models has led to the establishment of an active international research effort on this topic. No practical means of disposing of carbon dioxide has yet been found and present work is therefore concentrated almost entirely on predicting more accurately the implications of continued fossil fuel use.

7.5 Trace species

In addition to the major products of coal combustion and conversion, a large number of substances are emitted in minute amounts. Concern about the possible effects of these emissions is comparatively recent and, in many cases, more data are still required. Although some of the substances are known to be toxic, carcinogenic or mutagenic in high dosages, the possibility of health hazards arising from the low concentrations produced by coal use appears generally to be small.

Trace elements

Almost every element in the periodic table can be found in coal at some concentration. For convenience, the following categories may be defined.

Major elements

These are the main constituents of the organic coal substance: carbon, hydrogen and oxygen.

Minor elements

These comprise the minor elements in the organic coal substance (sulphur and nitrogen) and the main inorganic elements in the mineral matter (silicon, aluminium, iron, calcium, sodium, potassium and magnesium).

Trace elements

The term 'trace elements' usually refers to those elements which are not included as major or minor elements.

Average concentrations of trace elements in coal are generally of the same order as their occurrence in the earth's crust. The main exceptions are selenium, which is highly enriched, and arsenic, antimony, boron, cadmium, chlorine and molybdenum, which are slightly enriched. The concentrations are therefore small, as indicated in terms of orders of magnitudes for some of the elements in table 7.4.

Although the average concentrations of trace elements in coal are similar to their crustal abundance, the range of variation between coals can be large. In general, only two-thirds of coals are likely to have concentrations of a trace element lying within a factor of two of the mean value.

Trace elements appear to have been incorporated within the coal during and after the initial deposition of the seam. Most of the elements are present both in organic and inorganic form. The inorganic material occurs mainly as sulphides, aluminosilicates and carbonates and may be removed partially by washing the coal. Little is known about the organic chemical structures containing trace elements. The principal forms in which some of the trace elements occur are given in table 7.4.

Concern about the possible environmental impact of trace elements in coal is associated mainly with possible health hazards arising from the disposal and

Table 7.4 Occurrence of trace elements in coal[a]

Element	Average concentration (order of magnitude, in ppm)	Main form of occurrence	Emission characteristics
Antimony	1	Sulphide	Enrichment
Arsenic	10	Sulphide	Enrichment
Beryllium	1	Organic	
Boron	100	Organic	Enrichment
Cadmium	1	Sulphide	Enrichment
Chlorine	1000	Organic and inorganic	Vapour
Chromium	10		Enrichment
Cobalt	10	Organic and inorganic	Refractory
Copper	10	Sulphide	Enrichment
Fluorine	100		Vapour
Lead	10	Sulphide	Enrichment
Manganese	100	Carbonate	Refractory
Mercury	0.1	Sulphide	Vapour
Molybdenum	1	Organic and inorganic	Enrichment
Nickel	10	Organic and inorganic	Enrichment
Selenium	1	Sulphide	Vapour
Thallium	1		Enrichment
Titanium	1000	Aluminosilicate	Refractory
Thorium	1		Refractory
Tungsten	1		
Uranium	1		Enrichment
Vanadium	10	Organic	Enrichment
Zinc	100	Sulphide	Enrichment

[a]Sources: *Trace Elements from Coal Combustion Atmospheric Emissions*, Technical Information Service report ICTIS/TR05, IEA Coal Research, London; *National Coal Board submission to the Commission on Energy and the Environment*, Topic 4, 1980.

dispersion of combustion residues. So far, research has concentrated on the elemental composition of these residues although the main elements of interest are unlikely to be emitted in significant amounts in elemental form.

Although it is known that the extent of the health hazard from such elements (for example, toxicity) depends critically on the type of chemical compounds formed, for most elements little is yet known about this aspect.

The residues arising from combustion may leave the system in two ways: as solids for disposal or as emissions to the atmosphere.

The possible environmental impact of the trace elements present in ash for disposal will be considered in section 7.7. The remainder of the present section will therefore concentrate on atmospheric emissions where possible effects on human health could occur by inhalation or ingestion via the food chain.

An important factor in assessing the potential health hazard is the partitioning of the trace elements between the above two categories (that is, solid waste and atmospheric emissions). This varies from element to element and depends on the properties of the compounds formed during combustion. Three possibilities may be identified.

Refractory

For elements that are not volatilised during combustion, the distribution between the two methods of removal is in proportion to the mass flows (on a carbon-free basis). Emissions to the atmosphere are, therefore, comparatively low in this case. However, because the ash represents only a small proportion of the original coal, the final concentration of the refractory elements in the ash is from 3 to 20 times their concentration in the original coal (as, for example, given in table 7.4).

Enrichment

With an important class of elements, enrichment occurs on the fine particles that cannot be captured effectively by dust collection equipment and are consequently emitted to the atmosphere. This arises with elements that are partially or totally volatilised at combustion temperatures but condense as the combustion gases are cooled by heat exchangers. The condensation occurs preferentially on the fine particles entrained in the gas stream because of their relatively high surface area per unit mass.

Vapour

Some trace elements are emitted to the atmosphere partially or completely as a vapour phase component of the combustion gases.

The broad pattern of partitioning for some of the trace elements is indicated in terms of the above categories in table 7.4.

The degree of enrichment resulting from condensation on fine particulates and vapour-phase emissions may be assessed as the ratio of the concentration of an element to that of a standard refractory element (for example, aluminium or iron) in the atmospheric emissions, divided by the corresponding ratio for the original coal. For power stations, enrichment factors have been estimated to be generally less than 10 and are thought to be no greater for industrial and domestic systems.

Concern about trace element emissions from coal combustion arises from the known toxic, carcinogenic or mutagenic effects that have been observed either in individuals exposed to high doses of these elements as a result of industrial processes or in animal bioassay or cell culture tests.

The only reported health hazard associated with atmospheric trace element emissions from coal combustion appears to be the case of a power station in Czechoslovakia burning a coal with an arsenic content 100 times higher than the average given in table 7.4.

Apart from this, the general view is that ground-level atmospheric concentrations of trace elements from coal combustion are well below the levels at which there is a significant health hazard (for example, several orders of magnitude below current Threshold Limit Values–TLVs). There are, however, indications that some heavy metals, for example cadmium, may accumulate in the soil around a power plant under certain circumstances (for example, high initial concentrations in the coal coupled with poor particulate emission control). This aspect, together with enrichment, is the subject of a number of active research programmes.

Radioactivity

A special case of trace element emissions is that of radioactive elements. As noted in the previous section, almost every element in the periodic table can be found in coal and this includes the naturally occurring radioactive isotopes (principally uranium-238, thorium-232, radium-226, polonium-210, lead-210 and potassium-40).

Attention has been attracted to the topic of radioactive emissions by comparisons which show that, per unit of electricity produced, a coal-fired power station emits more radioactivity to the environment than a nuclear station.

In both cases, however, the emissions have to be seen in the context of their overall level. In the United Kingdom, for example, the maximum annual dose of radioactivity received by a member of the public living in the vicinity of a large coal-fired power station (the so-called 'critical group') is likely to be about 230 microsieverts per year or about 13 per cent of the average natural background radiation dose of 1860 microsieverts per year. This is less than the variation in natural background radiation from one part of the United Kingdom to another and is less than 5 per cent of the dose limit of 5000 microsieverts per year recommended by the International Commission on Radiological Protection (ICRP) for critical groups of the general public. The average annual dose to the entire United Kingdom population from all coal-fired power stations in the United Kingdom is even lower at about 4 microsieverts per year or 0.2 per cent of natural background radiation. This is less than the year to year variation in the natural background radiation arising, for example, from changes in cosmic ray activity.

A number of other non-nuclear factors contribute radioactivity to the environment. Building materials, for example, contain the radioactive radon daughter isotopes polonium-218, lead-214 and bismuth-214. These can result in an in-house dose enhancement of 700 microsieverts per year under adverse conditions and therefore can be considerably more important than coal-fired power station emissions.

In terms of the comparison between emissions from coal-fired and nuclear stations, it may be noted that there are no abnormal operating conditions likely to occur at a coal-fired station that can result in a significant increase in the emission of radioactive elements. Furthermore, although radioactive emissions from a nuclear power station are low, significant amounts of radioactivity are also released elsewhere in the nuclear fuel cycle; overall, the contribution of nuclear electricity generation to radioactivity in the environment is likely to be higher than that from coal-fired power generation.

Trace gases

On combustion, the major and minor organic elements in coal are released mainly as carbon dioxide, steam, nitric oxide, nitrogen dioxide and sulphur dioxide. In addition to these compounds, there are indications or suspicions that a number of other compounds of the major and minor organic elements may be present in trace quantities. Such compounds are referred to as trace gases and include the following

methane
ethylene
higher hydrocarbons
carbon monoxide
nitrous oxide (see section 7.3)
hydrogen cyanide
hydrogen sulphide
carbonyl sulphide
carbon disulphide

Work on the identification and measurement of trace gases is at an early stage but the indications are that these compounds are generally present only in minute quantities in the gases resulting from industrial and power station coal combustion and represent no significant health hazard. Larger quantities may be contributed by road vehicle engines and, particularly in the case of hydrocarbons, industrial solvents. These sources certainly expose the population to greater ground-level concentrations because of the less effective dispersal of the gases.

Although the environmental impact of coal conversion plants has yet to be fully evaluated, there appears to be no reason for trace gas emissions to be higher than for present combustion applications.

Polycyclic aromatic hydrocarbons

Polycyclic aromatic hydrocarbons (abbreviated to PAH) are a class of organic compounds that are contained in the volatile matter evolved from coal above about 450 °C. They are of particular concern because of their known carcinogenic properties. If emitted to the atmosphere, they may condense to form droplets of a respirable size so that, in sufficient quantities, they may be a contributory cause of lung cancer.

Other classes of complex, coal-derived organic compounds such as aromatic amines, benzene, substituted benzenes and methyl sulphates are also known to be carcinogenic but are of more concern for occupational health in coal conversion industries than as environmental hazards.

Atmospheric PAH concentrations may be characterised in several ways; two of the most common are the concentration of benzene soluble material (BSM) and the concentration of the compound benzo-a-pyrene (BaP).

Concern about emissions of PAH lies in three main areas.

Occupational exposures at coke ovens

Fugitive emissions of PAH arise at coke ovens, principally from poor seals (particularly on oven doors) and during charging and pushing.

As a result, coke oven workers, especially those employed on the top of the ovens, may be exposed to high atmospheric concentrations of PAH, sometimes up to about 3000 $\mu g/m^3$ of benzene-soluble material. Statistical analyses indicate that such highly exposed workers have a substantially increased mortality rate arising from malignancies, particularly of the lungs.

The identification of this hazard has prompted the establishment of a threshold limit value (TLV) of 200 $\mu g/m^3$ (BSM) in the United States. This is thought to represent the level below which workers may be exposed as part of their daily duties without significant adverse effects.

Techniques that may be adopted for reducing emissions to a safe level include improvements in the sealing of doors and lids and careful control of charging and pushing.

Urban atmospheric concentrations

The smoke produced by burning bituminous coal on open fires contains PAH. Before controls on emissions from domestic fires were introduced, urban concentrations of benzene-soluble materials in the United Kingdom averaged 25 $\mu g/m^3$ with peaks of several hundred $\mu g/m^3$. These values have been reduced dramatically with smoke levels, and urban air concentrations of benzene-soluble materials now average about 3 $\mu g/m^3$. It is thought that less than half of the current concentrations arise from coal combustion with a major contributor now being motor vehicle exhausts. The risks of cancer induced by the emissions from coal combustion are, therefore, correspondingly small.

Implications for future coal liquefaction plants

The direct coal liquefaction processes involve the production of PAH as constituents of intermediate products in the hydrogenation of coal to finished fuels. This has led to concern for the safety of the workers who will be required to operate these plants in the future.

However, coal liquefaction plants differ markedly from coke ovens in that the processes take place within completely sealed high-pressure containment vessels. In this respect, coal liquefaction plants resemble oil-refineries which also handle PAH-rich streams. Exposure levels may, therefore, be expected to be minimal.

Concern has also been expressed about the effects of handling and use of the finished fuels produced by coal liquefaction. In general, these fuels are similar in nature to the liquid fuels produced from crude oil, as discussed in chapter 5. Their use should not, therefore, present any major new environmental problems.

7.6 Liquid effluents

Liquid effluents arise mainly from coal processes which include gas treatment, for example, coke ovens and gasification.

As discussed in chapter 4, the initial stage of gas treatment is generally aqueous scrubbing. This involves the use of a recycled aqueous liquor stream to cool the gas and to permit the removal of any particulates or tars that may be present. These materials are readily separated from the aqueous stream by density devices, for example, settling tanks. A purge stream is taken from the aqueous recycle loop for processing and disposal. Depending on the process, further aqueous liquor streams may be obtained as the gas is cooled after the initial aqueous scrubbing stage or after a subsequent shift reaction stage.

None of these aqueous liquor streams are now generally considered suitable for direct disposal because of the dissolved contaminants that they contain. The contaminants include the following.

Gases

The principal dissolved gas is ammonia but some hydrogen sulphide and hydrogen cyanide are also present.

Organic compounds

These are mainly present as phenols.

Inorganic salts

Dissolved inorganic salts include chlorides, sulphides, sulphates, thiosulphates, thiocyanates, cyanides and ammonium compounds.

Environmental impact

Formerly, it was common practice to discharge aqueous liquor streams from coke oven gas processing directly into natural water courses. The main environmental impacts were as follows.

Aquatic life

Effects detrimental to fish and plant life can arise for three reasons: toxicity, oxygen depletion and turbidity.

Some of the contaminants present in aqueous liquor streams are known to be highly toxic to fish. Concern centres on ammonia and phenols (of the minor constituents, cyanides and sulphides are also highly toxic). Since ammonia is toxic mainly as 'free ammonia' (that is, as a dissolved gas), rather than fixed as ammonium salts, effluent streams containing ammonia are more hazardous under alkaline conditions than under acidic conditions.

The dissolved organic material (principally phenols) and some of the inorganic components (including cyanides and sulphides) are oxidised by biological action if discharged into natural water courses. Although removing toxic materials from the environment, this process results in the waters becoming depleted of dissolved oxygen. The ability of such water courses to sustain aquatic life can be seriously impaired. Empirical indices such as the biological oxygen demand (BOD) and the chemical oxygen demand (COD) have been introduced to quantify the capacity of effluents for oxygen depletion.

The presence of contaminants can also reduce light transmission and aquatic plant growth may consequently be inhibited. Phenols are thought to be partly responsible for this effect, referred to as turbidity.

Effects on drinking water

The effects of dilution, biological oxidation in the natural environment and processing at waterworks combine so that, with present standards, no significant human health hazards via drinking water arise from aqueous effluent disposal.

The main concern, however, is that phenols present in the water combine with the chlorine added for water purification purposes to form chlorophenols. These compounds have an unpleasant taste and are detectable in extremely low concentrations (a few parts per billion).

Control

Methods of control for aqueous effluent pollutants are well developed and are generally employed in coke works. As the processes available for future coal conversion plants are likely to be based on those developed for coking, the remainder of this section concentrates on the main types of existing technology.

The two most important pollutants in aqueous liquor are usually ammonia and phenols. Traditionally, these species were recovered from coke oven effluent streams because of their value as chemical feedstocks rather than for reasons of pollution control.

Present conditions differ in two respects. The development of new chemical processes has made the economics of ammonia and phenol recovery in coke works less favourable and the efficiency of the recovery processes is not now generally regarded as adequate from an environmental viewpoint. Further stages of treatment based on destructive removal of the aqueous liquor contaminants have therefore been developed for use either in addition to, or instead of, the recovery processes.

For the coal conversion processes currently under development, design studies often provide for the recovery of ammonia and phenol although their destruction by recycle to a gasifier also appears attractive (see chapter 4).

The main recovery and removal technologies are summarised below in the order in which they would normally be employed.

Phenol extraction

Phenol may be recovered from aqueous liquor by scrubbing with a light aromatic solvent (for example, benzole). The 'phenolised' solvent is treated with caustic soda and the phenol (as sodium phenolate) is separated out by a density method. The regenerated solvent is recycled to the scrubber.

The Phenosolvan process is of this general type except that butyl acetates are used as solvents with the advantage of more effective removal of polyhydric phenols.

Although removal efficiencies of 99 per cent can be achieved by extraction, in general additional, more efficient removal processes are required to meet pollution controls.

Distillation

Dissolved ammonia gas can be removed by distillation in an ammonia still. An alkali, such as lime, is first added to the liquor to change the equilibrium from fixed ammonia towards free ammonia in order to maximise recovery.

The ammonia may be absorbed in sulphuric acid to produce ammonium sulphate fertiliser. Alternatively, the ammonia may be incinerated although this may create nitrogen oxide emission problems.

Biological treatment

Many of the pollutants present in aqueous liquor streams can be oxidised to a harmless form by the action of microbiological organisms. Work on this approach began in the first decade of the twentieth century and it is now established as a standard technology for treating aqueous liquors from coking plants.

The use of biological processes for this purpose arose as an application of the sanitary treatment of sewage. Indeed, many coke oven sites still use local municipal sewage treatment plants to process their aqueous effluent. However, better control and more efficient pollutant removal can be achieved with a purpose-designed biological plant on site.

The standard method of on-site biological treatment of aqueous liquor is the 'activated sludge' process. The liquor is mixed with dilution water and nutrients and fed to an agitated tank where it comes into contact with the activated sludge. The activated sludge contains several strains of bacteria capable of oxidising various constituents of the liquor. The bacteria are all aerobic and the tank is therefore aerated. A stream containing effluent and sludge is taken continuously from the tank to a gravity separation vessel where the treated effluent is decanted. The remaining sludge is recycled to the reaction tank with a purge stream being removed in order to maintain the activity of the sludge.

Present activated sludge processes are designed mainly to remove phenol although a significant reduction in thiosulphate, thiocyanates and cyanides can also be obtained.

Phenol removal efficiencies can be extremely high, values approaching 99.99 per cent having been achieved in commercial-scale plant.

Conventional activated sludge processes do not reduce the ammonia content of the liquor. However, bacteria capable of converting ammonia into nitrogen in two stages (aerobic oxidation to nitrates, followed by anaerobic reduction to nitrogen) are known to exist and commercial processes based on this approach may be developed in the future.

Tertiary processes

The treated effluent stream from an activated sludge process, while satisfying existing requirements, still contains small quantities of pollutants. If further improvements in effluent quality are required, tertiary 'polishing' processes will have to be introduced. For example, active carbon clean-up has been suggested for this purpose.

7.7 Solid residues

When coal is used in any combustion or conversion process, a solid residue is obtained. In many cases, this is derived mainly from the mineral matter in the original coal although unconverted carbon and sulphur acceptor materials may also be present.

Currently, nearly all of the solid residues resulting from coal utilisation are ash produced by conventional coal combustion systems. For this reason, the environmental impact of this material has been studied extensively. In the future, however, increasing quantities of solid residues may arise from other processes, principally flue gas desulphurisation, fluidised bed combustion and gasification.

The main features of the formation, utilisation and disposal of combustion ash are outlined below together with the differences between combustion ash and the solid residues from future technologies.

Formation of combustion ash

Although, to a first approximation, the mineral matter in coal may be regarded as an inert material in coal processes, chemical and physical changes do occur at combustion and gasification temperatures. The properties of the resulting ash depend on the type of process and can affect the uses or methods of disposal suitable for the material.

Under conventional combustion conditions, the main chemical changes are

removal of the water of crystallisation
calcination of the carbonates
oxidation of the sulphides

Ash comprises mainly silicates, oxides and sulphates, the above reactions being essentially complete by 750 °C.

The 'physical' changes depend on the temperature and residence time of the material and do not begin until about 1000 °C. At this temperature most ashes sinter (that is, become sticky). Above about 1400 °C the particles become completely molten (that is, form a slag) and, when cooled, solidify into a hard, fused glassy material.

Although, in pulverised fuel furnaces, the ash is heated to a high temperature (in excess of 1500 °C), the residence time is short and only partial melting of the ash occurs. The resulting material is therefore a mixture of fused and unfused particles.

The ash produced by mechanical stokers varies according to the design from an almost completely fused to a largely unfused product.

Utilisation of combustion ash

In most industrial countries, the main source of ash at present is power stations. This reflects not only the importance of power generation as a market for coal but also the relatively low ash content of the coal grades supplied for other applications. Most of the ash is therefore produced by pulverised fuel combustion and is referred to as pulverised fuel ash (PFA).

There exist a number of beneficial uses for PFA as a construction material. In the United Kingdom, over one-third of the PFA produced is employed in this way and, in the United States, the proportion is about one-quarter. The principal applications are as a specialised road fill and in the manufacture of building blocks. Smaller quantities are used as a light-weight aggregate for concrete, as cement substitute for concrete and as a fill for building sites.

Markets similar to those for PFA exist for the ash from industrial combustors and, in some cases, the ash can be sold thereby providing a source of income.

Disposal of combustion ash

The ash that is not utilised for construction purposes is disposed of by dumping. Two methods are used:

(1) The creation of a lagoon to which the ash is transferred by a water-slurry pipeline (usually 60 per cent water and 40 per cent ash). This method is often employed where ash disposal is required in the vicinity of a power station.
(2) The transportation, tipping and levelling of dry ash using conventional civil engineering machinery. The 'dry ash' would normally be conditioned with about 15 per cent water before leaving the site to improve handling and to reduce the lifting of dust in windy weather.

The main environmental concern that arises in connection with the disposal of ash is the possibility that trace elements (see section 7.5) may be leached out as soluble salts by rainfall. This could result in the contamination of surface and underground watercourses and, ultimately, of the food chain or drinking water.

Fortunately, only about 2 per cent of the ash is readily soluble and, of this, only about 1 per cent comprises hazardous trace elements. The alkalinity of the leachate inhibits the solution of many of the heavy metals.

Where disposal takes place as 'dry ash', it may require several years of rainfall to provide sufficient water to saturate the dump and for leachate to appear. If, however, the ash is stored in a lagoon, it is saturated by the transport water and leachate appears quickly. In this case, the leachate also has a lower concentration of dissolved solids.

The overall concentration of dissolved solids in the leachate is usually referred to as the total dissolved solids (TDS). This is highest in the initial leachate, decreasing by an order of magnitude after two bed volumes of water have passed through the ash. Only a few toxic elements occur in the initial leachate at concentrations greater than the World Health Organisation standards for drinking water. The concentrations are further reduced by absorption when passing through subsoil strata (for example, gravel), and by dilution in the aquifer.

Provided that hydrological data are available and appropriate management strategies and dumping rates are observed, the hazard from ash leaching is thought to be low.

A secondary environmental concern associated with ash disposal is the creation of a dust nuisance. This problem can be minimised by careful management. For dry ash disposal, surface treatment (for example, the addition of topsoil) should follow the ash-tipping phase closely so that the exposure of unprotected ash is minimised. In lagoons, plants such as bullrushes may be grown in the floating ash layer.

Flue gas desulphurisation (FGD)

The technology of flue gas desulphurisation is described in section 6.7.

Conventional lime/limestone scrubbers produce a sludge comprising calcium sulphite and sulphate. The amount of sludge produced depends on the sulphur content of the coal; for a 4 per cent sulphur/15 per cent ash coal, the amount of sludge (on a dry basis) is similar to the amount of ash.

The sludge is often simply dumped in lagoons where it remains as a thixotropic gel and does not become structurally stable with time. The main environmental concerns associated with FGD sludge disposal are as follows.

(1) Since the sludge is not structurally stable, the large areas of land required for the lagoons cannot be returned to agricultural or other uses when disposal is completed.

(2) Leaching of the sludge may contaminate surface or underground waterways.

For these reasons, it is now widely accepted that stabilisation of FGD sludge before disposal is desirable. This is possible by mixing the sludge, after thickening or filtration, with dry fly ash, lime or fine blast furnace slag. The resulting material hardens sufficiently in time to permit disposal as landfill and also has acceptable leaching characteristics.

Other FGD systems have been developed (see section 6.7) to produce materials that are more easily disposed of (such as dry acceptor waste) or saleable (such as gypsum, sulphur, sulphuric acid or liquid sulphur dioxide). The widespread adoption of systems that produce saleable products might, in some regions, provide greater quantities than the market is capable of absorbing. It is thought, however, that a major proportion of the coal-fired power stations in Europe could be equipped with this type of FGD without meeting a marketing problem.

Fluidised bed combustion

The composition of the solid residue obtained from fluidised bed combustion depends on the design and operating conditions. The main cases are summarised below (see also chapter 3).

No sulphur retention

Fluidised bed combustors operate at a relatively low temperature so that the solid residue in this case is a non-sintered ash.

Atmospheric pressure combustion with sulphur retention

With atmospheric pressure combustion, limestone would usually be employed as a sulphur acceptor material. The solid residue is, therefore, a mixture of ash, calcined stone and sulphated stone.

Pressurised combustion with sulphur retention

Dolomite is the preferred acceptor material for operation at elevated pressure and results in a solid residue comprising ash, partially calcined stone and sulphated stone.

The main opportunities for using non-sulphated residues are as a construction material in markets similar to those for PFA and stoker ash (for example, in road construction and as an aggregate). However, little work has yet been carried out on these potential applications.

Sulphated fluidised bed combustion residues are generally inferior to PFA as construction materials and are unlikely to find outlets in most of the traditional PFA applications. Owing to their free lime content, sulphated residues might be used in agriculture as a neutralising additive for acidic soils.

Since fluidised bed combustion residues may differ markedly from PFA in their physical and chemical characteristics, their leaching characteristics may also differ. However, leaching studies indicate that the concentrations of toxic elements are generally acceptable although the leachate can be quite alkaline.

Gasification

The reducing conditions in a gasifier may be expected to produce an ash product with properties different from those obtained from combustion processes. However, gasifiers generally contain a high-temperature combustion zone that provides oxidising conditions similar to those in a combustor for some of the reaction time.

Although more work has to be done in this area, on present evidence it appears that, in general, gasification residue may be utilised as a construction material or disposed of by dumping in the same way as for PFA. The leaching of trace elements from gasification ash is expected to be within acceptable limits. For gasifiers that operate under slagging conditions and require the addition of limestone as a fluxing agent, utilisation of the residue is unlikely to be attractive and an alkaline leachate may be expected.

7.8 Other environmental impacts

The environmental concerns discussed in the previous sections all relate specifically to possible forms of pollution from coal combustion and conversion processes. There are, however, a number of general impacts on the environment that are similar to those of many other major industrial activities such as oil refining and power generation. These are noted briefly in the present section for completeness.

Catalysts

In common with other chemical processes, coal conversion plants generally produce small quantities of spent catalyst. These may either be treated before disposal to reduce their toxicity or returned for reprocessing into new catalyst material.

Thermal pollution

Almost all energy consumption is ultimately rejected as heat to the environment. Normally, such heat is dissipated over a wide area. However, the tendency for large energy conversion or consumption facilities to be concentrated in a small area has given rise to concern that the local climate might be affected.

The amount of heat rejected from a conversion process depends on its thermal efficiency. The thermal efficiency of coal conversion processes is usually 50 to 70 per cent so that 30 to 50 per cent of the chemical energy in the coal is rejected as heat to the environment. As shown in table 7.5, although this amount of heat is greater than that produced by an oil-refinery (80 to 90 per cent efficiency), it is less than that from power stations (35 to 40 per cent efficiency).

Table 7.5 Typical thermal pollution, land and water requirements[a]

Process	Heat rejected (%)	Land (ha/GW input)	Water	
			(1/GJ)	(t/t, coal-equivalent input)
Coal-fired power station	30–50	*c.* 25	*c.* 300	*c.* 10
Oil-refinery	10–20	*c.* 10	*c.* 50	*c.* 1.5
Coal conversion process (SNG/liquefaction)	60–65	15–30	15–150	1.5–5

[a]Sources: *Liquid Fuels from Coal*, National Coal Board, Hobart House, London, 1978; *Constraints on Liquefaction in the U.K. to the Year 2020*, Coal Research Establishment, Cheltenham, 1981.

Land

In table 7.5 typical data on land utilisation for coal conversion processes are compared with those for coal-fired power stations and oil-refineries. Expressed in terms of energy input, the land requirements for coal conversion are similar to those for power generation but greater than those for oil refining.

However, considerably more land is used in coal production (100 to 200 ha/GW in the United Kingdom) than is likely to be used in the subsequent conversion. Land availability is not, therefore, expected to be a major constraint in the exploitation of coal utilisation technology.

Water

Typical water requirements for power generation, oil refining and coal conversion are also summarised in table 7.5. As with land requirements and thermal pollution, coal conversion processes may be regarded as intermediate between oil refining and power generation in terms of their water consumption.

Although it can be preferable to locate a coal conversion plant near the mine, if this is in a remote unindustrialised area, water supply may be a problem. Under such circumstances, the additional costs of transporting water to the plant, or of transporting the coal to a more favourable location, can be considerable.

Visual impact

Any large coal processing plant will, inevitably, have some visual impact. However, the new technologies described in earlier chapters are unlikely to appear substantially different from existing power stations or oil-refineries. The main items contributing to the visual appearance of conversion plants are

- the plant structure
- coal storage and handling facilities
- product storage or distribution facilities
- cooling towers
- miscellaneous buildings (for example, offices)

It is expected that large coal conversion plants will be situated either near a coal mine or in an existing industrial or port area. None of the above facilities are, therefore, likely to create a visual impact that is out of place in such an environment.

CHAPTER 8

ECONOMICS

Ultimately, the requirements for new coal utilisation technologies will be determined by their economics and economic assessments have accordingly played a prominent role in research and development programmes. Although extensive, however, the literature on coal utilisation economics is often not particularly helpful because of differences in the economic conventions and design ground rules used by different authors.

It is, in any case, generally acknowledged to be extremely difficult to obtain agreement on capital cost estimates from independent sources to within 10 per cent even with conventional technologies such as pulverised fuel power generation. For technologies still being developed, definitive estimates are not available and realistic error bands must be wider still, perhaps plus or minus 30 per cent. Although efficiencies and yields can be calculated more accurately, significant differences can also occur with these parameters.

In view of these uncertainties and difficulties, it may be tempting to conclude that economic studies of developing technologies are of little assistance in assessing their future prospects. Nevertheless, using the methods of analysis outlined in sections 8.2 and 8.3, some general features do emerge, as discussed qualitatively in subsequent sections. Specific data on investment and efficiencies are given in appendix 2.

8.1 Design considerations

The first stage in carrying out an economic assessment is to obtain a design for the plant. If the assessment is preliminary, then the design might be assumed to be similar to that for an earlier study or an existing plant. For a more accurate and detailed estimate, a specific design study is required.

Since economic assessments are generally an aid to decision-making, most studies are comparative. One of the difficulties in such work is in ensuring that the information obtained is on a consistent basis. In particular, for the economic comparisons to be valid, the underlying plant designs have to be carried out using a standard set of assumptions and ground rules. This applies especially to assessments that are not site-specific.

Some of the more important economic implications of the design basis are discussed below. Throughout this chapter, capital costs are expressed in terms of the useful energy output ($/kW), unless otherwise stated.

Location

The plant location can have a significant effect on costs; existing industrial areas are generally more favourable than remote, non-industrial areas. The main factors are as follows.

The site

If the site is undeveloped and in a non-industrial area (commonly referred to as a 'greenfield' site), it may be necessary to provide roads, a water supply, power and houses for the workforce. In an industrial area, however, many or all of these facilities will already be available and the considerable additional investment on infrastructure is avoided.

The country or region

Surveys have shown that the general level of plant investment varies from country to country. Typical 1979 cost ratios for a range of countries and regions are shown in table 8.1. In broad terms, industrialised regions have lower costs than developing regions. Western Europe, the United States and Japan, for example, have lower costs than Third World countries. The reasons are thought to be differences in factors such as labour rates, productivity, raw material

Table 8.1 Relative costs for process plant construction[a]

Country/region	Relative cost
Australia	1.3
Canada	1.15
Central Africa	2.0
France	0.95
West Germany	1.0
India (imported element)	1.8
(indigenous element)	0.65
Italy	0.9
Japan	0.9
Middle East	1.1
South America (north)	1.15
South America (south)	1.35
United Kingdom	0.9
United States	1.0

[a]Source: A. V. Bridgwater, International Construction Cost Location Factors, *Chemical Engineering*, **86**, No. 24 (1979) 119.

costs, the strength of the engineering industries, methods of financing and the degree to which it is necessary to import components and skilled labour. The differences may therefore also change with time.

By-product values

By-product values may be expected to vary according to a pattern similar to that of investment with the more remote areas tending to offer lower values than industrial areas. Indeed, in some remote locations, by-products that would find ready markets in industrial regions may be unmarketable and the plant design may have to be modified accordingly. This applies particularly where a fuel gas is produced as a by-product and inevitably leads to a reduction in the overall process efficiency (often by 10 percentage points, or more).

Scale

Although the capital cost of a plant increases with size, it is generally found that the increase is not proportional. Almost invariably, the investment required per unit of throughput is less for a large plant than for a small plant. This effect of the 'economies of scale' is usually expressed by an equation of the form:

$$K = K_0 \ (S/S_0)^{J-1} \qquad \$/\text{kW} \qquad (8.1)$$

where K, K_0 are the specific investments (\$/kW) for the plant of interest and a standard plant respectively
S, S_0 are the sizes (throughputs) of the plant of interest and a standard plant respectively
and J is the 'scale factor'.

The value of the scale factor varies with the type of process but typically lies in the range 0.6 to 0.9. A value of 0.75, for example, indicates that trebling the plant size reduces the specific capital cost by about 25 per cent.

Equation 8.1 may be applied to the total plant cost but is more accurate if used for each of the constituent units as some items are more affected by the economies of scale than others. In a large plant, for example, some units may be used at their maximum practical size. Further increases in scale therefore require multiple streams for such units with correspondingly reduced economies of scale.

As a result of the above effects, comparative economic assessments of competing technologies are valid only at the specified scale.

Development status

For developing technologies, a distinction can be drawn between cost estimates that refer to a 'pioneer plant' and those that anticipate a 'mature technology'. A pioneer plant is always relatively expensive, for example, because of the need for generous design factors, built-in redundancy to ensure a workable plant, higher design/engineering costs for a non-standard specification and, possibly, research and development. As further plants are built and operating experience accumulated, costs can be reduced and the process optimisation further refined.

The effect of experience on unit costs has been studied extensively, particularly in relation to manufacturing processes. The work of the Boston Consulting Group, for example, indicates that a power law of a form similar to that used for the economies of scale may be widely applicable

$$K = k\,(D/d)^{j-1} \qquad \$/\mathrm{kW} \tag{8.2}$$

where K, k are the specific investments (\$/kW) for the plant with and without the benefit of additional experience respectively
D, d are corresponding indices of the accumulated experience (for example, the number of plants built)
and j is a constant.

A value commonly taken for the constant j is 0.9 (the so-called 'nine-tenths rule'), suggesting a 20 per cent reduction in unit costs for every order of magnitude increase in experience. In practice, the initial part of the learning curve may be rather steeper than implied by this rule as a pioneer plant for coal utilisation processes is often taken to be 50 to 100 per cent more expensive than mature technology.

In the research and development phase of a project, the opposite of the above trend may be observed. As more is learned about a technology, performance shortfalls and problems become apparent and, although technical solutions may be identified, these generally introduce additional costs. During this period, the costs can therefore increase with experience (for example, it has been noted that the research and development project that appears the most attractive always involves the technology that one knows least about). Owing to this effect, it can be misleading to compare the economics of a well-established commercial technology with those of one at an early stage of development.

Ground rules

Some of the general considerations that arise in defining design ground rules for comparative economic assessments are as follows.

Technical conventions

The technical conventions for the plant design should define the design standards and performance targets for all the major items of equipment (for example, pressure vessel standards should be consistent, as should the efficiencies specified for compressors). Also, a common set of ambient conditions should be adopted throughout a study.

Coal type

In general, the characteristics of the coal (for example, composition, swelling index, grade, reactivity, ash content and ash composition) affect the performance and economics of a coal utilisation process either through the choice of operating conditions or by influencing the equipment design. Combustion processes, notably pulverised fuel and fluidised bed combustion, are generally more tolerant to coal type than gasification and liquefaction where the coal specification can often determine the choice of technology.

Environmental control standards

The cost of meeting environmental control standards can be substantial, and it is therefore important in comparative economic studies that the plants are designed to operate under the same regulatory conditions.

Plant boundaries

The plant boundaries need to be carefully defined, particularly the treatment of energy and product flows 'across the perimeter fence'. The cost and performance of a plant designed to be self-sufficient in electricity, for example, may be significantly different from one that imports electricity from the grid (or exports electricity to the grid).

Technical optimism

As far as possible, the plants considered in a comparative study should be designed with the same degree of technical optimism, so that each has a comparable probability of success. This is particularly relevant to the performance of materials and catalysts under unconventional conditions.

Design optimisation

The extent and nature of the plant optimisation should be consistent, particularly for the trade-off between capital costs and efficiency. For example, in a plant designed for expensive coal, there is a strong incentive to achieve a high efficiency, even if this incurs greater capital expenditure. With cheap coal, however, the overall economic optimum will be a less efficient, lower capital cost plant.

8.2 Economic assessment methods

As background for the subsequent discussions of the economics of coal utilisation processes, the present section describes the main features of economic assessment methods and, in particular, presents an example of a simple economic model.

Cost components

The major costs associated with a coal utilisation project may be summarised as follows.

Capital costs

These are basically the costs of building the plant and are usually taken to include

process plant cost
general facilities
engineering fees
royalties
contingencies

In addition to the costs of constructing the process plant, general facilities (for example, roads, workshops, offices, laboratories, canteen) have to be provided where they do not already exist. These costs generally lie in the range 5 to 20 per cent of the process plant cost. Engineering fees may be contained in the process plant and general facility costs and are usually equivalent to 10 to 15 per cent of the process plant cost. Royalties vary from plant to plant, depending on the design and process technologies selected; a typical value for coal utilisation technologies is 2 per cent of the total process plant cost (taking into account that some units may incur higher percentage payments but that not all the process units will be subject to royalties). The allowance for contingencies should also depend on the design, reflecting the degree of novelty or risk involved. Values range from 10 per cent to 40 per cent; an allowance of the order of 15 per cent is common in assessments of new coal processes assuming mature technology although higher values are used for pioneer plants.

The capital costs are usually spread over a construction period of between 4 and 6 years for large plants although the period is usually much shorter for industrial-scale installations. Within this period, the expenditure is not spread evenly but is concentrated in the middle of the construction period so that the cumulative expenditure follows an 'S'-shaped curve. Typical distributions of costs are given for various construction periods in table 8.2 and figure 8.1 assuming that the yearly costs are corrected for inflation to bring them to constant money values. The total capital cost is usually expressed in terms of constant money values for the reference year at the end of the construction period. This is sometimes referred to as the 'instantaneous' or 'overnight' capital cost to distinguish it from the 'out-turn' cost which represents the total expenditure in current (inflating) money values.

Table 8.2 Percentage distributions of construction costs[a]

Year	Construction period (years)			
	3	4	5	6
1	12	6	4	3
2	52	25	14	9
3	36	48	32	19
4		21	36	33
5			14	25
6				11

[a]Based on data contained in *Economic and Technical Criteria for Coal Utilisation Plant, Part 1: Economic and Financial Conventions*, Economic Assessment Service report A1/77, IEA Coal Research, London.

Energy costs

The purchase of the coal is usually a major item in the cost of operating coal utilisation equipment. The actual cost depends on

the coal price ($/GJ)
the efficiency
the load factor

The coal price varies considerably with location. Prices in Western Europe, for example, are generally 2 to 3 times higher than in the United States. The efficiency and the load factor (that is, the fraction of the year for which the equipment is operated) depend on the design and application. The efficiency is important because it determines the amount of coal required to meet the specified output.

Energy costs other than coal may also be incurred. For example, some plants import substantial amounts of electricity to meet compression and pumping requirements.

Operation and maintenance costs

Operation and maintenance costs are usually divided between fixed costs and variable costs.

Fixed costs are independent of the plant load factor and include operating labour, maintenance labour, insurance and local taxes (rates). The labour costs should contain allowances for sickness, supervision and social costs.

Variable costs are proportional to the plant load factor, and include maintenance materials and consumables (water, chemicals, catalysts, etc.).

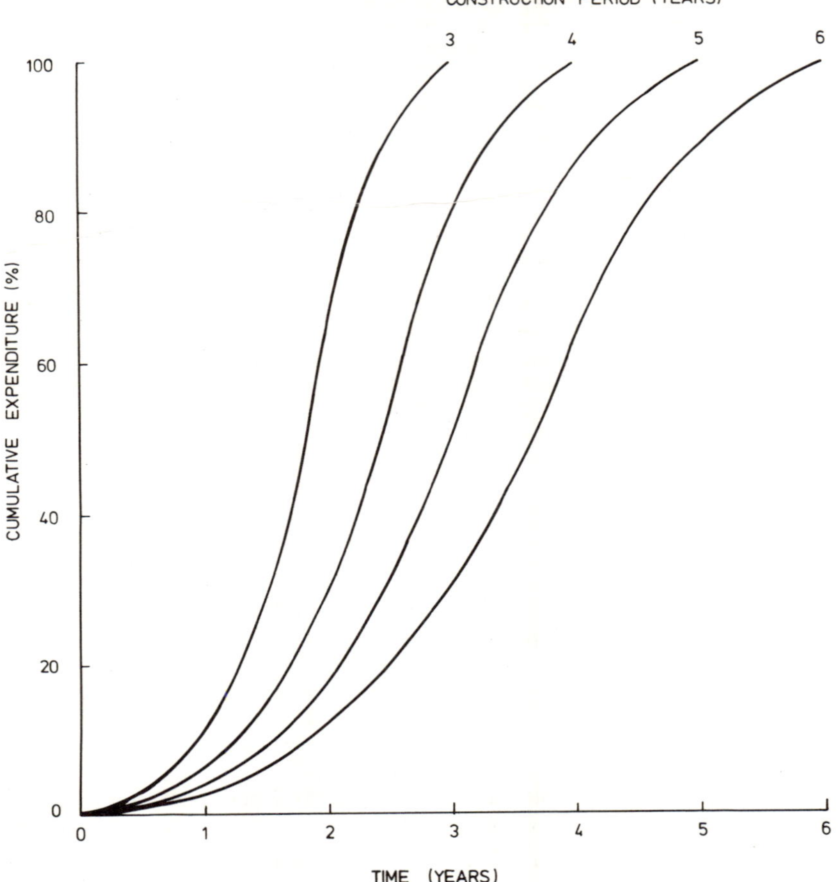

Figure 8.1 Cumulative expenditure during plant construction. Based on data contained in *Economic and Technical Criteria for Coal Utilisation Plant, Part 1: Economic and Financial Conventions*, Economic Assessment Service report A1/77, IEA Coal Research, London

Initial costs

The initial period of plant operation is a special case, because additional, once-only costs are incurred. These arise from three sources

working capital
start-up costs
reduced availability

The detailed definitions of these terms and the assumptions adopted for assessment purposes can vary significantly from organisation to organisation. Working capital normally covers the cost of stocks to provide an adequate inventory for operation and to meet obligations as they become due. Items usually included are coal, maintenance materials, water, catalysts, chemicals and products. Start-up costs represent the additional expenses for labour, fuel and materials incurred during the initial period of operation. It is, for example, necessary to allow for the labour costs associated with 'trouble-shooting' and staff training and for the fuel and materials consumed when no satisfactory product is produced. In addition, during the initial period of operation, the load factor of the plant is usually low because of 'teething problems' which introduces a further penalty in terms of the capital and fixed operating charges. Typically, operation at half of the normal load factor may be assumed for the first year of operation.

The above costs are recovered, over the lifetime of the plant, by the revenue from the sale of the products. For economic assessments, a lifetime of between 15 and 30 years is usually assumed for coal utilisation plant. The actual lifetime may well be longer and a shorter lifetime may be assumed in calculating depreciation for taxation purposes.

A Discounted Cash Flow (DCF) model

In order to illustrate the main features of economic assessment models, a simple example is considered in the present section. The approach is, however, general and the model can readily be modified to apply to more complex situations.

The process of interest is assumed to have only one energy input, coal, and one product. The equations and notation used to describe the major cost items are summarised in table 8.3 and the timing of the cash flows is illustrated in figure 8.2. The construction time for the plant is taken to be M years and it is assumed that the initial costs of working capital, start-up and reduced availability are all incurred in the first year of operation. The plant then operates normally for a further N years.

Since the expenditure and revenue occur at different times, an economic assessment must provide a method of comparing present and future cash flows that takes into account the time value of money. Probably the most widely used approach is Discounted Cash Flow (DCF).

The basis of this method is as follows. If a sum of money, h, is invested now at a rate of return r (fractional) then, in n years, the sum of money obtained is

$$g = h(1 + r)^n \tag{8.3}$$

Conversely, a sum of money, g, available in n years' time would be worth the same as a sum of money, h, invested now at a rate of return, r, where

$$h = g/(1 + r)^n \tag{8.4}$$

These equations can be regarded as establishing an equivalence between sums of money at different times. By selecting a rate of return and applying equations

Table 8.3 Equations and notation for the economic assessment model

Cost item	Equation	Notation	Units[a]
Capital costs	$X(I) = K.T(I)$	$X(I)$ = capital expenditure in year I	\$/kW
		K = total capital cost[b]	\$/kW
		$T(I)$ = proportion of capital expenditure occurring in year I	–
Energy costs	$Y = 31.5CL/e$	Y = annual coal costs	\$/kW
		C = coal price	\$/GJ
		L = load factor	–
		e = efficiency	–
Operation and maintenance costs	$Z = F + VL$	Z = annual operation and maintenance costs	\$/kW
		F = fixed costs	\$/kW
		V = variable costs	\$/kW
		L = load factor	–
Initial costs		I = net cash flow in first year of operation	\$/kW
Product revenue	$Q = 31.5PL$	Q = annual product revenue	\$/kW
		P = product price	\$/GJ
		L = load factor	–

[a]Units given as \$/kW refer to the plant product output capacity.
[b]The capital cost, K, includes royalties and contingencies.

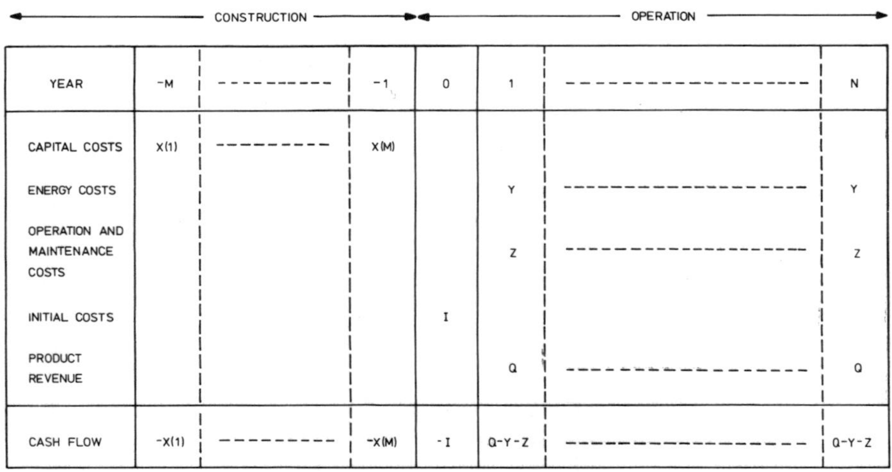

Figure 8.2 Pattern of expenditure and revenues in a coal utilisation project

8.3 and 8.4, the cash flows occurring at every year in a project can be brought to a common time. The sum of these 'Discounted Cash Flows' is termed the 'Net Present Value' (NPV) of the project. The rate of return that makes the net present value equal to zero is known as the 'DCF rate of return' for a project. Appropriate values for DCF rates of return are discussed later.

In the present model, no account is taken of taxation because the conventions vary from country to country. However, the analysis could be extended to incorporate this aspect, as discussed briefly later.

For simplicity, DCF models (including the present) are often carried out in 'real terms', that is, using constant money values for all the cash flows. This has the advantage of making the analysis independent of the inflation rate and estimates of its future trends over the plant lifetime. Using constant money values also makes the resource costs involved in the various factors more visible and comparable. The analysis can instead be conducted taking into account general inflation and, in this case, the appropriate rate of return would be higher by an amount approximately equal to the inflation rate. The higher rate of return used in the calculations cancels the effects of the inflationary increase in current money values and consequently the real rate of return is similar in both cases. The main advantage of operating in current money values is that it reveals the actual cash flow pattern that is expected to occur and this can then be budgeted for accordingly.

It may, however, be the case that the various components of the overall cost escalate at different rates. For this reason, some studies allow for 'real' increases in coal prices, labour costs and product revenue with time. The load factor may also be assumed to decrease with time. In the present model, such variations are excluded and all costs, revenues and performance parameters are assumed to remain constant throughout the operating period. The model could be extended to incorporate these effects if required.

With the additional assumptions of annual accounting periods and all payments and revenues occurring at the beginning of the year, equations 8.3 and 8.4 can be used to discount the cash flows illustrated in figure 8.2 to the first year of operation. Equating the Net Present Value to zero and substituting for the expressions given in table 8.3 leads to the following equation

$$P = C/e + 0.0317\ [V + (F + I/B + AK/B)/L] \quad \$/\text{GJ} \qquad (8.5)$$

Equation 8.5 may be used to calculate the product cost (P), as a function of the capital cost (K), coal price (C), load factor (L), efficiency (e), fixed costs (F), variable costs (V) and initial costs (I). The product cost also depends on the parameters A and B which may be calculated from the plant construction period, the operational lifetime and the DCF rate of return. Values for a range of these variables are given in tables 8.4 and 8.5 (the general equations defining the parameters are given as footnotes to the tables).

Table 8.4 Values of parameter A[a]

Construction period (years)	DCF rate of return (%)				
	0	5	10	15	20
3	1	1.090	1.185	1.284	1.388
4	1	1.112	1.232	1.361	1.500
5	1	1.136	1.285	1.449	1.629
6	1	1.159	1.339	1.542	1.770

[a]The values assume the patterns of expenditure given in table 8.2. The general equation for parameter A is

$$A = \sum_{I=1}^{M} T(I) \times (1 + r)^{M-I+1}$$

where r is the DCF rate of return (fractional)
$T(I)$ is the proportion of the capital spent in year I
and M is the construction period (years).

Table 8.5 Values of parameter B[a]

Plant lifetime (years)	DCF rate of return (%)				
	0	5	10	15	20
10	10	7.722	6.145	5.019	4.192
15	15	10.380	7.606	5.847	4.675
20	20	12.462	8.514	6.259	4.870
25	25	14.094	9.077	6.464	4.948
30	30	15.372	9.427	6.566	4.979

[a]The general equation for parameter B is

$$B = \sum_{I=1}^{N} 1/(1 + r)^{I} = [1 - 1/(1 + r)^{N}]/r$$

where r is the DCF rate of return (fractional)
and N is the plant lifetime (years).

For some conversion processes such as ammonia manufacture, the product cost may be required per tonne of output, rather than per gigajoule. In these cases, the following alternative form of equation 8.5 may be used

$$P' = C/\epsilon + (0.0317/\epsilon)\,(V' + (F'_{,} + I'/B + AK'/B)/L)\ \$/\mathrm{t} \qquad (8.6)$$

The meaning of the symbols in equation 8.6 is as follows. P' is the product cost (\$/t), ϵ is the product yield (t product/GJ coal input) and V', F', I', K' are, respectively, the variable costs, fixed costs, initial costs and capital cost, but expressed as \$/kW coal input, rather than per unit rate of product output.

Table 8.6 Approximate values for annual operation and maintenance costs[a]

Component	Costs (% of capital)			
	Power generation		Coal conversion	
	Fixed	Variable[b]	Fixed	Variable[b]
Operating labour	0.5		1	
Maintenance labour	1		2	
Maintenance materials		1		2
Consumables		1		1
Insurance, local taxes etc.	3		3	
Total	4.5	2	6	3

[a] Based on data contained in Economic Assessment Service reports A1/77, E1/79 E2/80, E3/81, IEA Coal Research, London.
[b] Variable costs are quoted at 100 per cent load factor.

Table 8.7 Approximate values for initial costs[a]

Component	Duration (months)		Cost[b] (\$/kW)	
	Power generation	Coal conversion	Power generation	Coal conversion
Working capital				
Coal	2	1	$(1/6)31.5C/e$	$(1/12)31.5C/e$
Maintenance materials	12	6	$0.01K$	$(1/12)0.02K$
Consumables	1	1	$(1/12)0.01K$	$(1/12)0.01K$
Product receivables	1	1	$(1/12)31.5P$	$(1.12)31.5P$
Start-up costs				
Coal	1	1	$(1/12)31.5C/e$	$(1/12)31.5C/e$
Consumables	1	1	$(1/12)0.01K$	$(1/12)0.01K$
Operating labour	12	12	$0.005K$	$0.01K$
First year costs				
Coal	4	5	$(1/3)31.5C/e$	$(5/12)31.5C/e$
Fixed O & M	12	12	$0.045K$	$0.06K$
Variable O & M	4	5	$(1/3)0.02K$	$(5/12)0.03K$
Product revenue	4	5	$-(1/3)31.5P$	$-(5/12)31.5P$
Total			$(7/12)31.5C/e$ $+(1/12)0.82K$ $-(1/4)31.5P$	$(7/12)31.5C/e$ $+(1/12)1.13K$ $-(1/3)31.5P$

[a] Based on data contained in Economic Assessment Service reports A1/77, E1/79, E2/80 and E3/81, IEA Coal Research, London.
[b] The units refer to the plant product output capacity.

Approximations

For preliminary estimates or calculations, it is possible that not all of the parameters required to use equations 8.5 and 8.6 will be known. In particular, the operation and maintenance costs and the initial costs may not have been evaluated. Under these circumstances, 'rule of thumb' estimates are often made for the unknown parameters in terms of items of cost that are known, for example by expressing them as a percentage of the capital cost. Approximate relationships for operation and maintenance costs and initial costs are given in tables 8.6 and 8.7, respectively. The values are typical for large plants, although the range of variation is considerable.

The relationships given in the tables may be combined with equation 8.5 to give the following 'order of magnitude' estimate for the product cost

$$P = C/(Ue) + K/W \quad \$/\text{GJ} \tag{8.7}$$

The parameters U and W depend on the DCF rate of return, load factor, plant lifetime and plant construction period. Values for a range of DCF rates are given

Table 8.8 Values of parameters U and W[a]

Parameter	Power generation	Coal conversion
Load factor (per cent)	65	85
Lifetime (years)	30	20
Construction period (years)	6	5
U		
0% DCF	0.983	0.986
5% DCF	0.968	0.978
10% DCF	0.950	0.968
15% DCF	0.931	0.958
20% DCF	0.913	0.947
W		
0% DCF	221.5	194.7
5% DCF	152.3	149.9
10% DCF	102.8	113.2
15% DCF	71.5	85.7
20% DCF	51.6	65.9

[a]The general equations for the parameters are as follows:

Power generation $U = (12BL + 3)/(12BL + 7)$
$W = 31.5(12BL + 3)/(0.24BL + 0.54B + 0.82 + 12A)$

Coal conversion $U = (12BL + 4)/(12BL + 7)$
$W = 31.5(12BL + 4)/(0.36BL + 0.72B + 1.13 + 12A)$

where L is the fractional load factor
A is the parameter given in table 8.4
and B is the parameter given in table 8.5.

in table 8.8, with assumptions for the other parameters representative of large, base-load plants. The general expressions for the parameters are given as footnotes to the tables.

Equation 8.7 may also be expressed in mass terms, as follows (see equation 8.6 for a definition of the notation)

$$P' = C/(U\epsilon) + K'/(W\epsilon) \quad \$/\text{t} \tag{8.8}$$

It is interesting to note that the product cost estimated by equations 8.7 and 8.8 varies linearly both with the coal price and the capital cost. For example, at a 5 per cent DCF rate of return, the product costs for power generation and coal conversion processes are given by equations 8.9 and 8.10, respectively

$$P = C/(0.968e) + K/152.3 \quad \$/\text{GJ} \tag{8.9}$$

$$P = C/(0.978e) + K/149.9 \quad \$/\text{GJ} \tag{8.10}$$

The variation of product cost (\$/GJ) with coal price (\$/GJ), capital cost (\$/kW) and efficiency is illustrated in figure 8.3 for the conditions of equations

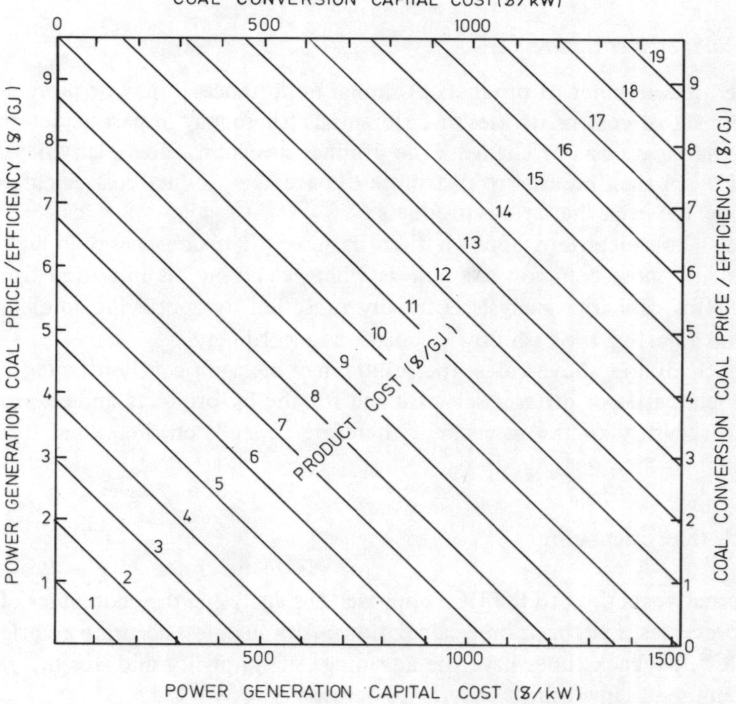

Figure 8.3 Variation of product cost with coal price, process efficiency and capital cost. The estimates refer to a 5 per cent DCF rate of return cent load factor

8.9 and 8.10. The predictions are general, and are subject only to the assumptions described above. For example, a coal conversion process having an efficiency of 60 per cent and a capital cost of 1000 $/kW would give mean product costs of 8.4 to 11.8 $/GJ for coal prices of 1 to 3 $/GJ.

Representative values of efficiencies and capital costs for a range of coal utilisation processes are given in appendix 2.

Multi-product systems

Although the model described in the previous sections assumed a single product, the analysis can be readily adapted to include the situation where by-products or co-products are produced. Two approaches are widely used.

By-product credits

Where there is one main product and some minor by-products, the simplest method is to allocate values to the by-products and to adjust the cash flow calculations to take into account the credit obtained from their sale.

Co-product cost structure

If there are a number of products of similar importance, it may be preferable to assume a set of cost relativities or differentials (these may in part be determined by the marginal costs of changing the product spectrum). The relativities or differentials can then be used to distribute the average product cost calculated by the model between the various products.

Similar considerations apply if there is more than one energy input to the process; the most common example is where electricity is imported from the grid. In this case, the analysis is usually modified by specifying an electricity price and adjusting the cash flow calculations accordingly.

In each of the above cases, the results may be significantly affected by the values, relativities or differentials assumed for the by-products and co-products and the accuracy of the assessment therefore depends on the extent to which these are realistic.

Payback time calculations

An alternative method to the DCF approach for analysing the economics of coal-based processes is payback time calculations. Although less accurate and rigorous than DCF, payback times have the advantage of simplicity and, for this reason, are often used as investment criteria by industry.

The payback time, usually expressed in years, is defined as the ratio of the capital investment to the net annual revenue. It is implicitly assumed that the capital cost is an 'instantaneous' value (that is, including an allowance for interest during construction) and that there are no initial costs (so that the first year of operation is the same as subsequent years).

For example, consider a project for which the capital cost is K (\$/kW) and the product value is P (\$/GJ). Then, using the notation in table 8.3, the payback time t (years) is given by

$$t = K/(31.5PL - 31.5CL/e - F - VL) \tag{8.11}$$

Payback times may be related to DCF rates of return as follows. Applying the DCF model of equation 8.5 to the project considered above (setting the initial costs, I, to zero and the parameter A to unity) shows that the payback time, t, is identical to the parameter B. The payback time may therefore be related to the DCF rate of return and the project lifetime, the relationship being given by table 8.5.

Cost comparisons

As noted earlier, it is often not the absolute cost of a process but its performance relative to other processes that is of most interest in an economic assessment.

Consider, for example, the comparison between two processes producing an identical product. Where one process has both a lower capital cost and lower running costs (fuel and operating costs) then it would normally be more economic. If, however, one process has a higher capital cost but lower running costs than the other, then there is a trade-off between capital and running costs.

In the latter case, the economics can be analysed by assuming that the product price is the same for the two processes. This is equivalent to considering only the cash flows corresponding to the differences in the capital costs and the running costs. If a DCF model is used, the Net Present Value of these cash flows can be calculated or, more usually, set to zero in order to determine another parameter such as the DCF rate of return. Alternatively, the payback time can be calculated simply by dividing the difference in the capital costs by the difference in the running costs.

The results are often presented in terms of 'break-even' conditions for the main parameters, usually coal price, load factor and DCF rate of return (or, equivalently, payback time). The break-even conditions correspond to the values of the parameters above which one process is more economic and below which the other is more economic. The break-even values are interrelated so that, for example, at a fixed DCF rate of return, there is a relationship between coal price and load factor representing the conditions of indifference between the processes.

The approach described above applies both to comparisons between different coal-based processes and to comparisons between a coal-based process and a process using another energy source (such as oil). In the latter case, the price of the other energy source is an additional variable in the analysis, and break-even conditions between this and the other variables may be calculated. In particular, the break-even relationship between coal price and the alternative feedstock price is often investigated and is normally linear.

Rate of return

One of the most important assumptions in an economic assessment study is the required real rate of return. The values used vary significantly from organisation to organisation and from country to country. In general, however, much depends on the type of organisation and the method of financing employed.

In the case of 'public sector' financing, a relatively low rate of return is usually assumed, a typical value being 5 per cent per annum in real terms. The detailed arguments on public sector discount rates are complex, the values used partly reflecting a judgement on the social benefit of present, as opposed to future, consumption (the 'Social Time Preference' rate) and partly on the appropriate distribution of investment between the public and private sectors (the 'Social Opportunity Cost' rate).

For 'private sector' financing, a higher rate of return, often 10 per cent per annum or more in real terms, is employed. This represents the return required to finance the project by a combination of borrowing at market rates (debt) and the investment of private risk capital (equity). In order to attract equity participation, significantly higher real rates of return have to be offered than those corresponding to debt. The overall required rate of return for the project may be calculated as the weighted average of the rates of return for the debt and equity components and therefore depends on their relative sizes. This aspect of project financing is usually expressed as the 'debt/equity ratio'; a ratio of 100/0 corresponds to all debt, no equity, whereas 0/100 corresponds to no debt, all equity. In practice, the debt/equity ratio will depend on the type of organisation and the extent of financial risk involved in the project.

The effective rate of return required may also depend on taxation. The rules are complex and vary considerably from country to country. Generally, however, tax is payable only from the 'profits' arising from equity participation, typically at a rate of about 50 per cent. In calculating 'profits', depreciation of the capital sum is usually allowed to offset the yearly operating surpluses and may take place over a shorter period than the plant lifetime used in the DCF calculations. Depreciation has a direct effect only on the calculation of taxation. In the DCF analysis, the taxes payable are taken into account as a further cost item in the cash flow, varying from year to year throughout the operating lifetime of the plant.

A further complication is that, in some situations, a project may qualify for government grants, tax credits or tax 'holidays' made available to assist conversion from oil to coal, conservation, regional development, or for other purposes. Such grants would also affect the required rate of return.

8.3 Energy models

Energy supply systems

The energy supply systems of industrial economies involve a chain of activities that include energy production, conversion, storage, distribution and use by consumers. In analysing such systems, a distinction can be drawn between an 'energy source' (the form in which the energy is produced) and an 'energy carrier' (the form in which the energy is distributed to consumers).

The main energy sources are coal (run of mine), natural gas, crude oil, nuclear energy and the alternative energy sources; the main energy carriers are coal (generally after washing), gases (including natural gas and manufactured gases), liquid fuels (including oil-refinery products), electricity and hot water. Coal and natural gas may therefore be considered to be both energy sources and energy carriers. In other cases, however, the link between an energy source and an energy carrier is a conversion process.

In general, conversion processes are required in order to produce a more suitable or convenient form of energy for distribution or use by consumers. At present, there are two main conversion processes: oil refining and electricity generation. In the future, it may be expected that the energy supply system will become considerably more complex as coal conversion processes are introduced and coal substitutes for declining supplies of oil and natural gas.

The main features of the energy supply system are illustrated in figure 8.4 together with options that may become widespread in the future. The figure shows both the conversion routes to the energy carriers and the ways in which these may be used to meet energy demands in the principal market sectors.

Energy modelling studies

Although the economic assessment methods described in the previous section can often provide a useful indication of the relative merits of various technologies, the size of the demand for which they may be required can usually be assessed only by considering the economic implications for the overall energy system. In particular, factors such as the existing investment both in distribution networks and by consumers, additional costs incurred elsewhere in the energy system and the effects on the markets for competing forms of energy may all affect the potential for a new technology. It may also be necessary to consider the overall energy system in order to identify the most effective competing route to the technology of interest.

In order to assist such studies, a number of large, computer-based energy supply models have been developed, including

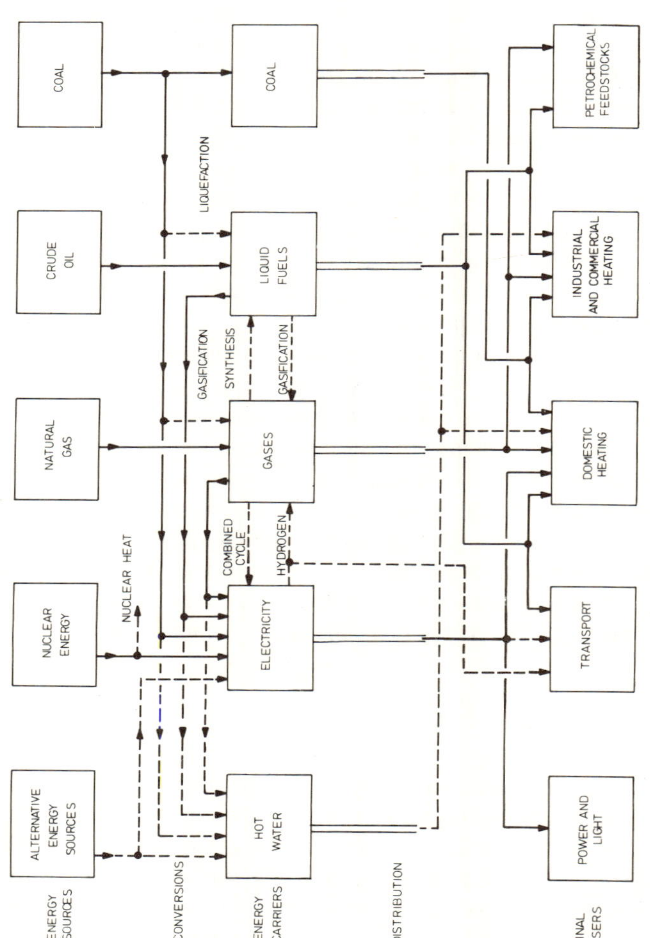

Figure 8.4 The structure of the energy supply system. Continuous lines represent a major current activity whereas broken lines represent a minor current activity or a possible future route

EFOM (European Economic Community)
MARKAL (International Energy Agency)
MESSAGE (International Institute for Applied Systems Analysis)
BESOM (Brookhaven National Laboratory)

In their simplest form, such models require information on the size of existing facilities for energy production, conversion, distribution and end-use, together with the costs and constraints associated with future extensions of these activities. Also required as input are estimates of imported energy prices and the size of the final energy demand by market sectors.

The models are based on a linear programming (LP) approach and predict the optimal allocation of energy sources and pattern of investment in production, conversion, distribution and end-use technologies. The criterion for the optimisation (the objective function) is usually minimum overall cost although other criteria can also be adopted.

The optimisation can be carried out at a single specified time in the future; such a model is termed a 'static model'. More commonly, however, the model considers simultaneously a series of 'snapshot' years so that the optimal development pattern for the energy system over a period of time can be investigated. Such models are referred to as 'time-phased' models and involve the replication of each variable for each of the snapshot years together with the introduction of additional equations and constraints to ensure sensible relationships between the situations in the snapshot years (for example, limits on plant construction rates). Alternatively (although less common), a series of static models can be applied to successive snapshot years, each using the results of the previous optimisations. This approach, referred to as a 'time-marching' model, does not provide inter-year optimisation but may reflect more accurately the nature of decision making.

An important aspect of the level of detail included in the model (referred to as the degree of 'disaggregation') is the number of demand sectors chosen. For example, domestic heating may be treated as a single sector or may be broken down into water heating, background-space heating and peak-space heating. Although a highly disaggregated model may be expected to give more accurate results, it will also be more expensive and difficult to use. Practical considerations of the size of the model impose more severe limits on the detail that can be included when modelling a group of nations or the global energy system than for a national energy supply model.

Often the estimates of energy demands by market sectors are made using a separate 'energy-demand' model. In such a model, energy demands are predicted using relationships that take into account the effects of economic growth (for example, more personal transport), energy prices (including present price/demand elasticities and the effects of conservation) and the existing socio-economic system (for example, the transport infrastructure and housing stock). The importance of economic growth is such that this variable may, in turn, be predicted using a further model (a 'macroeconomic' model) from information about capital, labour and resources in the economy.

Energy supply LP models have proved to be a valuable technique in providing an insight into future requirements for new technology, particularly because they ensure that a self-consistent set of energy prices are used in the comparisons. However, the models possess some important disadvantages and the results must be interpreted with caution.

For example, a widely noted feature of LP models is that a model designed to minimise cost will install the cheapest technology to the fullest extent even when only a small margin exists between the cheapest and next-cheapest technology. This can be misleading when a situation is finely balanced, especially when considerable uncertainties exist about the process economics, constraints and demand levels. Indeed, the large number of assumptions necessary in an LP model often makes it difficult to assess the validity of the results.

In practice, it is seldom possible to draw conclusions of wide generality on the relative merits of different technologies either from the results of a large energy supply model or from the simpler economic assessment methods discussed earlier. This is because of the uncertainties in the cost data, particularly for coal technologies still under development, and the diversity of situations that may arise. It is likely, for example, that one technology will be more economic than another under one scenario and vice versa.

8.4 Economics of coal conversion processes

Previous sections have outlined the main methods used in economic analyses of coal utilisation processes and considered some of the more important factors involved. The remaining two sections discuss the results of applying these methods to the technologies described in earlier chapters. In the interests of generality, the present text concentrates on the broad conclusions that can be drawn from such studies rather than on numerical examples that are, of necessity, limited to a specific time and type of location. For convenience, the discussion is subdivided into the economics of converting coal at large central installations into other energy carriers (the present section) and the economics of use by final consumers (section 8.5).

Electricity

Although at present almost all coal-fired power stations are based on pulverised fuel combustion, as the technologies discussed in earlier chapters become available, the options for new power stations in the future may be extended to comprise the following systems

pulverised fuel combustion
atmospheric pressure fluidised bed combustion
pressurised fluidised bed combustion
gasification combined cycle

Because of its dominant position, pulverised fuel combustion may be regarded as a convenient standard compared with which the new technologies have to show a significant economic advantage in order to justify their development.

In the absence of sulphur-emission control regulations, atmospheric pressure fluidised bed combustion may offer a small reduction in capital cost but no increase in efficiency. The combined cycle systems (pressurised fluidised bed combustion and gasification combined cycle) have capital costs that are of the same order as those for pulverised fuel systems but may benefit from increased efficiencies. For large central plants, any net generating cost reduction is unlikely to be sufficient to attract utilities to introduce the new technologies in view of the uncertainties involved. Indeed, advanced pulverised fuel combustion designs using supercritical steam cycles and 1000 MW turbine sets may be a preferable line of development. There may, however, be opportunities for the new technologies in small modular plant where such installations are favoured by siting or distribution system restrictions.

Where sulphur emissions have to be controlled, pulverised fuel combustion generally requires the use of flue gas desulphurisation plant (see chapter 6). This increases the capital cost by about 14 per cent and reduces the efficiency by about 1.9 percentage points for a regenerable process (see appendix 2). In this case, all the new technologies appear to offer lower generating costs than pulverised fuel combustion, typically by more than 10 per cent.

Comparisons between the new technologies are difficult because the costs of each are of a similar order and the preferred choice will vary from location to location. In the long term, gasification combined cycle systems may be expected to gain an advantage over the other systems as future increases in gas turbine inlet temperatures result in higher efficiencies and lower capital costs for this route.

The savings projected for the new technologies, if achieved, seem sufficient for them to be preferred to pulverised fuel combustion with flue gas desulphurisation where sulphur retention is required. However, the savings are not so large that the prospects of coal in competition with other energy sources for the power generation market would be greatly affected.

At present, the main energy sources for power generation, other than coal, are oil, nuclear power and hydropower, with natural gas also being used in some countries. The alternative energy sources (for example, wind, tides, solar, geothermal, ocean thermal gradients and waves) could make an important contribution to electricity supplies in the future but it may be several decades before this occurs and the extent of the contribution is likely to be dependent on local climatic and geographic factors.

The balance of energy sources used by a utility is determined by two main considerations: the way in which the system is operated and decisions on investment in new capacity.

Since the storage possibilities for electricity are limited, the variations in demand that occur diurnally and seasonally are met mainly by reserving some power stations for periods of high and intermediate demand ('peaking' and 'mid-merit' duties respectively) while other stations run for the maximum amount of time ('base-load' duty). In order to minimise system operating costs, the duties of the stations are determined by a 'merit order'; the stations cheapest to run are used for base-load duty while the more expensive stations to run provide mid-merit and peaking duties. The initial investments may be regarded as 'sunk' costs and do not affect the merit order of operation.

The operation of the merit-order system can be illustrated by a 'load–duration' curve which shows the relationship between the generation capacity in a utility system and the load factor at which it operates. An example is given in figure 8.5 for a system in which nuclear, oil-fired and coal-fired power stations have been installed. The nuclear stations have the lowest running costs and therefore are preferentially used for base-load operation. Even with advanced power generation cycles and cheap coal, it appears unlikely that coal-fired stations would displace nuclear stations in the merit order. The relative positions of coal-fired and oil-fired stations in the merit order are determined mainly by the fuel prices. During the 1960s, when oil was cheap, many oil-fired power stations were built and introduced as base-load plant. The oil price rises of the 1970s have increased the running costs of these stations dramatically so that now all but the least efficient coal-fired stations have displaced oil-fired stations to low load factor duties.

The primary energy sources used for power generation in the future will be affected by the pattern of investment in new plant. Although, in suitable locations, hydropower will continue to expand and some of the alternative energy sources may be introduced, power generation is widely expected to remain based predominantly on thermal systems for the foreseeable future, particularly in industrialised areas. In view of the increasing prices and diminishing reserves of oil and natural gas, it seems likely that the main options for new thermal plant will be coal and nuclear power.

Coal-fired power stations require less capital investment than nuclear stations but, as noted above, have higher running costs. The relative economics depend on the load factor; at low load factors, capital charges dominate whereas, at higher load factors, running costs become more important. As shown in figure 8.6, there can exist a load factor ('the break-even load factor') at which the additional capital charges of the nuclear plant are just offset by the lower running costs so that the overall generating cost is the same as for coal. Provided that the break-even load factor is less than the maximum availability of the plants (assumed to be the same), the optimum strategy would be to install sufficient nuclear capacity to meet the requirements for plant operating at higher load factors than the break-even value with the lower load factor duties being met by coal.

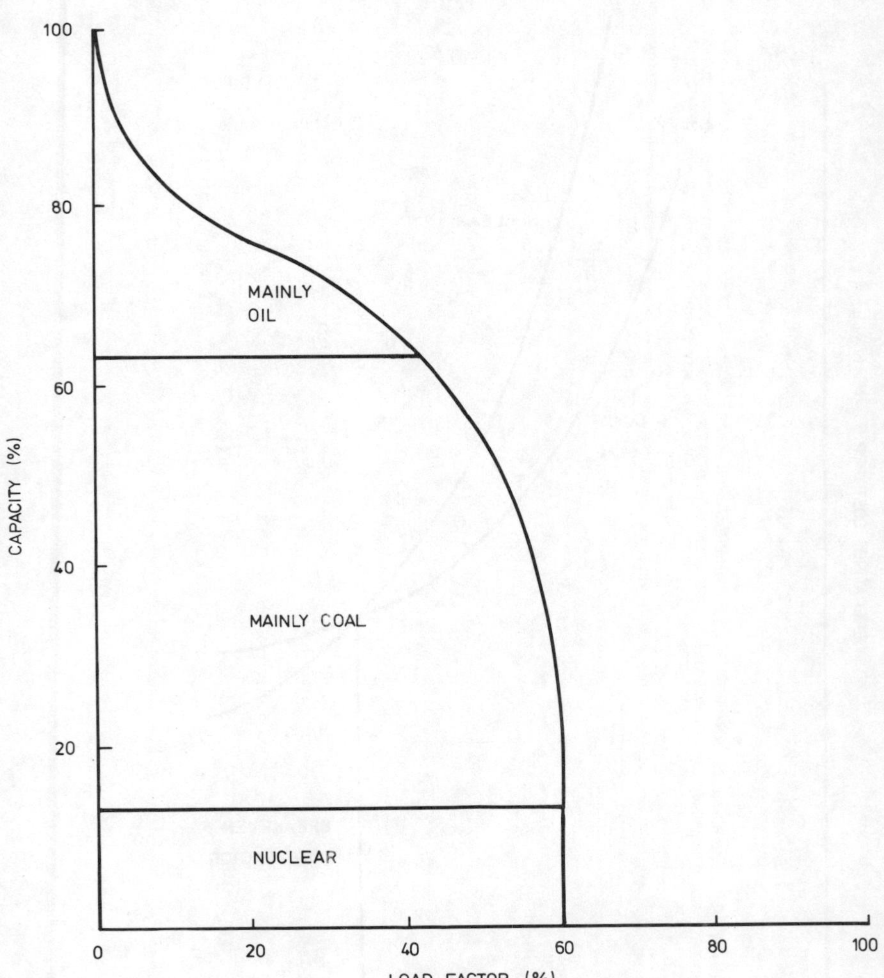

Figure 8.5 Load–duration curve for an electricity system containing nuclear, coal-fired and oil-fired stations

The choice of numerical values for use in analyses of the type described above is the subject of intense controversy. The uncertainties are compounded by large variations in estimates for nuclear power station construction costs from country to country and substantial real increases with time. Although the government view in many of the industrial nations appears to favour the installation of further nuclear capacity, the issues are not solely economic. An important consideration for utilities is the diversification of their energy supplies and, in particular, the present low penetration of nuclear power creates

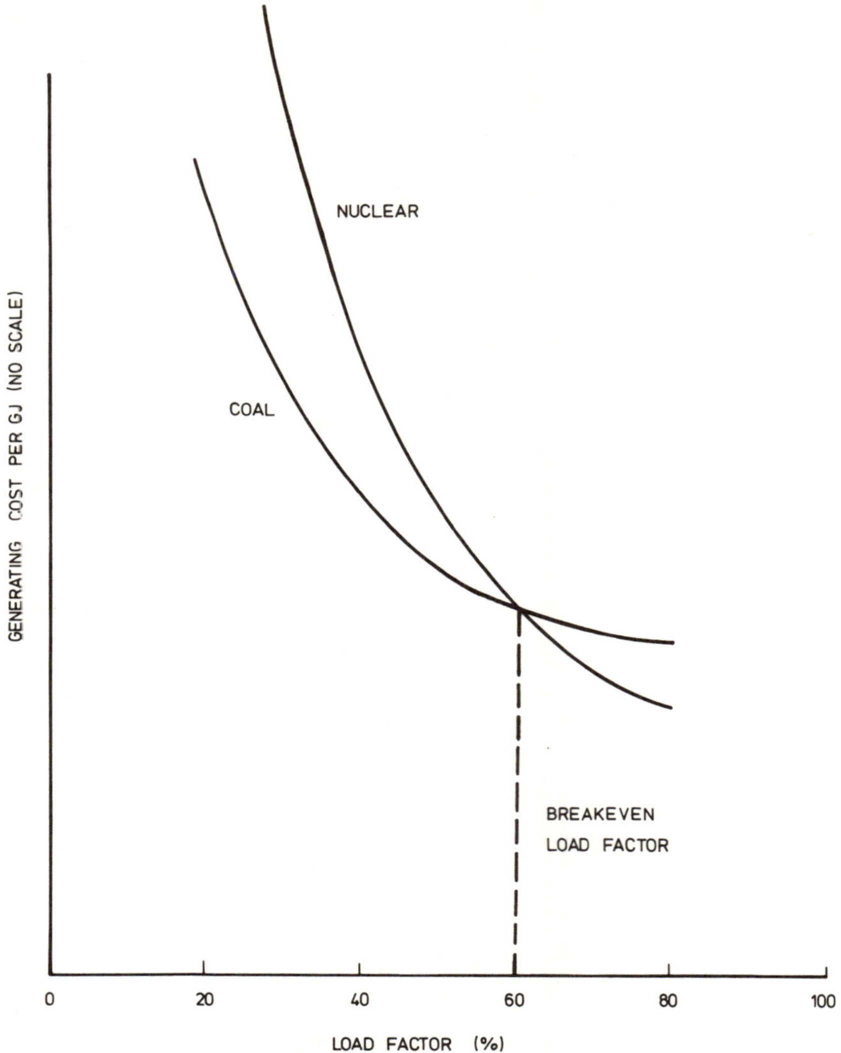

Figure 8.6 Effects of load factor on nuclear and coal-fired power station generating costs

an incentive to increase the size of the nuclear contribution. However, at least in the short term, it seems probable that the prospects for new nuclear capacity will depend as much on considerations of strategic importance, safety, environmental effects and nuclear proliferation as on other factors.

Future construction programmes for coal-fired power stations are critically dependent on those for nuclear stations. If nuclear capacity increases with the growth in electricity demand, the need for new coal-fired stations, and therefore for the new power generation technologies of atmospheric fluidised bed combustion, pressurised fluidised bed combustion and gasification combined cycle, is reduced accordingly. A moratorium on nuclear power, however, would mean that requirements for new capacity would probably be met largely by new coal-fired stations, at least for the time being. In this case, the number of new stations built would depend on the growth in electricity demand (which, in turn, is closely related to the overall level of economic activity) and the retirement rate for existing stations.

In addition to the technologies appropriate to new stations, modifications to existing stations can also have a significant impact on the contribution of coal to power generation. Interest centres on three main types of modification.

Refurbishing

When present pulverised fuel power stations are due for retirement, refurbishing to provide an additional period of useful life (see section 6.7) may be more economic than retirement as planned, together with the construction of new replacement capacity.

Sulphur-emission control

Almost all existing pulverised fuel power stations were constructed without flue gas desulphurisation plants. Where it is necessary to reduce sulphur emissions from existing stations, the addition of flue gas desulphurisation plant (possibly in conjunction with a special coal preparation system) appears to be more economic than changing the method of firing to fluidised bed combustion or retrofitting with gasification.

Conversion of oil-fired stations

The conversion of oil-fired stations to coal firing using pulverised fuel combustion is not usually practical because oil-fired boilers are designed for higher convective section entry temperatures and gas velocities (see also section 3.8). Technologies that might permit the conversion of oil-fired stations to coal-firing include coal–oil mixtures, coal–water mixtures and gasification.

At present oil and coal prices, the economics of coal–water mixtures generally appear to be more attractive than those of coal–oil mixtures, although coal-water mixture technology is at an early stage of development. The economics of conversion using gasification depend on the design of the boiler. If the amount of derating required for operation with coal–oil or coal–water mixtures is small, then these options are probably preferable to gasification. Where a substantial derating would be necessary, the direct-firing methods become unattractive and gasification may be economic.

Hot water

Hot water produced centrally may be distributed as an energy carrier for residential and industrial heating in district heating schemes. Hot water is also produced by some industries for process and space heating applications (this option is considered in the following section as an 'end-use' technology). Two main systems may be used for district heating: heat-only boilers and combined heat and power.

The relative economics of these approaches depend on local factors. In general, however, heat-only boilers appear to be restricted mainly to small schemes with combined heat and power predominating in large-scale applications. Indeed, an important stage in the construction of large district heating networks is often the creation of small, local district heating schemes that are run from temporary heat-only boilers until they can be linked to the central combined heat and power station. Heat-only boilers may also be used in large combined heat and power schemes to assist in meeting peak heating demands.

Most combined heat and power systems for district heating are based on steam turbines. As noted in section 6.7, these suffer a reduction in electrical output compared with similar electricity-only stations. The preferred systems are usually those that offer the highest ratio of electrical to heat output.

The use of hot water as an energy carrier is an attractive option for coal as it provides a means of access to the premium heating market and offers a high level of amenity to the consumer. Also, coal handling and environmental controls are easier and cheaper in large central combustion units than in small units at the points of consumption.

For large schemes, the main energy sources competing with coal are oil and nuclear power. As with power generation, coal is now substantially cheaper than oil and, where available, would normally be preferred. Compared with nuclear power, coal may possess a locational advantage in that coal-fired power stations can often be sited nearer to centres of population, and therefore nearer to the heat demand, than is generally considered acceptable for nuclear plant.

For small schemes of less than 20 MW (electrical), the economies of scale generally make steam turbine systems less competitive so that coal is effectively limited to heat-only boilers. Oil and natural gas can, however, be used in combined heat and power systems employing diesel engines or gas turbines. Technologies under development that may assist coal to compete in the small combined heat and power market include

fast fluidised bed air heaters (for gas turbines)
pressurised fluidised bed combustion (for gas turbines)
low calorific value gas manufacture (for gas turbines)
coal-oil and coal-water mixtures (for diesels)

Little has yet been published on the relative costs of these options and the economics are therefore difficult to assess.

As an energy carrier, hot water competes mainly with natural gas, electricity and distillate fuels distributed directly to final consumers. In comparison with these options, hot water requires a high level of investment in the distribution system, although this may be offset by the lower running costs associated with the use of a non-premium fuel. Since distribution costs are high, district heating can usually be justified only where the demand density is high implying either an area of high population density, an area of concentrated industrial demand or a cold climate so that space heating demands per dwelling are large.

In regions with a low demand density, hot water distribution is unlikely to be economic compared with the other energy carriers.

Gasification

The main types of coal-derived gas that may be considered for long-distance transmission as energy carriers are substitute natural gas, medium calorific value gas and hydrogen.

The economics of using these gases as heating fuels mainly depend on the following factors.

Production costs

The cheapest of the gases to manufacture is medium calorific value gas as this is an intermediate in the manufacture of both substitute natural gas and hydrogen. The production cost of hydrogen is probably broadly similar to that of substitute natural gas if a methane co-product is produced although it may be somewhat higher (see appendix 2) if hydrogren is the only product.

Storage and distribution costs

These are related to the calorific value of the gases (that is, the energy content per unit volume, see chapter 4). Substitute natural gas has the highest calorific value (approximately 35 MJ/m^3) and therefore the lowest storage and distribution costs. Medium calorific value gas (9 to 13 MJ/m^3) and hydrogen (12 MJ/m^3) have similar storage and distribution costs; in each case they are substantially higher than those of substitute natural gas. The storage and distribution costs of low calorific value gas (5 to 7 MJ/m^3) are such that this gas is considered to be economic only for on-site use and is therefore excluded from the present discussion.

The relative costs of the three gases are illustrated in figure 8.7. As shown in the figure, hydrogen is more expensive than both substitute natural gas (because of higher storage and distribution costs) and medium calorific value gas (because of higher production costs). Although hydrogen has been widely studied as an energy carrier, particularly in the context of the ‘hydrogen economy’, its manufacture from coal for distribution as a heating fuel does not appear to be econ-

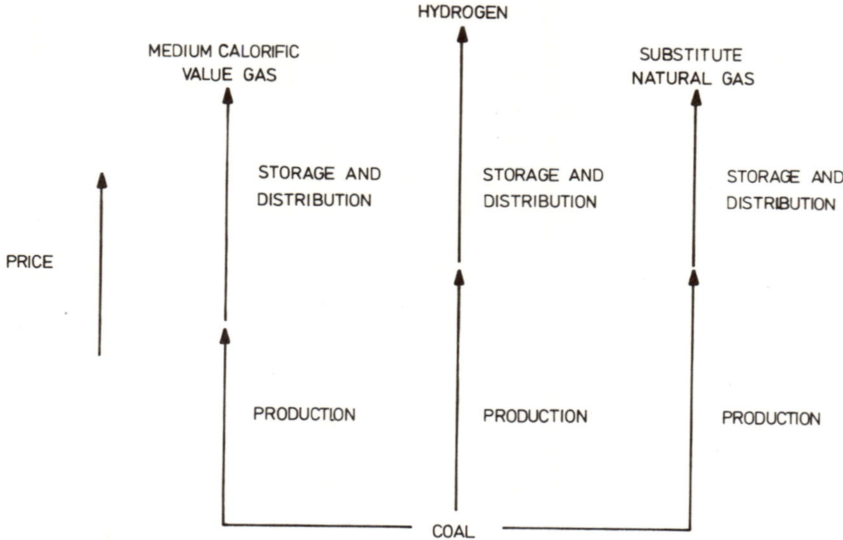

Figure 8.7 Relative costs for coal–derived gases (not to scale)

omic compared with the other coal-derived gases. Hydrogen may of course be manufactured from coal for use in markets where its special properties are required, for example, as a chemical feedstock (see next section). Furthermore, hydrogen manufactured from other energy sources (for example, by electrolysis using low-cost electricity) might be competitive with coal-derived gases for heating applications.

However, the two main options for coal-derived gases in the heating sector are substitute natural gas and medium calorific value gas. Depending on the type of consumers and the distances involved, the overall costs are probably similar, medium calorific value gas being cheaper to produce but more expensive to store and distribute than substitute natural gas. However, substitute natural gas has the important advantage of compatibility with natural gas supplies and the existing stock of consumer appliances. Also, the acceptability of distributing a toxic (high carbon monoxide content) gas, particularly for domestic consumers, is now doubtful. For these reasons, it is generally expected that substitute natural gas will be the preferred coal-derived gas for national distribution. Under suitable conditions, there may be a role for medium calorific value gas for distribution to bulk industrial consumers using a special pipeline system provided that a sufficiently large demand exists within the region.

It is interesting to note that coal-based substitute natural gas plants will probably operate on base-load. This is because storage costs, even for seasonal demand variations, are generally less than the capital charge penalty of low load factor operation of the plant. The main method of storage is likely to be underground caverns including, perhaps, depleted natural gas fields. The situation is less clear for medium calorific value gas and it may be that the relatively high storage costs for this gas restrict its economic use to demands that have a high load factor.

The choice of gasification technology for various applications has already been considered in chapter 4. In general terms, the high conversion efficiency of coal into product gas achieved with the counter current contacting system of fixed bed designs is an advantage for manufacturing substitute natural gas or medium calorific value gas. The high direct yield of methane is of particular benefit if substitute natural gas is the desired product. However, fixed bed gasifiers are at present limited in the range of coals that can be used and entrained phase gasifiers or the gasifiers under development using fluidised bed or molten bath systems may be more suitable for some types of coal.

If natural gas is regarded as competing mainly with distillate heating fuels in the premium heating market then, in the long term, the price of natural gas may be expected to be similar to that of such fuels. Indeed, in recent contracts for internationally traded gas, the price is generally fixed at oil-related levels. At these prices, the manufacture of substitute natural gas from coal is generally uneconomic at present. With the exception of demonstration plants, it therefore seems unlikely that substitute natural gas will be manufactured from coal on a large scale until natural gas supplies decline and become insufficient to meet the demand, possibly from the 1990s onwards.

The manufacture of substitute natural gas from petroleum fractions is an alternative to its manufacture from coal. The conversion of naphtha into substitute natural gas by the Catalytic Rich Gas (CRG) process is an established commercial technology. However, on the view that gas prices will be similar to those of distillate oil fractions, such an approach may not be economic in the future except possibly for easing peak supply problems on the gas grid. Alternatively, fuel oil could be converted into substitute natural gas by gasification and synthesis (for example, using a Texaco gasifier), or perhaps by hydrogenation. However, the cost of these conversions will probably be greater than the cost of upgrading fuel oil to distillates and, as before, if gas prices are similar to those of distillate oil fractions, this route may also be uneconomic.

Although the above discussion contains a number of simplifications (for example, storage and transport costs are excluded), it serves to illustrate some of the economic factors relevant to substitute natural gas manufacture. The relative prices for the scenario described are shown in figure 8.8.

Liquefaction

The two main approaches to converting coal into liquid fuels are as follows.

The direct route

This involves the decomposition of coal, usually in the presence of a solvent, and the hydrogenation of the resulting products to liquid fuels.

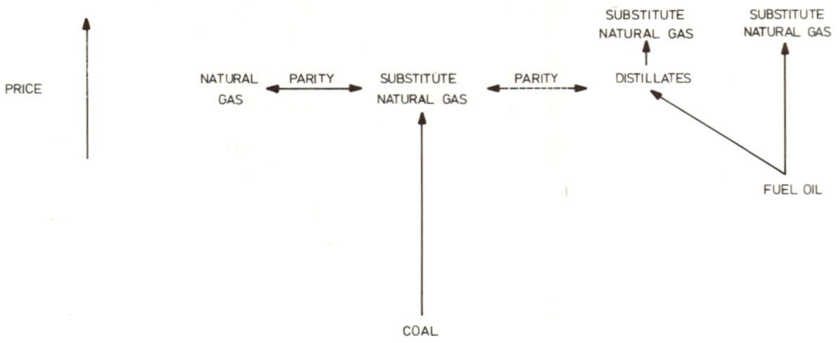

Figure 8.8 Possible relative prices for substitute natural gas manufacture (not to scale)

The indirect route

This involves the gasification of the coal to produce a synthesis gas which is then converted catalytically into liquid fuels.

These approaches are compared technically in section 5.11. Economic comparisons are made difficult by the different states of development of the technologies. The only commercially available coal liquefaction technology is the Fischer–Tropsch process which is acknowledged to be expensive and inefficient. The new technologies, both direct and indirect, offer the potential of considerable economic improvements. Those direct liquefaction processes at the most advanced stage of development (Exxon Donor Solvent, H-Coal and Ruhrkohle) are generally somewhat more efficient than the new indirect processes (such as methanol synthesis followed by MTG conversion). However, they also produce a wider, less valuable product slate. The overall effect is that the economics of the two technologies are probably quite similar at present.

Improvements may be expected with both approaches. Two-stage direct liquefaction processes are thought to offer a more attractive product slate than the single-stage processes on which development efforts have so far been concentrated. Also, further breakthroughs with improved catalysts for the indirect route are possible and could result in increases in efficiency and reductions in capital cost (for example, by eliminating some of the conversion stages).

The main factor affecting the economics of coal liquefaction (by whichever route) is the price of crude oil. Under present conditions, coal liquefaction is generally uneconomic by a significant margin. Plants built in the immediate future will therefore have an importance which is strategic rather than economic (as is probably the case with the Sasol plants in South Africa).

In the longer term, however, world economic growth and the associated increase in energy consumption, together with the depletion of oil reserves, will change the balance of oil supply and demand and result in an upward pressure on oil prices being maintained. If oil producers were competing in a classical free market, the oil price would equal the long-run marginal cost of supply (that is, the cost of oil produced from the most expensive source). In this case, as demand increased, so the next 'tranche' of production capacity, slightly more expensive than previously, would be brought into operation. With the existence of the OPEC cartel, however, the oil markets depart substantially from this well-ordered ideal. For example, oil prices are no longer related to production costs so that even marginal oil fields have costs that are only a fraction of the current price. Consequently, as the large and rapid changes of the 1970s have demonstrated, oil prices can be extremely volatile.

Although the changes may not be smooth, there is a widely held view that oil prices will continue to increase over the next few decades, possibly doubling in real terms by the beginning of the next century. Although coal prices may also increase, the trend will eventually lead to coal liquefaction becoming economic. Coal would then be the marginal source of liquid fuels and the cost of coal-derived liquids may become the price-setter for liquid fuels generally.

The time-scale on which coal liquefaction will become economic is naturally difficult to estimate but widespread commercialisation by the early decades of the next century is often assumed for the purpose of research and development programme planning. The factors affecting the timescale include

- world economic growth
- the effect of price increases on substitution for liquid fuels by other energy sources (such as the direct use of coal)
- the effect of price increases on conservation in the premium markets for liquid fuels
- the degree to which enhanced recovery techniques will extend the effective reserve base for oil
- the discovery of new oilfields
- the extent to which the large reserves of oil present in oil shales and tar sands can be recovered economically and in an environmentally acceptable way

It is also uncertain whether coal liquefaction plant would best be located in coal exporting or importing regions. Coal is about three times more expensive to transport than oil, on an energy basis, and therefore the cost of transporting the coal required to produce liquid fuels is about five times greater than that of transporting the liquids (assuming a conversion efficiency of 60 per cent). On the basis of transport costs alone, it may therefore be expected that coal liquefaction would be more economic near the point of coal production than in a remote, energy consuming region (see figure 8.9). However, the transport cost advantage may be offset by higher costs in building and operating a liquefaction plant in a non-industrialised area (see section 8.2). The liquefaction plant may, therefore, instead be best sited in regions with the necessary industrial infrastructure and with the coal feedstock being imported if necessary. In prac-

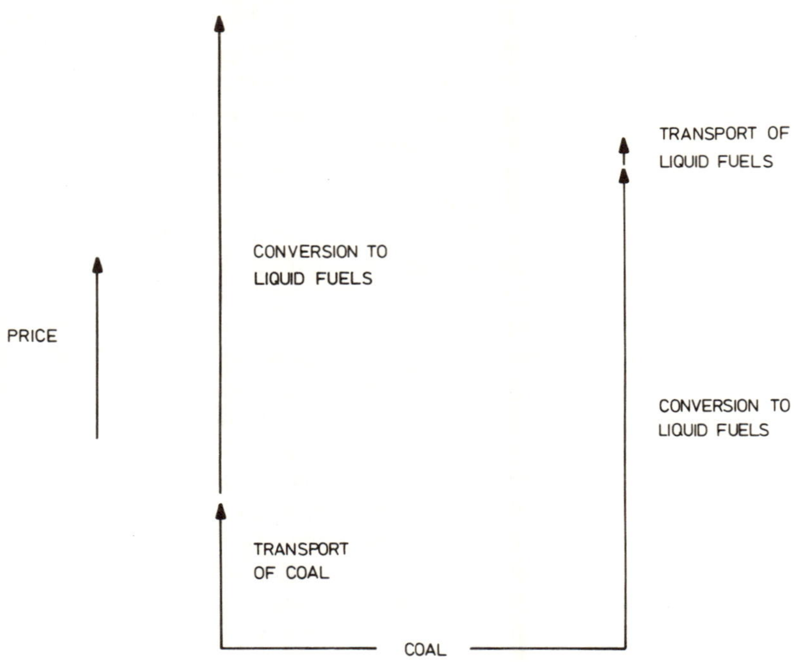

Figure 8.9 Effect of transport costs on the location of coal liquefaction plant (not to scale)

tice, policy considerations such as the growth of an industrial base for coal exporting regions or diversity and security of energy supplies for consumers may be more important than small cost differentials.

8.5 Economics for final consumers

The conversion processes discussed in the previous section provide final consumers with a range of energy carriers capable of satisfying their demands. For each energy carrier, there may be several utilisation technologies that could be employed. The economics of the options available may be analysed using models such as those described in section 8.2.

An important factor in such analyses is the DCF rate of return. Both domestic and industrial consumers generally require considerably higher rates of return for end-use equipment than do public or private enterprise for large central conversion plants. Pay-back times of three years or less are often required for investment by final consumers.

The discrepancy in the rates of return used by utilities and consumers introduces distortions into the pattern of investment in energy utilisation equipment. In particular, routes that have a low investment at the point of consumption are favoured even though a high investment may be required in central conversion plants. For this reason, the linear programming energy models discussed earlier are often operated with a uniform DCF rate of return in order to identify the optimum pattern of investment. The results of such studies do not, however, represent a realistic prediction of events under present conditions.

In some sectors of energy consumption, the visible economic factors may be less important than considerations such as amenity and space requirements. Although these aspects are difficult to quantify in economic studies, they may have a significant impact on the prospects for coal, particularly in the domestic and premium industrial sectors.

Power and light

Domestic requirements for power and light are almost always met by electricity from a utility grid. In the case of industrial consumers, however, the option of self-generation is practical and may be economic.

Although industrial self-generation capacity has increased steadily over recent decades, the rate of increase has been slower than the growth in the utility networks. Its relative importance has therefore declined to an average of about 8 per cent of total capacity for the major western industrial economies with individual values ranging from 3 per cent for the United States to 19 per cent for West Germany.

The principal economic factors favouring utility power generation rather than industrial self-generation are

the economies of scale that can be achieved in large central power stations
the security of supply of an integrated grid system
the availability of cheap fuels (including coal and nuclear power) for large power stations as compared with the need to use expensive premium fuels for small industrial schemes based on diesel engines or gas turbines

These factors are offset by the possibility of providing both power and process steam in a combined heat and power system. Although overall efficiencies of up to 75 per cent are possible, the economic incentives for industry to make the substantial investments required may not be sufficient to stop the decline in the relative importance of self-generation.

As noted in section 8.4, coal can be used in combined heat and power schemes larger than about 20 MW (electrical) with steam turbines but is effectively excluded from the market for the smaller industrial combined heat and power systems unless the new technologies which would permit coal to be used with gas turbines or diesels are developed successfully.

Transport

At present, liquid fuels dominate transport energy requirements and any change in feedstock would require a substantial change in end-use technology. Arguably, the simplest way in which coal could substitute for oil in this sector is therefore by coal liquefaction, the economics of which are discussed in section 8.4. However, a number of other options involving changes in transport technology also exist and are considered below. With the exception of rail electrification and possibly coal-fired ships, it appears likely that transport will remain largely dependent on conventional liquid fuels in the immediate future. As supplies of crude oil decline, substitution using coal may be expected to be by coal liquefaction with transport becoming one of the main markets for synthetic liquid fuels from coal.

Rail

Electrification of railway networks allows almost any energy source to be used to provide rail transport. In particular, coal can contribute to rail transport energy requirements through coal-fired power stations. This approach probably provides a higher overall efficiency than supplying diesel trains with synthetic fuel manufactured from coal or returning to coal-fired steam trains (even of modern design). The economics of electrification depend on the traffic volume.

Ships

Initially, coal-fired ships will be powered by steam turbines and fired with conventional stokers although pulverised fuel, coal-liquid mixtures or fluidised bed combustion may also be used in the future (see section 6.6). Although the efficiency of such systems is about 25 per cent, compared with over 40 per cent for a modern marine diesel, fuel prices are such that a saving in operating costs can still often be obtained.

However, the cost of building a coal-fired ship is greater than that of a diesel-powered ship of the same size because the propulsion plant is more expensive and additional facilities are required for coal and ash handling. The difference in capital costs becomes proportionately less for larger ships.

Since freight payment is made per tonne of cargo-carrying capacity, the utilisation of the ship is also affected by a change to coal firing. Storage requirements for coal are 2.5 to 3 times more by weight and 3 to 3.5 times more by volume than for oil (taking into account the above differences in end-use efficiency) and can have a considerable effect on the net cargo capacity of a ship of a given size. A compensating factor is that the optimum sailing speed is higher for a new coal-fired ship than for a new optimally designed oil-fired ship (although about the same as for existing oil-fired ships). This is because fuel consumption increases with sailing speed so that on economic grounds a lower speed is preferred with a more expensive fuel.

Overall, the trade-off between the lower running costs of coal-fired ships and the lower capital costs of oil-fired ships is such that coal may already have some advantage for the larger ship sizes depending on fuel prices, the voyage distance and the type of cargo. A number of coal-fired ships have recently been ordered and the conversion of a few oil-fired ships to coal-firing is also taking place. In general, the most attractive opportunities for coal-fired ships appear to be bulk carriers, particularly for coal and grain.

A future possibility is for diesel-powered ships to be fuelled by synthetic liquids manufactured from coal. The overall efficiency is probably similar to that of a steam cycle with direct coal firing; the higher efficiency of the diesel engine is offset by the losses in the manufacture of synthetic fuel. However, the capital costs of coal conversion are relatively high and, for this reason, coal-fired steam cycle propulsion may be more economic at least for the larger vessel sizes.

A further option may be direct coal-firing of marine diesel engines (see section 6.6). Although the economics appear attractive, considerable technical problems have yet to be overcome.

Aircraft

The only alternative to liquid fuels that appears to have been considered for air transport is hydrogen. Its main advantage is a high energy content per unit mass (2.8 times that of conventional jet fuel). This enables a lower mass of fuel to be carried for a given journey and results in better fuel economy, expressed as kJ/seat-km. In addition, the specific heat of hydrogen is high, making it effective for cooling purposes.

For aviation applications, hydrogen would have to be stored as a cryogenic liquid. The energy content per unit volume of liquid hydrogen is low (0.25 of that of conventional jet fuel), so that much larger storage volumes are required. This would necessitate a radical redesign of the aircraft.

Studies carried out by Lockheed indicate that, for subsonic passenger aircraft, a 12 per cent reduction in fuel consumption could be achieved because of the lower payload with the capital costs being similar to those for conventional jets. Greater savings would be possible for supersonic passenger aircraft because of their higher fuel consumption; a 38 per cent fuel cost reduction and a 26 per cent capital cost reduction is estimated in this case.

For civilian applications, the main market is in subsonic passenger jets. Although the production costs of hydrogen from coal are probably somewhat less than those of hydrocarbon liquid fuels from coal, the economic incentives to change to hydrogen appear small compared with the costs and difficulties of storing and handling liquid hydrogen on a routine basis.

Road vehicles

Besides conventional petrol and diesel fuels, energy requirements for road transport could possibly be supplied by methanol (with internal combustion engines), hydrogen (with internal combustion engines), and electricity (with battery storage).

Methanol has attracted interest as a possible coal-derived road transport fuel because it is cheaper to manufacture from coal than petrol using present synthesis technology. It has an octane number (RON) of 112 so that a high compression ratio can be employed resulting in a high thermal efficiency. A further advantage is that pollutant emissions are lower than with gasoline; methanol fuel is virtually sulphur-free and the formation of nitrogen oxides can be controlled by operating at leaner fuel to air ratios than is possible with gasoline.

A major obstacle to the widespread use of methanol as a road transport fuel is that extensive vehicle design changes would be necessary. These include increasing the compression ratio, heating the inlet manifold (because of the high latent heat of vaporisation) and increasing the size of the fuel tank (because of the low energy content per unit volume). It is doubtful whether such changes would be economic or practical for existing vehicles. Although new vehicles designed for methanol use could be manufactured, changes to present practice would also be required in storage and distribution. These arise principally from the attack of plastic and rubber components, the solubility of water in methanol and associated corrosion problems, toxicity and the wide range of methanol-air mixtures that are inflammable. For such reasons, it is possible that methanol would be more attractive as a feedstock for conversion to conventional gasoline by MTG processes than for direct use as a vehicle fuel.

Methanol may, however, find applications as a blending agent for gasoline. Up to about 15 per cent (by volume) of methanol can be added to gasoline without the need for major changes to the vehicle design. Although the advantages of neat methanol would not be obtained, an improvement in octane number of about 2 points may be possible. Disadvantages of the blended fuel include tendencies toward vapour lock formation and phase separation, the latter occurring in the presence of water at concentrations as low as 0.1 per cent. These problems will need to be overcome before such fuels would be acceptable.

Hydrogen has been considered as a fuel for internal combustion engines from the beginning of their development. The early work of Otto, for example, included the use of hydrogen-rich gaseous fuels. The advantages of hydrogen are similar in nature to those of methanol

> high engine efficiencies are possible because high compression ratios can be used
>
> emissions of pollutants are low—in particular, the release of carbon monoxide, sulphur dioxide, hydrocarbons, particulates and lead is negligible; also the formation of nitrogen oxides can be limited to values typically an order of magnitude less than for gasoline by the ability of hydrogen to burn in lean fuel to air ratios

The main development problem with hydrogen-fuelled vehicles is 'flashback' (pre-ignition of the fuel mixture while the inlet valve is open, leading to combustion propagating back into the inlet system). This can be controlled by injecting water vapour (or cooled, recycled exhaust gases) to act as a thermal diluent or by delaying the injection of the hydrogen until combustion is required. Flashback is more severe at high fuel to air ratios and, even with the above measures, operation will probably be limited to comparatively lean fuel mixtures. This leads to larger (and therefore heavier and more expensive) engines than with gasoline for the same power output.

Probably the most important disadvantages of hydrogen as a vehicle fuel are the problems of storage and handling. The two preferred methods available at present are metal hydrides and liquid hydrogen but each is expensive and has important limitations when compared with gasoline. Hydrides generally require a mass per unit energy stored of at least an order of magnitude greater than that for gasoline and liquid hydrogen storage can be subject to significant boil-off losses. In both cases, the large-scale distribution of hydrogen on a routine basis poses safety problems.

Since the cost of manufacturing hydrogen from coal is probably only marginally less than for liquid fuels, it seems unlikely that hydrogen from coal would be economic as a means of providing road transport energy requirements. If a cheaper source of hydrogen became available (for example, electrolysis using low-cost electricity), then hydrogen-fuelled vehicles may be more attractive.

Electric vehicles are already in widespread use as fork-lift trucks and, in the United Kingdom, for the local delivery of milk. However, using present lead-acid batteries, they are severely limited in range and performance. There therefore seems to be little prospect of significant growth in the use of electric vehicles unless advances are made in battery technology. New developments such as the sodium–sulphur battery will permit the speed and range of electric vehicles to be increased but, even so, their performance is unlikely to match that of internal combustion-engined vehicles.

A further disadvantage of electric vehicles is the comparatively slow charging rate. For domestic premises at least, this would be limited to a few kilowatts, whereas filling a tank from a gasoline pump corresponds to an energy transfer rate of several megawatts. A possible solution to this problem may be 'battery exchange' stations.

The energy efficiency of electric vehicles is compared with that of conventional petrol and diesel vehicles in table 8.9, assuming that coal is the primary energy source in each case. As shown in the table, electric vehicles offer a somewhat higher system efficiency than petrol-engined vehicles but are similar in efficiency to diesel-engined vehicles. Such comparisons are not, of course, relevant where another energy source (such as nuclear power) provides the marginal supply of electricity.

Although electric vehicles based on overhead wires (for example, trams and trolley buses) were widespread before the Second World War and still exist in some cities, their re-introduction on a significant scale is regarded as unlikely.

Overall, the best prospects for electric vehicles appear to be for limited-range battery-powered light goods vehicles and urban buses and the develop-

Table 8.9 Comparative efficiencies[a] of coal-based road transport systems[b]

Operation	Internal combustion engine		Electric vehicle (%)
	Petrol (%)	Diesel (%)	
Coal liquefaction and fuel distribution	50–60	50–60	
Internal combustion engine	25	35	
Transmission	90	90	
Power generation			35–40
Electricity distribution			93
Battery charge/discharge			80
Electric motor, controls, transmission			70
Overall efficiency	11–14	16–19	18–21

[a]The efficiencies for the electric vehicle exclude regenerative braking and vehicle heating in cold weather.
[b]Sources: Appendix 2 (conversion efficiencies); J. Porter *et al.*, *A Review of Market Prospects for Battery Electric Road Vehicles*, Transport and Road Research Laboratory report 630, 1974.

ment of improved batteries may be expected to lead to some increase in use for these purposes. The inconvenience of reduced range, poor performance and long recharging times may make a substantial penetration of the personal transport and heavy goods vehicle sectors of the market extremely difficult in the foreseeable future.

Domestic heating

The main energy requirement in this sector is for space heating, other demands such as water heating and cooling being of secondary importance. For this reason, the present discussion is directed specifically towards space heating, although many of the considerations also apply to other parts of the domestic heating sector.

At present, the two major sources of domestic heating energy in the OECD are oil and natural gas with relatively small amounts of electricity and solid fuels also being used. However, the pattern varies considerably from country to country.

As noted earlier, a number of special factors affect consumer choice in the domestic sector. Pay-back times tend to be short, favouring energy supply systems that involve little capital expenditure at the point of consumption even if running costs are high. In particular, this militates against the installation of pollution control equipment by consumers so that relatively 'clean' energy carriers have to be used. Domestic consumers also show a marked preference for high degrees of amenity and convenience.

Domestic heating is one of the most complex energy demand sectors to analyse because of the wide range of potential options for energy carriers and end-use technologies. Amongst the future methods by which coal could replace oil and natural gas from this market are the following

direct coal combustion
district heating
electric resistance heating
electric heat pumps
substitute natural gas
synthetic liquid fuels

The direct use of coal or smokeless fuel in modern domestic appliances (see section 6.5) can give high efficiencies without the smoke emissions previously associated with domestic coal combustion. Where supplies of solid fuel are available, the economics of such appliances are often already competitive with those of other heating systems and may be expected to become increasingly so in the future. However, the extent to which domestic coal consumption grows in the future may be limited by consumer resistance to returning to a lower level of amenity than that offered by present heating systems (for example, based on oil or gas) coupled with the significant investment required. Environmental considerations may also become important. If controls on low-level sulphur emissions were introduced or existing regulations on particulate emissions were made more stringent, the direct use of coal in domestic premises may become prohibitively expensive.

Many of the disadvantages of domestic coal combustion can be overcome by district heating schemes in which coal is burned in large central boilers or combined heat and power systems with the heat being distributed to consumers as hot water. As discussed in section 8.4, such schemes require a substantial capital investment in the distribution network and are economic only where the demand density is high.

It is widely thought that, for much of the market, the two main competing energy carriers will be electricity and substitute natural gas. The relative merits of these options have been the subject of a number of analyses whose main results are summarised below.

Since electricity and gas offer similar levels of amenity, the competition between them may be expected to be in terms of the cost to the consumer. This can be sub-divided into three components: the price of the delivered energy, the appliance efficiency and the appliance cost.

Substitute natural gas is significantly cheaper to manufacture from coal than electricity per unit of thermal energy, partly because the conversion efficiency is approximately twice that for electricity. As distribution costs (using the existing systems) are similar, the delivered price of substitute natural gas would be less than that of electricity. However, this advantage is offset by a lower efficiency in use (about 70 per cent design efficiency compared with an efficiency of direct-acting electric resistance heating approaching 100 per cent) and significantly higher investment requirements by the consumer. The net effect is that the overall costs to the consumer are probably quite similar in the two cases. For new dwellings, there will therefore be little to choose between the two systems and, for existing dwellings, there will be no incentive to change to the other system. Substitute natural gas may be expected to retain most of the market for heating houses presently using gas although, outside the gas supply areas, electricity consumption for domestic heating may grow.

For heating purposes, electricity may be used more effectively than with resistance devices by employing a heat pump. Depending on the temperatures involved, heat pumps can have 'efficiencies' of 200 to 400 per cent (that is, deliver as heat twice to four times the heat that would be produced by electric resistance heating). However, heat pumps are comparatively expensive to install and the overall economics for the consumer may show little or no advantage over those for electric resistance heating. Heat pumps designed to be capable of operating in the reverse mode as air conditioners in the summer may be attractive for some applications.

An additional complication is the variation of electricity demand with time and the availability of off-peak electricity. Unlike gas, electricity has only limited storage possibilities and the diurnal and seasonal demand variations have to be met by installing sufficient generating capacity to meet peak demands. There therefore exists the opportunity to supply electricity during off-peak periods at a cheap tariff corresponding to the marginal production and distribution costs (broadly, this comprises operating costs but not capital charges). The amount of off-peak electricity available depends on the characteristics of the electricity demand. In systems that are summer-peaking (for example, those parts of the United States that have a heavy summer air-conditioning load), a considerable amount of winter heating could be supplied in this way (although winter heating demands in such regions tend to be light).

Elsewhere (Europe for example), electricity demand is generally winter-peaking and off-peak heating can take place only during winter nights; the amount of off-peak electricity available is limited accordingly. In this case, storage heaters are necessary so that electricity can be consumed overnight and heat released during the daytime. The advantage of storage heaters in being able to consume cheap off-peak electricity is offset by the following factors

capital costs for the consumer are somewhat higher than for direct-acting electric heating

the effective efficiency is less than for direct-acting electric heating (this is partly because storage heaters tend to heat the house to a higher temperature than needed early in the day and partly because their lack of response tends to lead to overheating on unexpectedly warm days)

some on-peak direct-acting electric heating is required as back-up because of the difficulty in forecasting in advance the amount of storage heating required

In general, however, the economics of off-peak electric heating appear attractive when compared with direct-acting electric heating and substitute natural gas.

Although the above discussion has concentrated on coal-based systems for domestic heating, in practice electricity systems contain a mixture of plants. The economics of electric heating are determined by the type of power station operated to meet the demand. In the case of on-peak heating (that is, direct-acting electric heating in a winter-peaking system), this will be low load factor plant, probably coal or oil for most utilities both now and for some time in the future. Off-peak heating, however, uses plants operating at higher load factors, possibly including nuclear stations in the future if nuclear capacity were to grow to provide all of the base-load power generation duty. Since the operating costs of nuclear stations are low (most of the generating cost is capital charges), off-peak electricity from nuclear stations charged at marginal generating costs would be substantially cheaper than from coal-fired stations. The amount would, however, be limited by the characteristics of the system, as discussed above.

Coal could substitute for oil in the domestic heating market by synthetic liquid fuels from coal liquefaction plants. This option would have the advantage that the existing distribution system and stock of consumer appliances could be utilised. Although liquid fuels from coal may be somewhat more expensive than substitute natural gas, distribution costs are generally lower so that the overall cost to the consumer is probably similar. There would therefore be no incentive for existing gas consumers to change to oil firing based on synfuels. However, liquid fuels from coal might retain a proportion of the convenience heating market outside gas supply areas in competition with electricity.

Industrial (and commercial) heating

It is widely felt that the industrial market is the most promising area for an expansion of coal sales in the immediate future. At present, coal provides only about 35 per cent of industrial heating in OECD countries; the remainder is supplied mainly by oil and natural gas. Furthermore, coal now has a substantial price advantage over oil and natural gas on a thermal basis and its direct use to replace these fuels may therefore result in significant running cost savings. Such savings are, however, offset by the greater investment required for coal-fired equipment (see appendix 2).

Four situations exist in which coal may be used instead of oil or natural gas:

(1) On a green-field site (where new equipment has to be installed). This is the most attractive situation for introducing coal.

(2) On an existing site where oil or gas-fired equipment is to be retired or replaced. This is often less economic for the introduction of coal than a green-field site because of constraints on layout and space.

(3) On an existing site with operational oil or gas-fired equipment where conversion (retrofitting) for coal firing is possible, perhaps with some derating. This is not always practical, but can be attractive provided that the amount of derating is not too large.
(4) On an existing site with operational oil or gas-fired equipment, where the only way of introducing coal firing is by early retirement of the present installations. This is the least attractive situation for introducing coal and is seldom economic with present fuel prices.

The term 'industrial heating' encompasses a wide range of end-uses and, for convenience, these may be sub-divided into boilers and furnaces. An approximate breakdown of the market for the countries of the OECD is given in table 8.10. Boilers supply about 55 per cent of industrial heating requirements and furnaces 45 per cent; other industrial demands for fossil fuels are accounted for mainly by feedstocks.

Table 8.10 Approximate breakdown of the industrial heating market for OECD countries[a]

Sector	Coal		Oil and natural gas (%)	Total (%)
	Direct (%)	As coke (%)		
Boilers	8		47	55
Furnaces	3	24	18	45
Total	11	24	65	100

[a]The data given are on an energy content (higher heating value) basis with the values for coke being expressed in terms of the coal input to the coke ovens. Source: *The Use of Coal in Industry*, The Coal Industry Advisory Board, International Energy Agency, OECD, Paris, 1982.

The boiler sector of the industrial heating market is particularly important as a possible area of expansion for coal because of coal's relatively low market penetration (about 15 per cent) at present. As discussed in chapter 3, industrial boilers are of two types, water-tube boilers and firetube boilers. Water-tube boilers are considerably more expensive than fire-tube boilers but have to be used for large-scale applications or where high-pressure steam is required (for example, for power generation). Water-tube boilers account for about 60 per cent of boiler capacity in OECD countries but, because they operate at a higher load factor (on average 35 per cent) than fire-tube boilers (on average 20 per cent), they account for approximately 75 per cent of the fuel consumed in industrial boilers.

As a result of this dominant position, increased coal use in water-tube boilers is essential if coal is to displace oil and natural gas significantly as industrial fuels. However, water-tube boilers last much longer than firetube boilers; typical lifetimes are 30 to 50 years and 15 to 20 years respectively so that a correspondingly low replacement rate applies. Indeed, much of the existing water-tube boiler stock in OECD countries was built during the 1960s and 1970s (over half of the total capacity in the case of the United Kingdom) and this would not normally therefore be replaced before the 1990s.

At first sight, coal would appear to be better placed to substitute for oil on economic grounds in the water-tube boiler market than in firetube boilers. The price advantage enjoyed by coal over oil is emphasised by the relatively high load factor of water-tube boilers and their larger unit sizes bring economies of scale in coal distribution and handling. However, these factors are offset by the relatively high capital cost differential between coal and oil firing for water-tube boilers compared with firetube boilers (see appendix 2). This leads to longer pay-back times for the same price differential between coal and oil. Furthermore, environmental controls may be less stringent for firetube boilers than for water-tube boilers. The overall situation is therefore unclear and coal may be regarded as competitive with oil in both sectors within the wide range of local conditions that exist.

The main technologies available for coal-fired boilers in new industrial installations (either at green-field or existing sites) are pulverised fuel, stokers and fluidised bed combustion.

In general, for boilers larger than about 30 MW (thermal) the method of firing is pulverised fuel; below this size, various types of stoker are used (see chapter 3). Atmospheric pressure fluidised bed combustion may be regarded as an alternative to each of these technologies throughout their respective boiler size ranges. However, as for power generation, the direct economic incentives to depart from the conventional technologies are not large if sulphur retention is not required although fluidised bed combustion may offer an attractively high degree of amenity for industrial consumers (for example, better automatic control, more compact design, less maintenance and better availability). If sulphur retention is required, fluidised bed combustion has a significant advantage but coal (and heavy fuel oil) then become relatively less attractive when compared with natural gas or low-sulphur distillates.

For converting existing oil-fired or gas-fired boilers to coal firing, it may be possible to employ the combustion technologies given above, particularly for water-tube boilers that were originally designed for coal-firing but were subsequently converted to use oil or gas. However, such situations are comparatively rare and, in most cases, direct conversion by these methods is likely to be impractical or prohibitively expensive. For this reason, the conversion of oil and gas-fired boilers to use coal as coal–liquid mixtures has attracted interest recently. The main options are coal–oil mixtures and coal–water mixtures.

The economics of using these fuels depends on trade-offs between running costs and capital costs. Two comparisons arise.

Competition with oil

Coal–liquid mixtures are cheaper than oil but involve some additional capital investment for plant modifications (see section 6.6). Furthermore, the boiler may have to be derated on conversion although, where there exists surplus boiler capacity on-site, derating is less important.

Competition with conventional coal firing

Coal–liquid mixtures are more expensive than coal but the capital investment required for the plant modifications on conversion is less than that for the installation of conventional coal combustion equipment, particularly if the early retirement of existing boilers would otherwise be necessary.

Because some of the factors that favour changing from oil-firing to coal–liquid mixtures (for example, high oil prices, low coal prices and high load factors) also favour direct coal combustion, the use of coal–liquid mixtures is limited to a range of economic conditions for which it is preferable to change from oil-firing to coal–liquid mixtures but not from coal–liquid mixtures to conventional coal combustion. These conditions are determined by the fuel cost savings that can be achieved by installing a new method of firing in relation to the capital expenditure necessary. In each case, the fuel cost savings depend mainly on the oil/coal price relativity and the load factor whereas the capital expenditure depends on the boiler design and the importance of the derating incurred when firing coal–liquid mixtures. These factors vary considerably from site to site and from country to country so that it is difficult to generalise about the conditions for which conversion to coal–liquid mixtures is economic.

Since oil prices are now much higher than coal prices, coal–water mixtures should be cheaper per unit of energy than coal–oil mixtures. As other differences between the two fuels are generally of secondary importance, coal–water mixtures may be expected to become the preferred technology on economic grounds if current research and development programmes are successful. One of the most attractive sectors for conversion appears to be water-tube boilers because their relatively high capital cost militates against early retirement.

In addition to their potential role in the conversion of existing oil or gas-fired boilers, coal–liquid mixtures may also prove to be an attractive fuel for new installations where convenience is of paramount importance. For these applications, coal–liquid mixtures offer the advantages of a low-ash fuel that can be delivered by tanker, stored, pumped and burned in a way similar to conventional liquid fuels. Such opportunities may be expected to occur mainly in the commercial and small-scale industrial sectors.

Although new developments in conventional coal firing and coal–liquid mixtures are making possible significant improvements in amenity, coal will always be less convenient than the premium fuels and there may therefore be a part of the market that remains inaccessible to coal. Some consumers will prefer to continue burning premium fuels and may eventually use substitute natural gas or synthetic liquids. The size of this demand is difficult to estimate although it is likely to be concentrated in small boiler applications and would increase if stringent environmental controls were introduced.

In the furnace sector of the industrial heating market, coal supplies about 60 per cent of the energy demand (see table 8.10). However, most of this is accounted for by coke used in the iron and steel industry. For other applications, coal supplies approximately 15 per cent of furnace energy requirements.

Furnace applications are diverse and furnace designs are often specific to an industry or even to a process. In general, however, the types of furnace demand may be classified by the following parameters: gas purity, gas or flame temperature and unit size.

Depending on the furnace conditions, coal may be used either directly or by conversion to a clean combustible gas. The characteristics of the various technologies available are summarised in figure 8.10.

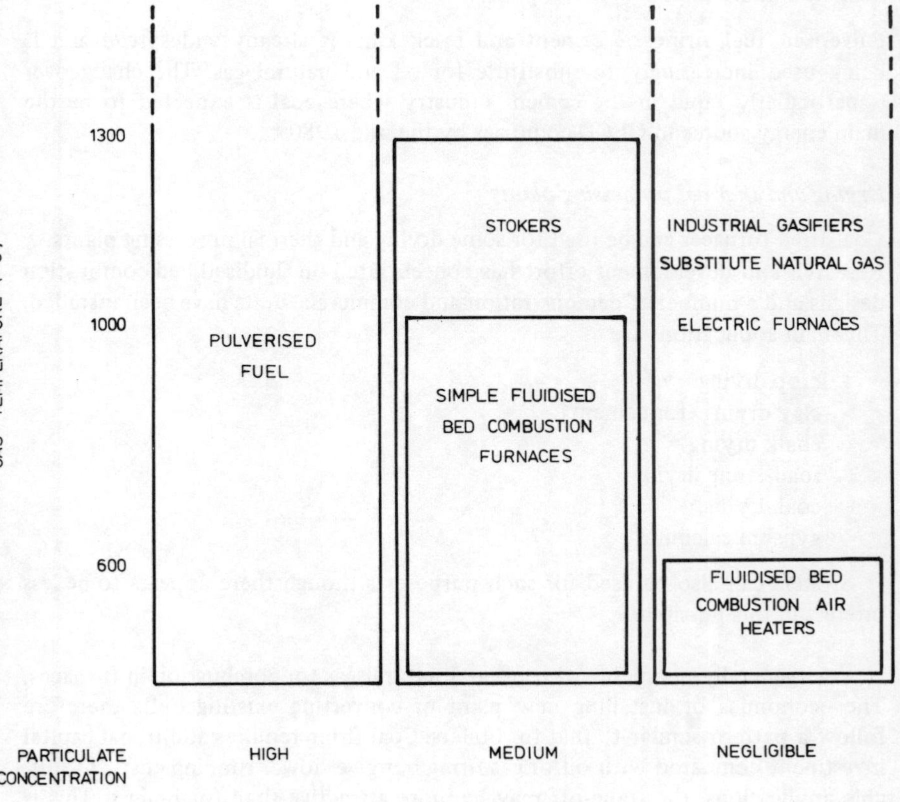

Figure 8.10 Comparison of coal-based furnace technologies

Where the presence of particles of coal ash in the gas can be tolerated, direct coal combustion can be employed. Indeed, almost all of the coal at present used in industrial furnaces is burned directly. Each of the main combustion technologies (pulverised fuel, stokers and fluidised bed combustion) may be applied in furnaces, as indicated in figure 8.10. Some of the more important opportunities for the direct use of coal in this sector include the following.

Blast furnaces

During the 1960s, low oil prices led to the widespread use of fuel oil for injection at the tuyeres of blast furnaces in order to reduce coke consumption and increase throughput. With the oil price rises of the 1970s, this practice has become less attractive and fuel oil consumption in blast furnaces has declined rapidly. Recently, the injection of pulverised fuel into commercial blast furnaces has begun and tests are also being carried out on coal–liquid mixtures.

Cement and brick kilns

Pulverised fuel firing of cement and brick kilns is already widespread and is being used increasingly to substitute for oil and natural gas. The changeover is particularly rapid in the cement industry where coal is expected to be the main energy source in OECD countries by the mid-1980s.

Drying and thermal processing plants

Coal-fired furnaces can be used for some drying and thermal processing plants. Research and development effort has concentrated on fluidised bed combustion designs and a number of demonstration and commercial units have been installed. The main applications are

crop drying
clay drying (for cement)
chalk drying
road-stone drying
coal drying
gypsum calcining

Stokers can also be used for such purposes although there appears to be less interest in this possibility.

Coal generally competes with oil and natural gas for combustion in furnaces. The economics of installing new plant or converting existing units therefore follow a pattern similar to that for boilers. Coal firing requires additional capital investment compared with oil or gas firing but gives lower running costs. In suitable applications, the trade-off may be more attractive than for boilers. This is because the difference in capital cost between coal firing and oil or gas firing is often less than for boilers; the kiln or furnace is usually the same with the only change being in the hot gas generator which is less expensive than an equivalent boiler. The fuel price advantage of coal over oil or gas therefore results in a comparatively short pay-back period.

The injection of pulverised fuel or coal–liquid mixtures into a blast furnace is a special case because this may be regarded as displacing coke rather than fuel oil. The economics then depend principally on the relative prices of coking coal and steam coal, the cost of coke manufacture and the replacement ratio of coal for coke.

For many types of furnace a clean hot gas is required and direct coal combustion cannot be used. At present, oil and natural gas dominate this sector of the market. The main technologies that may enable coal to compete (see also figure 8.10) are clean air heated indirectly in a fluidised bed, industrial gasifiers and substitute natural gas manufacture.

Systems in which clean air is heated indirectly in a fluidised bed combustor (for example, the Encomech furnace) are restricted to a maximum air temperature of about 600 °C. Applications available within this temperature limitation include the processing of foodstuffs, detergents and pharmaceuticals.

Where the gas must be both clean and at a high temperature, coal gasification may be used. Some of the major energy consuming industries in this case are glass manufacture, pottery and ceramics manufacture and metal melting and reheating.

The two main relevant coal-based technologies are on-site industrial gasifiers and remote substitute natural gas plants. Superficially, the economics appear to favour the use of on-site industrial gasifiers, as this approach avoids the costs of the more complex manufacturing process for substitute natural gas, and of the transmission and distribution network. However, there are a number of factors which offset these advantages of industrial gasification.

Economies of scale

Compared with small industrial gasifiers, large substitute natural gas plants benefit from the economies of scale.

Coal distribution costs

The cost of distributing substitute natural gas from a central plant is offset by the cost of distributing coal to numerous small industrial gasifiers.

Load factor

A substitute natural gas plant can operate at a higher load factor than a dedicated industrial gasifier, partly because of the effect of load sharing between consumers in an integrated distribution system and partly because storage is comparatively cheap for substitute natural gas.

Environment

It is generally thought to be easier and cheaper to meet strict environmental controls in large plants rather than in small plants.

Rate of return

A gas utility operating a substitute natural gas plant is likely to require a lower Discounted Cash Flow rate of return on the capital than an industrial company investing in an on-site gasifier.

In view of the above factors, it is difficult to assess the relative economics of the two routes and it may be that local conditions (such as low interest loans and environmental legislation) or application-dependent considerations predominate.

Coal competes with oil and natural gas for furnace use less effectively if gasification is necessary rather than if direct combustion is acceptable. This is because gasification inevitably involves a higher investment and a lower efficiency than combustion. It may therefore be expected that oil and gas will withdraw from clean, high-temperature furnace applications more slowly than from other sectors of the industrial heating market. Although, in principle, industrial gasifiers or substitute natural gas could be used for furnaces that can operate at a low temperature or with particulates present, it is likely to be more economic to employ direct combustion wherever this is practical.

A further option for clean high-temperature furnaces is electric heating and this may also be regarded as an indirect method by which coal can serve this sector. The economics of electric heating are strongly application-dependent although they can be attractive in terms of good control and high efficiency in use. It is widely expected that electric heating will increase its penetration of the furnace market in the future both with the well-established electric arc and induction furnaces and with the newer technologies of microwave heating and heat pumps.

Petrochemical feedstocks

The main feedstocks required by the petrochemicals industry are aromatics, olefins and synthesis gas.

At present, aromatics are manufactured mainly by the catalytic reforming of naphtha. Olefins are produced by the steam cracking of naphtha, gas oil and ethane extracted from natural gas (where available) with further amounts of aromatics also being obtained from some of these processes. Synthesis gas is manufactured principally by steam reforming natural gas with small amounts being obtained by steam reforming naphtha and the partial oxidation of heavy oil.

Processes for manufacturing aromatics and olefins from coal are essentially the same as those for manufacturing liquid fuels and it is to be expected that coal liquefaction plants would be designed to produce both types of product. The conversion of coal into aromatic and olefinic chemical feedstocks will therefore become economic at the same time as coal liquefaction generally (see section 8.4).

The economics of synthesis gas manufacture are, however, different. Where synthesis gas is at present manufactured from natural gas, the capital and operating costs of the conversion result in the cost of the synthesis gas being significantly higher than the price of the natural gas feedstock. For a coal-based process, the situation is reversed; the cost of synthesis gas production is lower than that of substitute natural gas because synthesis gas is an intermediate product in its manufacture. The price relativities are illustrated in figure 8.11. As shown in the figure, coal competes more effectively with natural gas for synthesis gas production than directly through substitute natural gas. A similar argument holds for the manufacture of synthesis gas from naphtha or heavier feedstocks.

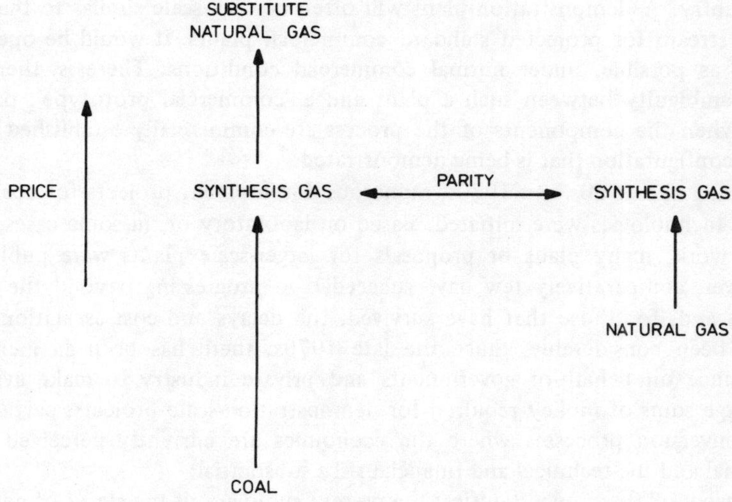

Figure 8.11 Relative costs of synthesis gas manufacture (not to scale)

It follows that the use of coal as a feedstock for synthesis chemicals may be expected to be economic before the conversion of coal into substitute natural gas or liquid fuels. Already, a number of coal gasification plants are in operation for synthesis gas manufacture, and more are under construction (see appendix 1).

APPENDIX 1

DEVELOPMENTS IN COAL UTILISATION TECHNOLOGY

In most fields of coal utilisation, the development of a new technology requires several stages of scale-up before being offered with normal commercial guarantees. Scale-up factors between stages of 100 (on throughput) are possible although values are generally less than this. The terms 'laboratory scale', 'pilot plant scale' and 'demonstration scale' are commonly used to indicate the size of the plant (in order of increasing throughput) but usage varies considerably from country to country. A demonstration plant will often be of a scale similar to that of a single stream for projected standard commercial plants. It would be operated, as far as possible, under normal commercial conditions. There is, therefore, some ambiguity between such a plant and a 'commercial prototype', particularly when the components of the process are commercially established but it is the configuration that is being demonstrated.

During the 1960s and 1970s, numerous development projects for coal utilisation technologies were initiated. Based on laboratory or, in some cases, pilot-plant work, many plans or proposals for larger-scale plants were publicised. However, comparatively few have succeeded in progressing beyond the initial phases and, for those that have survived, the delays and cost escalations have often been considerable. Since the late 1970s, there has been an increasing reluctance on behalf of governments and private industry to make available the large sums of money required for demonstration-scale projects, particularly for conversion processes where the economics are currently perceived to be marginal and the technical and financial risks substantial.

In view of these uncertainties, the present summary of the status of new coal utilisation technologies concentrates on developments that are approaching a commercial scale and for which completion is reasonably well assured. It follows that some promising projects are omitted where these are at present operating only on a small scale.

One of the most well-developed of the new coal technologies is atmospheric pressure fluidised bed combustion. Already, some boilers and furnaces are being offered with normal commercial guarantees, and others are at a commercial prototype stage. A large number of manufacturers are involved, mainly operating within their 'home' country. The current status of the technology is indicated in table A1.1.

Atmospheric pressure fast bed combustion and pressurised fluidised bed combustion are at earlier stages of development. Relatively few organisations are involved and only a small number of pilot plants have so far been operated (see tables A1.2 and A1.3).

Coal gasification is already well established commercially as indicated by the large number of plants that have been installed (see table A1.4). However, not all of these are currently in use and the main large-scale coal gasification plants in commercial operation are given in table A1.5.

Of the second-generation gasification processes, the greatest development activity is concentrated on Lurgi-based gasifiers and the Texaco gasifier. The main projects in the construction or operation phases involving these gasifiers are described in tables A1.6 and A1.7, respectively. Some of the other major developments of coal gasification systems (and related technologies) are shown in table A1.8.

Direct liquefaction is not at such an advanced stage of development as fluidised bed combustion or coal gasification. The relatively small number of large-scale pilot plants that have been constructed are summarised in table A1.9. All of these plants are single-stage processes; two-stage processes are at present operational only on a small scale.

Table A1.1 Industrial atmospheric pressure fluidised bed boilers[a] and furnaces in operation (1982)[b]

Number of units[c]	Countries[d]	Maximum plant size[e] (t/d coal equivalent input)
Over 50	China[f]	250
11-50	United Kingdom	100
	United States	250
Up to 10	Canada	40
	Finland	75
	Japan	60
	Sweden	120
	West Germany	90

[a] Fast fluidised bed combustors have been excluded from the table (see table A1.2).

[b] Sources: Jason Makansi and Bob Schwieger, Fluidized Bed Boilers, *Power*, August 1982; W. G. Kaye, Fluidised Bed Combustion, paper to *Symposium on Modern Coal Burning Technology for Industrial Boilers*, S. Wales Inst. Engrs., Cardiff, June 1982.

[c] Units smaller than 3 t/d coal input have been excluded from the table, such plants generally being experimental.

[d] Fluidised bed combustion units have also been built in other countries, including Australia, Denmark, France, India, Israel, Italy, Netherlands, South Africa and Switzerland.

[e] Not all of the units burn coal; the thermal input has therefore been expressed as coal equivalent using the conversion 29.3 GJ equals 1 tonne coal equivalent.

[f] It is thought that more than 2000 fluidised bed combustion units are in operation in China.

Table A1.2 Fast fluidised bed combustor developments[a]

Organisation(s)	Location	Plant size[b] (t/d coal equivalent input)
Lurgi/Vereinigte Aluminium Werke AG	Frankfurt, West Germany	4
	Lünen, West Germany	250
Battelle	Columbus, USA	5
Conoco/Struthers Wells Corp.	South Texas, USA	45
Gulf Oil/Pyropower Corp.	Bakersfield, USA	40
A. Ahlström Oy	Karhula, Finland	5
A. Ahlström Oy	Pihlava, Finland	45
Savon Voima Oy/A. Ahlström Oy	Suonenjoki, Finland	20
A. Ahlström Oy	Kauttua, Finland	200
Studsvik Energiteknik AB/ Generator Industri AB	Nykoping, Sweden	7

[a] Sources: W. Wein, L. Plass and K-H Maintok, Circulating fluid bed technology for efficient coal combustion, paper to *Conf. Coal Technology Europe*, Cologne, June 1981; Fluidized-bed Combustion–Industrial Application Demonstration Projects, Battelle's Multisolid Fluidized Bed Combustion Process, *US DOE report DOE/FE/2472-42*, Battelle Columbus Laboratories, October 1979; O. Jones and E. C. Seber, Initial operating experience at Conoco's South Texas multisolids FBC steam generator, paper to the *Seventh International Conference on Fluidized Bed Combustion*, Philadelphia, Pa., October 1982; J. O. Reardon, Pyroflow–commercial recirculating fluidised bed combustion, *Modern Power Systems*, November 1981, p. 37; F. Engström and H. H. Yip, Operating experience of commercial scale Pyroflow circulating fluidized bed combustion boilers, paper to the *Seventh International Conference on Fluidized Bed Combustion*, Philadelphia, Pa., October 1982; L. Strömberg and P. Wickberg, Fast fluidized bed combustion of coal, paper to the *Seventh International Conference on Fluidized Bed Combustion*, Philadelphia, Pa., October, 1982.
[b] Not all of these units burn coal; the thermal input has therefore been expressed as coal equivalent using the conversion 29.3 GJ equals 1 tonne coal equivalent.

Table A1.3 Pressurised fluidised bed combustion pilot plants[a]

Organisation(s)	Location	Plant size(s) (t/d coal input)
National Coal Board (CURL)	Leatherhead, UK	20, 5, 0.5
Argonne National Laboratory	Argonne, USA	0.2
New York University	New York, USA	20
General Electric	Malta, USA	1.5
Warsaw University	Warsaw, Poland	10
Curtiss Wright	Wood Ridge, USA	100, 5
Natal University	Durban, South Africa	5
NCB (IEA Grimethorpe) Ltd./ UK, West German, USA governments	Grimethorpe, UK	250
American Electric Power/STAL-Laval/ Deutsche Babcock	Malmo, Sweden	50

[a]Source: S. A. Miller *et al.*, *Technical Evaluation: Pressurised Fluidized Bed Combustion Technology*, Argonne National Laboratory, Prepared for US DOE under contract no. W-31-109-ENG-38, Report no. ANL/FE-81-65, 1982.

Table A1.4 Commercial coal gasifiers installed[a]

Type	Number of units	Size range (t/d coal input per unit)
Wellman	Over 600	0.3–75
Wilputte	Over 50	3–15
Woodall–Duckham/GI	Over 100	60–100
Wellman Incandescent	Approx. 30	15–100
Foster–Wheeler Stoic	3	30–60
Lurgi	142	60–600
Winkler	70	150–1000
Koppers–Totzek	54	300–600

[a]The data refer to gasifiers installed; many have now been taken out of service. Source: Robert A. Meyers (ed.), *Coal Handbook*, Marcel Dekker, New York, 1981.

Table A1.5 Large-scale commercial coal gasification plants in operation[a]

Type	Number of plants	Location(s)	Size (total coal consumption, t/d)
Lurgi/Fischer–Tropsch	3	South Africa	50000
Lurgi/ammonia	2	Pakistan Korea	450
Winkler/ammonia	3	India Turkey Yugoslavia	1000
Koppers–Totzek/ammonia	12	Greece India South Africa Thailand Turkey Yugoslavia Zambia	10000
Koppers–Totzek/methanol	1	South Africa	70

[a]Sources: *Chemistry of Coal Utilisation*, second supplementary volume, Wiley, New York, 1981; L. Grainger and J. Gibson, *Coal Utilisation: Technology, Economics and Policy*, Graham and Trotman, London, 1981.

Table A1.6 Lurgi-based gasification system developments[a]

Type	Organisation	Location	Plant size (t/d coal input)
Air-blown Lurgi gasifiers (5)/ combined cycle	Steag	Lünen, West Germany	1700
British Gas–Lurgi slagging gasifier	British Gas	Westfield, UK	350
British Gas–Lurgi slagging gasifier	British Gas	Westfield, UK	600
Ruhr 100 gasifier (high-pressure Lurgi)	Ruhrkohle/ Ruhrgas/Steag	Dorsten, West Germany	170
Lurgi gasifiers (14)/SNG	American Natural Resources	Great Plains, USA	14000[b]

[a]Sources: A. Baker and M. Teper, Synfuels Development Outside North America, *American Nuclear Society Conference*, Los Angeles, 1982; T. Wett, Synfuels Offer Challenging Future, *Oil and Gas Journal*, June 1981; L. Grainger and J. Gibson, *Coal Utilisation: Technology, Economics and Policy*, Graham and Trotman, London, 1981.
[b]The coal for this plant is a lignite.

Table A1.7 Texaco-based gasification system developments[a]

Type	Organisation(s)	Location	Plant size (t/d coal input)
Texaco gasifier	Texaco	Montebello, USA	15
Texaco gasifier	Ruhrkohle/ Ruhrchemie	Oberhausen-Holten, West Germany	150
Texaco gasifier/ ammonia synthesis	Tennessee Valley Authority	Muscle Shoals, USA	200
Texaco gasifier/ power generation (air blown)	Dow Chemical	Plaquemines, USA	400
Texaco gasifier/ acetic anhydride synthesis	Tennessee Eastman	Kingsport, USA	800
Texaco gasifier/ combined cycle	S. California Edison and others	Cool Water, USA	1000

[a] Sources: A. Baker and M. Teper, Synfuels Development Outside North America, *American Nuclear Society Conference*, Los Angeles, 1982; D. R. Simbeck and A. J. Moll, Synthetic Fuels from Coal: An Important Source of Energy, *World Coal*, Jan./Feb. 1982; T. Wett, Synfuels Offer Challenging Future, *Oil and Gas Journal*, June 1981.

Table A1.8 Other major gasification-related developments[a]

Type	Organisation	Location	Plant size (t/d coal input)
Kilngas gasifier retrofit power generation	Illinois Power Co./ Allis Chalmers	Wood River, USA	600
High-temperature Winkler gasifier	Rheinbraun	Frechen, West Germany	70[b]
High-temperature Winkler gasifier/ methanol	Rheinbraun	Cologne, West Germany	800[b]
U-gas gasifier	Institute of Gas Technology	Chicago, USA	6
Westinghouse gasifier	Westinghouse	Waltz Mill, USA	30
Molten iron gasifier	Sumitomo/Kamasaki	Kashima, Japan	60
Humboldt gasifier	Kloeckner	Maxhuette, West Germany	10
Saarberg-Otto gasifier	Saarbergwerke	Völkingen, West Germany	260
Shell gasifier	Shell	Harburg, West Germany	150
Mobil synthesis process	New Zealand Synthetic Fuels Corp.	New Zealand	*c*

[a]Sources: A. Baker and M. Teper, Synfuels Development Outside North America, *American Nuclear Society Conference*, Los Angeles, 1982; D. R. Simbeck and A. J. Moll, Synthetic Fuels from Coal: An Important Source of Energy, *World Coal*, Jan./Feb. 1982; T. Wett, Synfuels Offer Challenging Future, *Oil and Gas Journal*, June 1981.
[b]The size of the high-temperature Winkler plants is based on a lignite feedstock quoted on a dry basis.
[c]The Mobil demonstration project in New Zealand uses natural gas as the feedstock; the design production capacity is 1650 t/d of gasoline.

Table A1.9 Main direct liquefaction pilot plants[a]

Process type	Organisation(s)	Location	Plant size (t/d coal input)
SRC 1	Gulf Oil/ Southern Services	Wilsonville, USA	6
SRC 1	Mitsui	Omita, Japan	5
SRC 2	Gulf Oil/ Pittsburg & Midway	Fort Lewis, USA	50
Exxon Donor Solvent	Exxon	Baytown, USA	250
H-Coal	Hydrocarbon Research Inc.	Cattletsburg, USA	250
Ruhrkohle	Ruhrkohle/Veba	Bottrop, West Germany	200
Brown Coal Liquefaction	Nippon Brown Coal Liquefaction Co.	Latrobe Valley, Australia[b]	50

[a] Sources: A. Baker and M. Teper, Synfuels Development Outside North America, *American Nuclear Society Conference*, Los Angeles, 1982; D. R. Simbeck and A. J. Moll, Synthetic Fuels from Coal: An Important Source of Energy, *World Coal*, Jan./Feb. 1982.
[b] The coal input to the Latrobe Valley plant refers to dried brown coal.

APPENDIX 2

ECONOMIC DATA FOR SELECTED PROCESSES

The uncertainties in economic data for coal-based processes have already been discussed in chapter 8. While acknowledging the difficulties, an attempt is made here to present economic data, drawn largely from the literature, on a comparative basis for two aspects of coal utilisation technology.

In table A2.1, data on a representative selection of coal conversion processes are given. The costs assume mature technologies with plants located at North American greenfield sites. The designs are based on bituminous coal and this is the only energy input crossing the plant boundary (except during start-up). There are no energy co-products other than as indicated. The scale of the plants lies in the range 0.6 to 10 GW (thermal) of coal input.

The efficiencies are based on higher heating values. For the processes other than Fischer-Tropsch and Exxon Donor Solvent, the values were calculated using the NCB's ARACHNE process flowsheeting system according to a consistent set of ground-rules.

The capital costs are expressed in mid-1982 United States dollars and have been revised from the literature data using the *Chemical Engineering* plant cost index as necessary. The capital costs include a 15 per cent contingency and royalties at 2 per cent of the basic capital cost. Interest during construction, working capital and start-up costs are excluded. The costs are quoted per kW of rated output capacity.

Data on industrial boiler costs are given in table A2.2. The values refer to United Kingdom sites and include ancillaries such as fuel (and ash) handling and the stack. The costs have been revised to mid-1982 money values from the literature data using the *Chemical Engineering* plant cost index.

Table A2.1 Approximate costs and efficiencies of coal conversion processes[a]

Process	Efficiency (%)	Capital cost ($/kW output)
Power generation		
Pulverised fuel (no flue gas desulphurisation)	38.6	880
Pulverised fuel (flue gas desulphurisation, regenerable)	36.7	1000
Pressurised fluidised bed combustion (supercharged boiler cycle)	40.6	800
Gasification combined cycle (British Gas–Lurgi slagging gasifier, 1425°C, water-cooled gas turbine)	42.5	850
Gasification		
Medium calorific value gas (British Gas–Lurgi slagging gasifier)	79.4	430
Substitute natural gas (British Gas–Lurgi slagging gasifier, HCM methanation)	70.9	600
Hydrogen–fuel grade (Texaco gasifier, low temperature shift)	63.2	720
Liquid/fuels		
Methanol–fuel grade (Texaco gasifier, ICI methanol synthesis)	57.9	910
MTG (Texaco gasifier, ICI methanol synthesis, Mobil MTG)	52.1	1190
Fischer–Tropsch (Texaco gasifier, Synthol)	43.7	1070
H-Coal	62.8	835

[a] Sources: Process flowsheeting studies using the NCB's ARACHNE system; *The Cost of Liquid Fuels from Coal*, Economic Assessment Service report E3, IEA Coal Research, London; *The Economics of Gas from Coal*, Economic Assessment Service report E2/80, IEA Coal Research, London; *Technical Assessment Guide*, Electric Power Research Institute report EPRI P-2410-SR, May 1982; *Preliminary Assessment of Alternative PFBC Power Plant Systems*, Electric Power Research Institute report EPRI CS-1451, July 1980; *Coal-Fired Power-Plant Capital-Cost Estimates*, Electric Power Research Institute report EPRI PE-1865, May 1981.

Table A2.2 Typical costs for industrial boilers[a]

Fuel	Boiler type	Cost ($/kW)				
		1 to 5 MW	5 to 10 MW	10 to 20 MW	20 to 40 MW	40 MW and above
Coal	Fire-tube boilers (one combustion chamber)	66				
	Fire-tube boilers (two combustion chambers)		75			
	Water-tube boilers (packaged)			242		
	Water-tube boilers (site-fabricated)				313	280
Oil	Fire-tube boilers (one combustion chamber)	25	17			
	Fire-tube boilers (two combustion chambers)			17		
	Water-tube boilers (packaged)			120	106	
	Water-tube boilers (site-fabricated)			123	110	95

[a]Source: *The Use of Coal in Industry*, The Coal Industry Advisory Board, International Energy Agency, OECD, Paris, 1982.

APPENDIX 3

ABBREVIATIONS

Some common abbreviations used here and elsewhere in the literature on coal utilisation technology are given below.

BTX – Benzene, Toluene and Xylene
CHP – Combined Heat and Power
CLM – Coal–Liquid Mixtures
COM – Coal–Oil Mixtures
CWM – Coal–Water Mixtures
daf – dry ash-free
DCF – Discounted Cash Flow
dmmf – dry mineral-matter-free
FBC – Fluidised Bed Combustion
FGD – Flue Gas Desulphurisation
LCG – Low Calorific Value Gas
LPG – Liquefied Petroleum Gas
MCG – Medium Calorific Value Gas
MHD – Magnetohydrodynamics
PAH – Polycyclic Aromatic Hydrocarbons
PF – Pulverised Fuel
PFA – Pulverised Fuel Ash
SNG – Substitute Natural Gas
SRC – Solvent Refined Coal

APPENDIX 4

UNITS

In general, the units used in the present text have been limited strictly to those of the SI system although, for convenience, derived units such as bar (10^5 N/m^2) for pressure and °C for temperature are also employed where appropriate.

The fundamental units of mass, energy and power are the kilogram (kg), joule (J) and watt (W), respectively. In the present context of fossil fuel consumption, however, the tonne (t) is often used as the unit of mass with mass flow rates expressed as tonnes per hour (t/h), tonnes per day (t/d) or tonnes per year (t/a). The prefixes kilo (k), mega (M) and giga (G) are used to indicate multiples of 10^3, 10^6 and 10^9 respectively (for example, million tonnes per year may be denoted by Mt/a).

The following approximate equivalents may be used for 'off-the-cuff' calculations.

Energy

0.1055	GJ	=	1 therm
1.055	GJ	=	1 million BTUs
1.024	GJ	=	1000 lb steam (from and at 100 °C)
29.3	GJ	=	1 tonne coal (equivalent)
45	GJ	=	1 tonne petroleum (products)
6	GJ	=	1 barrel petroleum (products)
39	GJ	=	1000 m^3 natural gas
1.1	GJ	=	1000 ft^3 natural gas

Power

Note that in each case the MW values are thermal—*not* the electrical equivalent.

105.5	MW	=	1 therm per second
0.2931	MW	=	1 million BTUs per hour
0.2844	MW	=	1000 lb per hour of steam (from and at 100 °C)
8.1	MW	=	1 tonne per hour coal (equivalent)
0.34	MW	=	1 tonne per day coal (equivalent)
930	MW	=	1 million tonnes per year coal (equivalent)
12.5	MW	=	1 tonne per hour petroleum (products)
0.52	MW	=	1 tonne per day petroleum (products)
1430	MW	=	1 million tonnes per year petroleum (products)
69	MW	=	1000 barrels per day petroleum (products)

190 MW = 1 million barrels per year petroleum (products)
450 MW = 1 million m^3 per day natural gas
1240 MW = 10^9 m^3 per year natural gas
12.8 MW = 1 million ft^3 per day natural gas
35 MW = 10^9 ft^3 per year natural gas

The conversion factors have been derived mainly from the *Digest of United Kingdom Energy Statistics 1981* (H.M.S.O., London).

BIBLIOGRAPHY

H. H. Lowry (ed.), *Chemistry of Coal Utilisation*, vols. 1 and 2, Wiley, New York, 1945

H. H. Lowry (ed), *Chemistry of Coal Utilisation*, Supplementary Volume, Wiley, New York, 1963

Martin A. Elliot (ed.), *Chemistry of Coal Utilisation*, Second Supplementary Volume, Wiley, New York, 1981

D. W. van Krevelen, *Coal: Typology Chemistry Physics and Constitution*, Elsevier Publishing Co., Amsterdam, 1961

G. J. Pitt and G. R. Millward (eds.), *Coal and Modern Coal Processing: An Introduction*, Academic Press, New York, 1979

L. Grainger and J. Gibson, *Coal Utilisation: Technology, Economics and Policy*, Graham and Trotman, London, 1981

N. Berkowitz, *An Introduction to Coal Technology*, Academic Press, New York, 1979

Robert A. Meyers (ed.), *Coal Handbook*, Marcel Dekker, New York, 1981

Douglas M. Considine (ed.), *Energy Technology Handbook*, McGraw-Hill, New York, 1977

C. Y. Wen and E. Stanley Lee (eds.), *Coal Conversion Technology*, Addison Wesley, Reading, Massachusetts, 1979

I. Howard-Smith and G. J. Werner, *Coal Conversion Technology*, Noyes Data Corp., Park Ridge, New Jersey, 1976

I. G. C. Dryden (ed.), *The Efficient Use of Energy*, IPC Science and Technology Press, Guildford, Surrey, 1975

Carroll L. Wilson (ed.), *Coal: Bridge to the Future*, Report of the World Coal Study (WOCOL), Ballinger Publishing Co., Cambridge, Massachusetts, 1980

The Coal Industry Advisory Board, *The Use of Coal in Industry*, International Energy Agency, OECD, Paris, 1982

Larry L. Anderson and David A. Tillman, *Synthetic Fuels from Coal: Overview and Assessment*, Wiley, New York, 1979

J. R. Howard (ed.), *Thermal Applications of Fluidised-bed Technology*, Applied Science Publishers Ltd., Barking, Essex, 1982

D. G. Skinner, *The Fluidised Bed Combustion of Coal*, Mills and Boon, London, 1971

Fluidised Bed Combustion of Coal, National Coal Board, Hobart House, London, 1980

D. Anson, Fluidised Bed Combustion of Coal for Power Generation, *Progress in Energy Combustion Science*, vol. 2, 1976, p. 61

W. C. Patterson and R. Griffin, *Fluidised-bed Energy Technology: Coming to a Boil*, Inform Inc., New York, 1978

Joseph Yerushalmi, Circulating Fluidised Bed Boilers, *Fuel Processing Technology*, **5** (1981) 25

Gas Making and Natural Gas, BP Trading Ltd., Brittanic House, Moor Lane, London, 1972

P. Nowacki, *Coal Gasification Processes*, Noyes Data Corp., Park Ridge, New Jersey, 1981

H-D. Schilling, B. Bonn and U. Krauss, *Coal Gasification: Existing Processes and New Developments*, Graham and Trotman, London, 1981

P. Nowacki (ed.), *Coal Liquefaction Processes*, Noyes Data Corp., Park Ridge, New Jersey, 1979

Liquid Fuels from Coal, National Coal Board, Coal Research Establishment, Cheltenham, 1978

J. Gibson and D. H. Gregory, *Carbonisation of Coal*, Mills and Boon, London, 1971

O. K. Foo and E. M. Jamgochian, *Assessment of Coal–Liquid Mixtures in Co-operating IEA Countries*, Report MTR-83W87-04, The Mitre Corp, McLean, Virginia, June 1983

R. Brown, *Environmental Effects of Coal Technologies: Research Needs*, The Mitre Corp., Washington, D.C., 1981

Commission on Energy and the Environment, *Coal and the Environment*, H.M.S.O., London, 1981

A New Guide to Capital Cost Estimating, The Institution of Chemical Engineers, Rugby, 1977

Technical Assessment Guide, Report P-2410-SR, Electric Power Research Institute, Palo Alto, California, 1982

In addition, reports on a range of coal utilisation topics are published by the Technical Information Service and Economic Assessment Service of IEA Coal Research, Lower Grosvenor Place, London.

INDEX